T0130982

SYSTEMS ENGINEERING IN WIRELESS COMMUNICATIONS

Heikki Koivo

Helsinki University of Technology, Finland

Mohammed Elmusrati

University of Vaasa, Finland

A John Wiley and Sons, Ltd, Publication

This edition first published 2009
© 2009 John Wiley & Sons Ltd.,

Registered office

John Wiley & Sons Ltd, The Atrium, Southern Gate, Chichester, West Sussex, PO19 8SQ, United Kingdom

For details of our global editorial offices, for customer services and for information about how to apply for permission to reuse the copyright material in this book please see our website at www.wiley.com.

The right of the author to be identified as the author of this work has been asserted in accordance with the Copyright, Designs and Patents Act 1988.

All rights reserved. No part of this publication may be reproduced, stored in a retrieval system, or transmitted, in any form or by any means, electronic, mechanical, photocopying, recording or otherwise, except as permitted by the UK Copyright, Designs and Patents Act 1988, without the prior permission of the publisher.

Wiley also publishes its books in a variety of electronic formats. Some content that appears in print may not be available in electronic books.

Designations used by companies to distinguish their products are often claimed as trademarks. All brand names and product names used in this book are trade names, service marks, trademarks or registered trademarks of their respective owners. The publisher is not associated with any product or vendor mentioned in this book. This publication is designed to provide accurate and authoritative information in regard to the subject matter covered. It is sold on the understanding that the publisher is not engaged in rendering professional services. If professional advice or other expert assistance is required, the services of a competent professional should be sought.

Library of Congress Cataloging-in-Publication Data

Koivo, Heikki
    Systems engineering in wireless communications / Heikki Koivo, Mohammed Elmusrati.
        p. cm.
    Includes index.
    ISBN 978-0-470-02178-1 (cloth)
  1. Wireless communication systems. 2.   Systems engineering.   I. Elmusrati, Mohammed. II.
Title.
    TK5103.2.K644 2009
    621.384–dc22

                                                                            2009026965

A catalogue record for this book is available from the British Library.

ISBN 978-0-470-02178-1 (H/B)

Set in 10/12pt Times Roman by Thomson Digital, Noida, India.
Printed in Great Britain by CPI Antony Rowe, Chippenham, UK

Contents

Preface

In Helsinki during a visiting lecture, an internationally well-known professor in communications said, 'In the communications society we have managed to convert our proposals and ideas to real products, not like in the control engineering society. They have very nice papers and strong mathematics but most of the real systems still use the old PID controllers!'. As our background is mainly in control as well as communications engineering, we know that this thought is not very accurate. We agree that most of the practical controllers are analog and digital PID controllers, simply because they are very reliable and able to achieve the required control goals successfully. Most of the controllers can be explained in terms of PID. The reasons behind this impressive performance of PID will be explained in Chapter 2.

There is a hidden fact that many researchers and professionals do not pay enough attention to. This is that control engineering is one of the main pillars of modern communication systems in both hardware and algorithms. On the hardware front, we know the importance of the phase-locked loop (PLL) and its modified version, the delay-locked loop (DLL), in the implementation of communication receivers. We can say that phase-locked loops are an essential part of all kinds of today's communication receivers. These systems are simply small closed-loop control systems. From an algorithmic point of view, the multiple access control (MAC) layer is based on control theory concepts. Some examples are power control, rate control, scheduling, handover, and generally radio resource management algorithms. When there is feedback measurement and action based on that measurement, then there is control engineering. We may generalize this view to include all kinds of adaptive systems such as equalizers, beamforming antennas, adaptive modulation and coding. Therefore, if we think of the modern communications system in terms of the human body, then control algorithms are the heart and communication algorithms are the blood. This is a general view of the contribution of control engineering in telecommunications. Moreover, if we extend this vision to include system engineering, which means in addition to control engineering, system identification and modeling, the contribution is much more. Channel identification and modeling is one major requirement in the design of transmitters and receivers. However, this inherent integration between communication and control is not that visible for many people. This is one reason why you have this book in front of you now. We have tried to show some of the applications of control engineering in communication systems, but we did not want to just collect algorithms from literature and bind them together into a reference book. We wanted to generate one easy-to-read book which is accessible to a wide range of readers from senior undergraduate students up to researchers and site engineers. To achieve this target, we have included many numerical examples and exercises. A key approach to becoming more familiar and understanding the deep secrets of sophisticated control algorithms is through simulation. Therefore MATLAB®/ Simulink® is used as a tool to illustrate the concepts and shows promise in a diversity of ways to attack problems. We believe that Simulink® can provide a more transparent way of studying, for example in the simulation of self-tuning estimators and controls and Kalman filters.

We have many people to thank for this book.

Heikki Koivo wants to thank his students Jani Kaartinen, Juha Orivuori and Joonas Varso, who helped in setting up the numerous MATLAB®/Simulink® examples. He is eternally grateful to his wife, Sirpa, who suffered the pain caused by writing and the time spent apart. Special thanks go to Jenny and Antti Wihuri Foundation and the Nokia Foundation who have supported the author with stipends.

Mohammed Elmusrati wants to thank his friend Dr. Naser Tarhuni for his constructive comments about the book. He would also like to thank his wife Nagat and daughters Aia and Zakiya for their care, patience and support. Special thanks to his father Salem Elmusrati for his continuous moral support and encouragement.

The Co-author Mohammed Elmusrati wishes to make it clear that section 8.2 of this book is reproduced from his Ph. D. thesis entitled: *Radio Resource Scheduling and Smart Antennas in CDMA Cellular Communication Systems*, ISBN 951-22-7219-9 (published by Helsinki University of Technology - Finland), and not from a work entitled *Advanced Wireless Networks: 4G Technology* written by Savo Glisic which in section 12.5 reproduces part of Elmusrati Ph.D. Thesis.

As stated in this book, proper feedback is critical for system performance and efficiency. Hence, your feedback is very important to us. If you have any comments, corrections or proposals, please do not hesitate to contact us on the following email address: sys.eng.wireless. comm@gmail.com.

List of Abbreviations

AWGN	Additive White Gaussian Noise
AM	Amplitude Modulation
ASK	Amplitude Shift Keying
ARMA	Auto Regressive Moving Average
ARMAX	Auto Regressive Moving Average Exogenous
ARX	Auto Regressive Exogenous
BER	Bit Error Rate
BLER	Block Error Rate
BIBO	Bounded Input Bounded Output
BPF	Band-Pass Filter
BPSK	Binary Phase Shift Keying
CAC	Connection Admission Control
CDMA	Code Division Multiple Access
CIR	Carrier to Interference Ratio
CMTTP	Centralized Minimum Total Transmitted PowerCSMA
CSMA/CA	Carrier Sense Multiple Access with Collision Avoidance
CSOPC	Constrained Second-Order Power Control
DBA	Distributed Balancing Algorithm
DCPC	Distributed Constrained Power Control
DDMA	Demand Division Multiple Access
DL	Downlink
DoA	Direction of Arrival
DPC	Distributed Power Control
DPSS	Discrete Prolate Spheroidal Sequence
ESPC	Estimated Step Power Control
FDMA	Frequency Division Multiple Access
FEC	Forward Error Correction
FER	Frame Error Rate
FFT	Fast Fourier Transform
FLPC	Fuzzy Logic Power Control
FMA	Foschini and Miljanic Algorithm
FSPC	Fixed-Step Power Control
GMVDR	General Minimum Variance Distortionless Response
GUI	Graphical User Interface

HSDPA	High Speed Downlink Packet Access
HSPA	High Speed Packet Access
IFFT	Fast Fourier Transform
ISE	Integral Square Error
ISM	Industrial, Scientific and Medical
ITAE	Integral Time Absolute Error
LoS	Line of Sight
LS-DRMTA	Least Square Despread Respread Multitarget Array
LTV	Linear Time Variant
MIMO	Multiple-Input Multiple-Output
MISO	Multiple-Input Single-Input
MLPN	Multilayer Layer Perceptron Network
MMSE	Minimum Mean Square Error
MO	Multiple Objective
MODPC	Multiple Objective Distributed Power Control
MTPC	Maximum Throughput Power Control
MVDR	Minimum Variance Distortionless Response
OFDM	Orthogonal Frequency Division Multiplexing
OFDMA	Orthogonal Frequency Division Multiple Access
OLPC	Outer-Loop Power Control
PAN	Personal Area Network
PD	Proportional Derivative
PI	Proportional Integral
PID	Proportional Integral Derivative
PLL	Phase-Locked Loop
PSK	Phase Shift Keying
QAM	Quadrature Amplitude Modulation
QoS	Quality of Service
RBFN	Radial Base Function Network
RLS	Recursive Least Squares
RRM	Radio Resource Management
RRS	Radio Resource Scheduling
RX	Receiver
SDMA	Spatial Division Multiple Access
SDMPC	Statistical Distributed Multi-rate Power Control
SDR	Software-Defined Radio
SIMO	Single-Input Multiple-Output
SINR	Signal to Interference and Noise Ratio
SIR	Signal to Interference Ratio
SISO	Single-Input Single-Output
SMIRA	Stepwise Maximum Interference Removal Algorithm
SMS	Short Message Service
SNR	Signal to Noise Ratio
SOR	Successive Over-Relaxation
SPC	Selective Power Control
SRA	Stepwise Removal Algorithm

STPC	Self-Tuning Power Control
STPPC	Self-Tuning Predictive Power Control
TDD	Time Division Duplex
TDMA	Time Division Multiple Access
TX	Transmitter
UMTS	Universal Mobile Telecommunications System
VCO	Voice Controlled Oscillator
WCDMA	Wideband CDMA

1

Introduction

1.1 Introduction to Telecommunications

Telecommunication systems in general consist of three main parts: transmitter; transmission media or channel; and receiver (see Figure 1.1). A more detailed description for each part now follows.

1.1.1 Transmitter

This unit is responsible for processing the data signal in order to be valid for the transmission media (channel). In its basic form the transmitter consists of three subunits: source data processing; data robustness; and signal modulation (see Figure 1.2).

In *source data processing* there are filters to determine the signal bandwidth; for an analog source and digital communication system there will be a sampler, analog to digital converters, and possibly data compression.

The *data robustness* subunit represents all processes performed on the data signal to increase its immunity against the unpredictable behavior and noise of the transmission media, therefore, increasing the possibility of successful reception at the receiver side. In wireless communication we transmit our data as electromagnetic waves in space which contain an almost infinite number of different signals. The ability of the receiver to receive our intended signal and clean it from associated interferences and noises, as well as distortion, is a real challenge. In digital communication systems this subunit may consist of channel coding and a pre-filtering process.

The aim of channel coding is to increase the reliability of the data by adding some redundancy to the transmitted information. This makes the receiver able to detect errors and even to correct them (with a certain capability). In digital communication systems we convert the source analog signal to a finite number of symbols. For example in the binary system there are only two symbols S_0 and S_1, or more popular 0 and 1. Generally we may have any finite number M of symbols, so that it is called an M-ary digital communication system. Those symbols are also known precisely at the receiver. The information flow is determined by the arrival sequences of these symbols. This is one major strong feature of digital systems over analog.

Systems Engineering in Wireless Communications   Heikki Koivo and Mohammed Elmusrati
© 2009 John Wiley & Sons, Ltd

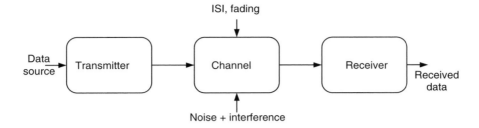

Figure 1.1   Simplified block diagram of communication system

Another possible way to increase the reliability of the transmitted signal is by making the symbols as unlike each other as possible. This increases the possibility of correct detection of the transmitted symbol. The pre-filtering process is a possible action done at the transmitter to minimize the effects of the channel distortion on the signal. Simply, if we can achieve pre-distortion of the signal, which has the reverse action of the channel distortion, then it is possible to receive the signal as distortion-free. Of course, this is only theoretical, but there are still several practical systems using this pre-filtering to reduce channel distortion effects.

Most of these systems require a feedback channel in order to adapt the characteristics of the pre-filter according to the channel behavior. One main type of signal distortion of the channel is due to the multipath reception of the transmitted signal.

The last part of this general presentation of the transmitter is the *modulation* subunit. We may express this subunit as the most important one for multiuser wireless communications systems. Usually the source data signal occupies a certain baseband bandwidth, for example for voice it could be between 20 Hz up to about 5 kHz. For the whole audible range the spectrum could be extended up to about 20 kHz. In wireless communications, it is not possible to connect this baseband signal directly to the transmission antenna. There are many reasons for this, such as lack of multi-usage of the channel, a very long antenna would be needed (antenna length is related to the signal wavelength), and non-suitable propagation behavior.

Modulation is the process of carrying information signals over a certain well-defined carrier signal. This carrier is usually a sinusoidal signal. Remember that the sinusoidal signal is the only signal which has zero bandwidth. In the spectrum domain, it is represented as a delta function located at the frequency value of the signal (and its mirror). Any sinusoidal signal is defined by three parameters: amplitude; frequency; and phase. It is possible to carry the data signal over any of those parameters, therefore we have three types of modulation techniques: amplitude modulation (or amplitude-shift keying in terms of digital communications

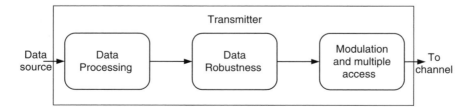

Figure 1.2   General representation of transmitters

terminologies); frequency modulation (or frequency-shift keying); and phase modulation (or phase-shift keying).

In digital wireless communications systems, we observe that phase-shift keying (PSK) is the most famous modulation technique. The reason is that PSK offers very high spectrum efficiency. For example, if the channel bandwidth is only 1 MHz, then after modulation (double sidebands) we can send a maximum rate of 1 Msymbol/s over this channel to be able to mitigate the intersymbol interference (ISI) problem. Using PSK, if the SINR is already very high at the receiver side, we can send large numbers of bits (i.e., more information) per symbol. For example, if we send 5 bits/symbol (i.e., 32-ary) then the transmitted data rate is 5 Mb/s over only 1 MHz bandwidth. However, as we increase the number of bits per symbol, we reduce the possibility of detecting the symbols correctly.

It is also possible to join different modulation methods together, for example amplitude and phase modulation. This is called quadrature amplitude modulation (QAM). These relations will be investigated more throughout the book.

Another important task of the modulation subunit is the multiple access possibility. Multiple access means the possible multi-usage of the channel between many agencies, operators, users, services, and so on. There are different kinds of multiple access methods such as: frequency division multiple access (FDMA); time division multiple access (TDMA); code division multiple access (CDMA); spatial division multiple access (SDMA); and carrier sense multiple access (CSMA). These techniques will be illustrated later.

1.1.2 Wireless Channels

A channel is the media where the signal will pass from transmitter to the receiver. Unfortunately, this transmission media is not friendly! Our signal will suffer from several kinds of problems during its propagation such as high power loss, where the received signal power is inversely proportional to the square of the distance in free space. In mobile communications, the received power may be inversely proportional with distance with power 4. For example, if the transmit average power is 1 watt, then after only 20 meters from the transmitter antenna, the average received power can be in order of $1/20^4 = 6.25\,\mu W$! By the way, this high-loss characteristic gives the advantage to reuse the same frequencies after some distance with negligible interference. This is one form of spatial division multiple access.

There is also the problem of shadowing where the received signal power may fall considerably because of buildings, inside cars, and so on. The large-scale power path loss is related to:

$$L \propto \frac{\aleph}{d^{-n}} \qquad (1.1)$$

where $\aleph$ is the shadowing effects, which can be represented as a random variable with log-normal distribution, d is the distance between the transmitter and the receivers, and n is the path-loss exponent which is usually between 2 and 4. Throughout the book, we sometimes refer to channel gain rather than channel loss. Channel gain is just the inverse of channel loss, i.e., $G = 1/L$. For mobile communications, the path-loss exponent 4 gives a good fit with practical measurements.

The signal itself reaches the receiver antenna in the form of multipaths due to reflections, diffractions, and scattering. The reason for this multipath is due to natural obstacles

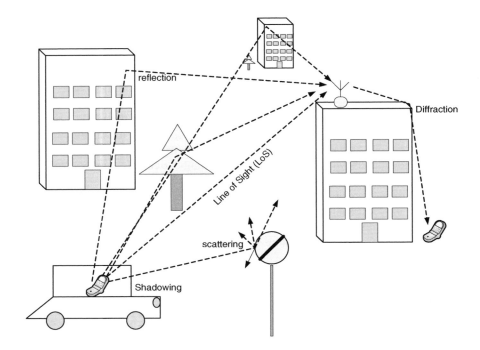

Figure 1.3 A simple picture illustrating the propagation conditions

(mountains, trees, buildings, etc.). Figure 1.3 shows a simple picture of the propagation conditions.

Most of the received signal comes through diffractions and reflections. It is rare that it follows a pure direct path, line of sight (LoS), except for point-to-point communication such as in microwave links. From Figure 1.3, assuming the transmitted signal in baseband to be $s(t)$, we may formulate the received signal as follows:

$$
\begin{aligned}
r(t) &= \text{Re}\left(\left\{\sum_{k=1}^{B}\alpha_k(t)s(t-\tau_k(t))\right\}e^{j2\pi f_c[t-\tau_k(t)]}\right) + n(t) \\
&= \text{Re}\left(\left\{\sum_{k=1}^{B}\alpha_k(t)e^{j2\pi f_c\tau_k(t)}s(t-\tau_k(t))\right\}e^{j2\pi f_c t}\right) + n(t)
\end{aligned}
\tag{1.2}
$$

where $\alpha_k(t)$ is related to the square-root of the path loss through path k and it is generally time varying, $\tau_k(t)$ is the delay of path k and it is also time varying in general, f_c is the carrier frequency, B is the number of paths, $n(t)$ is the additive noise and interference, and $s(t)$ is the transmitted signal in baseband. We will drop the additive noise part because we want to study the effects of channel propagations conditions.

Therefore, the equivalent received baseband is given by:

$$
z(t) = \sum_{k=1}^{B}\alpha_k(t)e^{j\varphi_k(t)}s(t-\tau_k(t))
\tag{1.3}
$$

where φ_k is the phase offset. If the transmitter and/or receiver are moving with some velocity then there will be a frequency shift, which is known as the Doppler frequency, and is given by:

$$f_d = \frac{V}{\lambda} \tag{1.4}$$

where V is the relative velocity and λ is the wavelength. Including this Doppler frequency shift, the baseband of the received signal becomes:

$$z(t) = \sum_{k=1}^{B} \alpha_k(t) e^{j[2\pi f_d t + \varphi_k(t)]} s(t - \tau_k(t)) \tag{1.5}$$

Actually the Doppler effect is observed even for fixed transmitters and receivers. In this case the time-varying dynamic behavior of the channel is the source of this Doppler shift.

The effects of mobile channels on the received signal can thus be classified into two main categories: large-scale effects, where the received signal strength changes slowly (this includes distance loss and shadowing); and small-scale effects, where the received signal may change significantly within very small displacement (in order of the wavelength), and even without moving at all (because of the dynamic behavior of the channel). These dramatic changes in the signal strength may cause signal losses and outages when the signal-to-noise ratio becomes less than the minimum required. This is known as signal fading. There are two main reasons for small-scale signal fading. The first is due to the multipath and the second is due to the time-varying properties of the channel and is characterized by the Doppler frequency. The multipath characteristic of the channel causes fading problems at band-pass signal and ISI problems at the baseband. The following example explains in a simple way how this multipath may cause fading for the received signal.

Example 1.1

Study the effect of the multipath environment by assuming the transmission of an unmodulated carrier signal, $A\sin(2\pi f_0 t)$, $f_0 = 1.8\,\text{GHz}$. Assume only two paths – one path causes a fixed delay and the other path a random delay. Plot the received signal power (in dB scale) using MATLAB® (a registered trademark of The Math-Works, Inc.) if the random delay has uniform distribution in the interval $[0,1]$ μs.

Solution

Name the two paths' signals as $x_1(t)$ and $x_2(t)$, and the received signal $y(t) = x_1(t) + x_2(t)$. Also let the received amplitude be fixed (usually it is also random). Assume the amplitude of the signal received via the first path to be 1 V and via the second path 0.9 V. The signal received via the first path is $x_1(t) = A_1\cos(2\pi f_0[t - \tau_0]) = A_1\cos(2\pi f_0 t - 2\pi f_0 \tau_0) = A_1\cos(2\pi f_0 t - \theta_0)$, and from the second path is $x_2(t) = A_2\cos(2\pi f_0[t - \tau_2(t)]) = A_2\cos(2\pi f_0 t - 2\pi f_0 \tau_2(t))$. The total received signal is $y(t) = x_1(t) + x_2(t) = \cos(2\pi f_0 t - \theta_0) + 0.9\cos(2\pi f_0 t - 2\pi f_0 \tau_2(t))$, which can be simplified to (assuming $\theta_0 = 0$) $y(t) = \sqrt{1.81 + 1.8\cos(2\pi f_0 \tau_2(t))}$ $\cos(2\pi f_0 t + \varphi_n)$. We can see clearly the random amplitude nature of the total received signal because of the random delay. The maximum amplitude is $\sqrt{1.81 + 1.8} = 1.9$ and the minimum is $\sqrt{1.81 - 1.8} = 0.1$. This indicates the high fluctuations of the received signal amplitude. The

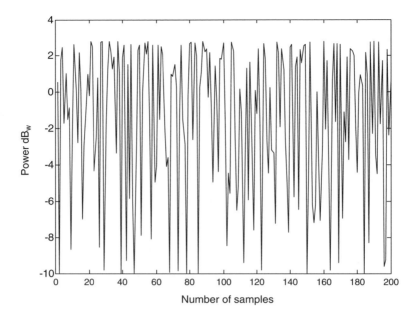

Figure 1.4   The effects of multipath on the received signal power

average received power can be expressed as $P_r(t) = [1.81 + 1.8\cos(2\pi f_0 \tau_2(t))]/2$. We may plot the received power in dBW as shown in Figure 1.4. The MATLAB® code is:

```
f0=1.8e9; % carrier frequency
tao=1e-6*rand(200,1); %generation of 200 random delay
Pr=(2+1.8*cos(2*pi*f0*tao))/2; % instantaneous power
Pr_db=10*log10(Pr); % expressing the power in dBw
plot(Pr_db) %plot the power
```

From Equation (1.5) we may deduce the impulse response of the wireless channel as follows:

$$h(t,\tau) = \sum_{k=1}^{B} \alpha_k(t) e^{j[2\pi f_d t + \varphi_k(t)]} \delta(t - \tau_k(t)) \tag{1.6}$$

If the symbol duration (T) is longer than the longest delay, i.e., $T > \max_{k}(\tau_k)$, we can approximate Equation (1.5) as

$$z(t) = s(t - \tau_0) \sum_{k=1}^{B} \alpha_k(t) e^{j[2\pi f_d t + \varphi_k(t)]} = \hat{\alpha}(t) s(t - \tau_0) \tag{1.7}$$

Therefore the received signal is simply the transmitted signal multiplied by the time-varying complex number representing the channel. Usually the relation between the received signal and the transmitted one is convolution because the channel is a linear time-varying system. Hence, the multiplicative relation shown in Equation (1.7) is a special case valid only when the signal duration is longer than the channel coherence time. This is known as the flat fading channel. The reason for this name is that when the channel loss is very high, i.e., $|\hat{\alpha}(t)|$ is very small, then the received symbol and all its replicas through the multipath will be faded at the same time.

The other type is called the frequency selective channel, which is obtained when signal duration is less than the channel delay profile. This leads to ISI between different symbols and causes large errors in the received signal. This problem sets an upper bound for the symbol rate that can be transmitted over wireless channels. However, it is interesting to note that this problem has been turned into an advantage by using CDMA.

With CDMA modulation, when the delay is larger than the duration of a single chip, the received replicas will have low correlation with each other. Using a Rake receiver, which consists of several correlators locked to each signal path, we obtain time diversity. In other words if one symbol is lost (because of deep fading), there is a chance of receiving the same symbol through another path.

Another way of mitigating the problem of ISI in frequency selective channels is by dividing the signal into many orthogonal parallel subbands and sending them simultaneously in the channel. Every subband will see the flat fading channel, and then at the receiver side we can recombine all subbands to their original band. A very efficient method to perform this process is called orthogonal frequency division multiplexing (OFDM).

The effects of time variations of the channel appear as a frequency deviation from the carrier frequency. For example if the transmitted signal is $\mathrm{Re}\{s(t)e^{j2\pi f_0 t}\}$, it will be received as $\mathrm{Re}\{s(t)e^{j2\pi(f_0 \pm f_d)t}\} = \mathrm{Re}\{s(t)e^{\pm j2\pi f_d t}e^{j2\pi f_0 t}\}$. We have a time-varying phase which will be difficult to track if f_d has large value. The effect of the time-varying behavior of channels is known as slow fading (if f_d is small) and fast fading when f_d is large. More analysis about channels is given in Chapters 3 and 4.

We have so far discussed the effects of the inherent properties of wireless channels on signals. Moreover, there is an almost infinitely number of signals in the wireless channel. Some of those are manmade and other are natural signals from the sun, stars, earth, and so on. Usually we consider only signals which fall within or close to our signal bandwidth. Other signals can be easily removed using front-end band-pass filters at the receiver side. Signals from other communication systems within our bandwidth are known as interferences. Other signals are known as noise. Actually, one major noise part is generated within the receiver circuits. It is known as thermal and shot noise. The noise generated by the receiver is characterized by the noise figure or the equivalent noise temperature of the receiver. Usually, the noise figure of the receiver is one of the parameters that determines receiver sensitivity.

1.1.3 Receiver

The receiver is an electronic circuit which is used to recover the original signal from the intentional modifications (modulations, coding, ADC, etc.) and the channel inherent distortions (channels, noise, interferences, etc.). The receiver can be described as reversing or undoing the operations of the transmitter. In this sense the receiver can be simplified into three blocks as shown in Figure 1.5.

On the receiver side, we have band-pass filters to reject all signals other than the interested band. In multiuser communications, the receiver should receive only its intended signal through the multiple access method. Next we have the demodulation process to move the signal from band-pass to its original baseband form. Data recovery is the block which uses all data processing made at the transmitter to enhance reception reliability. For example, if the transmitter uses

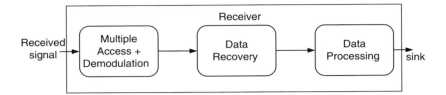

Figure 1.5   Simplified structure of wireless receiver

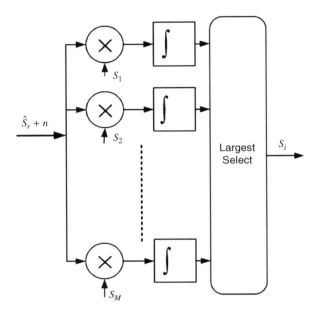

Figure 1.6   Correlator detection

forward error correction (FEC) coding, then in the data recovery block the decoding process will detect any errors and correct them if possible. Finally, in the data processing part we should retrieve the signal as close to its original form as possible. As mentioned before, in digital communications we send one of the known symbols at a time. The symbols are known precisely at the receiver as well. The information flow is in the transmission sequence of those symbols. From this, the optimum detector at the receiver is based on a correlation process. The received symbol will be distorted according to the channel and also corrupted by noise and interference. The receiver will measure the matching between the received symbols with all possible symbols using correlators and select the one which has the highest match score as shown in Figure 1.6. Finally, the detected signal is delivered to the sink (speaker, screen, computer input port, etc.).

1.2 The Quality of Service (QoS) Parameters

In wireless communications, it is important to define a set of parameters to measure the communication quality. The QoS parameters can be classified into two main parts. One part can be sensed by the end user such as voice quality, blocking and outage probability, and data rate. The second part is related to the security and privacy which usually cannot be sensed by the end user during the call. Operators should offer accepted quality of service and at the same time keep the security and the privacy of their customers. Feedback control can be applied to improve QoS. To assess the link quality, measures of performance are needed. Some of the QoS parameters can serve as such. One of the most important QoS parameters is the ratio of the average power of the desired signal - to - the average power of the interference and noise signals. It is commonly known as the signal to interference and noise ratio (SINR). Other names are possible such as signal-to-noise ratio (SNR), where noise is the dominant part; signal-to-interference ratio (SIR); where the interference is dominant; and carrier-to-interference ratio (CIR), which is used to indicate the ratio at band-pass signal. These different names will be used in this book interchangeably. There is one problem with the SINR indicator when it is applied to indicate the quality of digital modulated signals. The SINR does not consider the number of bits per symbol in multilevel modulation, which may lead to wrong conclusions. This is explained in the next example.

Example 1.2

Assume PSK modulation where the received signal is $x(t) = 3\cos(2\pi f_0 t + \phi_i(t)) + n(t)$ and the average noise power is 0.1 W. The input resistance of the receiver is 1 Ω. Compute the SNR for binary PSK (BPSK), where we have only two symbols and 32-PSK, where every symbol carries 5 bits.

Solution

We can calculate the SNR simply as the ratio of the average received power to the average noise power corrupting the signal such as $SNR = \frac{9/2}{0.1} = 45 \Rightarrow SNR_{dB} = 10\log_{10}(45) = 16.5\ dB$. This SNR value is the same for both cases of modulations (binary and 32-ary)! Since we do not increase the transmission power, it is clear that the probability of error will be greater when we send more data per symbol. This can be verified by studying the constellation map where all the symbols are positioned. The distance between the symbols becomes shorter as we increase the number of symbols without increasing the transmission power. The possibility of incorrect detection now becomes much greater because of noise. This fact can be observed in the bit error rate formulas of different modulation types.

We can make the SINR indicator more relative to digital communication systems by computing it at the bit level. In other words we compute the average received power per bit which can be done by multiplying the average received power with the bit duration. From physics we know that the product of power with time is energy. Therefore we can compute the received bit energy E_b as $E_b = ST_b$, where S is the average received signal power and T_b is time duration per bit. This can be made more understandable by forming a unitless expression dividing E_b with the noise energy, known as noise spectral density N_0, $N_0 = \delta_n^2/BW$, where δ_n^2 is average noise power and BW the bandwidth. The unitless expression is then E_b/N_0, which is useful when comparing different communication techniques having different modulation levels. It is related to SINR by the following expression $\frac{E_b}{N_0} = SINR\frac{BW}{R_b}$, where R_b is the bit rate and BW is the bandwidth, which can be the same as the symbol rate for some systems. Does the end user notice the value of the SINR? Actually, not directly! The end user notices more the block error rate (BLER). When BLER is large then there will be outage, voice degradation, low throughput (because of retransmissions), and so on. However, there is a direct relation between BLER and SINR. In AWGN channels the BLER decreases when the SINR increases. This mapping can be given in closed-form mathematical expression in some simple cases such as when noise is additive white noise or when the fading is modeled mathematically. However, in reality, these give only rough estimates of the real situation.

Exercise 1.1   What is the difference between bit error rate (BER) and block error rate (BLER), and what are the effects of coding in this case? *Hint*: see Chapter 8.

The SINR is used directly as an indicator for the channel quality. Since the channel condition is measured generally by the BER or BLER after coding, why do we not use them directly as channel indicators instead of SINR or E_b/N_0? The answer is that the required estimation time is too long. The SINR estimation can be done in a very short time, but a direct measurement of the BLER or BER may require a longer time, which leads to slow adaptation rate of the resources. For example, assume that the target BER is 10^{-3}, and the data rate is 10 kb/s, so that we will need to test about 10 000 bits to have robust estimation for the BER. This number of bits will require 1 s of monitoring time. This is a very long time in terms of resources adaptation. For example, in UMTS mobile systems, the power adaptation rate is 1500 Hz, i.e., the power is adapted once per 0.667 ms. Therefore we need a very fast channel quality estimator. This can be obtained with SINR (or E_b/N_0) measurements.

The throughput is another important parameter of the QoS in wireless communication systems. Throughput is an indicator for the actual information rate and it is different than the data rate. The data rate is the actual transferred data per unit time. This data contains (besides the required actual information) other types of bits such as coding bits, destination address header, and signaling protocol bits. Moreover, packets could be retransmitted if they were not correctly decoded. We can say that throughput is usually less than the data rate. Although the other bits are necessary to perform the communication between the transmitter and the receiver, the end user does not really care about them. We are much more interested in the actual information rate, for example how fast we can download certain files wirelessly through our mobiles.

Packet delay is another QoS parameter which each operator should specify accurately. When we look at the delay between data transmission and reception (end to end) we may call such delay the latency. In real-time applications maximum latency should be strictly specified, at least within a certain confidence interval, for example, 99% of the transmitted packets will have delay less than 30 ms. There are several reasons for packet delays in wireless communications such as protocol delay (e.g., in CSMA/CA), queuing delay (because of limited resources), retransmission delay (according to error reception), and propagation delay (significant in satellite links). When the delay exceeds some limit we will have packet loss which should be also specified in the QoS agreement. There is another problem with delays when they are not fixed, i.e., when the delay period is a random variable. This introduces something known as jitter, which is computed as the maximum delay minus the minimum delay. When the jitter is longer than the packet period, there is a high chance that the packets will not arrive in the correct order as well as interfere with others. Moreover, in certain applications such as wireless automation, it would be difficult to design high performance and stable control systems in the presence of a wide range of random delays.

Based on the above summary of the QoS parameters, we may state that the required QoS depends on the application. There are several classifications of applications such as real time and non-real time, fixed and elastic and so on. Let us take few simple examples for the required QoS parameters with different applications.

- **Voice communication** (GSM): fixed and low throughput ($\sim$13 kb/s), strict maximum delay, relaxed BER ($\sim 10^{-3}$), and relaxed packet loss probability (our brains are excellent at data fitting, so that the lost packets may not be missed if they are not too large). This is considered as a real-time application.
- **SMS messages** (non-real-time application): very low throughput, very relaxed delays, strict maximum allowed BER (we do not like to receive a different message than has been sent!).
- **Internet exploring** (large file download): the higher the throughput the better, relaxed delays, and strict BER. Here we are more interested in high throughput in the average sense. For example if a file can be downloaded within 30 s, it is much better than 3 min regardless of the continuity of the service. Let's explain this more. Assume that during this 30 s you get served only 1 s every 9 s (i.e., three times) with a data rate of 10 Mb/s. This means that during these three times you will download 30 Mb. On the other hand, if you have fixed continuous services but with a data rate of 30 kb/s, then you will need a total time of 1000 s to get this 30 Mb! Hence, in these types of applications, we do not care about the continuity of the service as long as we have a high throughput on average.
- **High secure applications** if we want to make a bank transfer, then we may relax the throughput as well as the delay, but we should have very low BER and the highest security and privacy.

From the above we see that the QoS requirements depend mainly on the applications.

1.3 Multiple Access Techniques

Multiple access methods are critically required for the multi-usage of the channel. In other words, how can several transmitter/receiver pairs use the same available channel with minimum interference between them. Consider a real situation which describes the multiple access need and methods.

Assume a classroom with many students (from different countries) and one teacher. When the teacher is talking and all students are listening, we have a broadcasting situation where we do not need to consider multiple access. Assume one student cannot hear clearly, then she may ask the teacher to increase his voice level. This can be considered as a feedback channel to improve the SNR at the receiver side. Suppose the teacher divides class into random pairs (not neighboring pairs), where each pair should talk to each other. What do you think will happen in this case? Every pair will start talking, but because of the interference from other pairs, each pair will try to talk more loudly (increasing the transmission power) and use the available degree of freedom to increase the possibility of correct reception such as beamforming (through adjusting the location of ears in the right direction) and image recognition (through looking at the mouth of the intended speaker). When the number of pairs is large within the limited size of the classroom, you can imagine that it will be chaos. What are the methods to make the multi-usage of the limited resources (bandwidth) more organized and successful?

For the first method, the teacher can divide the time between all pairs. Hence, during the first time slot the first pair talks freely without interference, and then they become silent during the next time slot, where the channel is given to the next pair and so on. In telecommunications this is called time division multiple access (TDMA). Can you suggest another way? Suppose every pair speaks a different language (e.g., Arabic, English, Finnish, French, etc.), which cannot be understood by other pairs. Therefore all pairs can talk simultaneously at just the proper voice level to minimize the interference to others. We can filter out the non-understandable languages. This is also used in telecommunications and known as code division multiple access (CDMA), where every terminal uses a special code which is almost orthogonal to other codes.

Another method is if we can reorganize the pair locations in the classroom so that every transmitter/receiver pair becomes very close to each other and each speaks with a very low voice, then all pairs can communicate simultaneously. In telecommunications, we call this spatial division multiple access (SDMA), where we reuse the same bandwidth keeping a proper distance between pairs.

And another way? Yes, our normal way of talking. Simply, the speaker who wants to say something to his partner, will listen to the channel. If no one is talking, then she can start talking. As soon as she starts talking, no one else will talk or interrupt until she completes the message (a polite environment!). Then it becomes free for others to talk in the same manner. If, at that moment two persons start talking simultaneously, then both will stop and be quiet for a while. One of them may start after some random time. This method is unorganized multiple access and it is widely used in ad hoc networks. In telecommunications it is known as carrier sense multiple access with collision avoidance (CSMA/CA).

The problem of CSMA/CA is that in general no target QoS can be guaranteed, however, there are more advanced forms which may guarantee a minimum QoS (based on user demand and channel situation) such as the multiple access methods proposed for WiMAX.

Are there other ways? From our side it is enough to show these examples of analogies between multiple access techniques in telecommunications and human communication. There are several other methods for multiple access methods. The frequency division multiple access (FDMA) may be the first technical method applied in telecommunications. In FDMA the available bandwidth is divided between users simultaneously. However, this method is not widely used in modern communication systems because of its lack of efficiency where we need to keep a guard band between different signals to minimize the interference. Its efficient version, OFDM, is much more popular and can be found in many recent standards such as WiFi, WiMAX, and cognitive radios. Furthermore for the spatial division multiple access, there are efficient methods using beamforming principles which can greatly enhance the system capacity as will be described in Chapter 9.

2

Feedback Control Basics

2.1 Introduction

Feedback control is a fundamental concept in nature and is applied in practically all technological fields. The simplest example of a feedback loop is a human being driving a car. The basic task is to stay on the road. If the road is straight, the steering wheel position could be fixed and the car would follow the road. This is called *open-loop control*. Neither visual sensing nor any control is needed. Since the road usually has turns, open-loop control in car driving is hardly possible. The driver serves as a way to close the loop, to form a *closed-loop control system*.

Let us analyze the car driving situation in more detail. From a systems point of view the car is the system to be controlled. The driver performs the duties of sensing, signal processing, control, and actuation. Figure 2.1 displays a block diagram of the overall feedback control system.

The driver uses his eyes to observe where the car is heading and compares this with the road direction. If the car does not follow the road, the difference between the desired and observed direction, the *error signal*, is not zero. This information is fed to the *controller*, which forms a control decision. In the case of a human being, brains act as a controller. The controller generates a *control command*, which is transmitted to the arm muscles turning the steering wheel so that the car follows the desired direction. This means that the error signal at the next time instant should be smaller than it was before. Muscles together with the steering mechanism perform the duties of an *actuator* in feedback control.

A typical application of feedback in a cellular network is fast power control. Once a base station admits a mobile user to the network, a sufficiently high SINR has to be achieved in order to keep the link valid. On the other hand, one overpowered user could block all the other users in the cell. The conventional strategy is to equalize the received power per bit (E_b) of all users. In addition, the radio channel is subject to fast fading. Taking all this into account, fast and accurate closed-loop power control is a necessity.

Fast *uplink* (from mobile to base station) power control is performed so that the base station forms an estimate of the received SINR and compares it with the desired SINR. If the SINR level is smaller than desired, the error is negative. In this case, the base station asks the mobile user to boost up its power by a fixed step. If the SINR level is larger than the desired SINR, the

Systems Engineering in Wireless Communications   Heikki Koivo and Mohammed Elmusrati
© 2009 John Wiley & Sons, Ltd

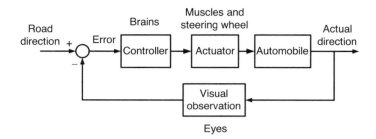

Figure 2.1 Block diagram of person driving a car. The person observes the road direction, the brain performs the error estimation and control decision, while the muscles together with the steering mechanism provide the actuation

error is positive. Now the command from the base station to the mobile user is to decrease power. This also saves the handset battery power. The control command is performed by one bit per time slot. In control systems literature, this type of simple controller is called a *relay* or *on-off controller*.

Another important application of feedback control is the phase-locked loop (PLL). The feedback loop in PLL contains a voltage-controlled oscillator, the output of which is designed to synchronize with the phase of the input signal. PLLs are used in billions of communication applications such as cellular phones, radio transmitters and receivers, wire and optical digital communications, and in a number of other applications. PLLs will be discussed in more detail in Section 2.13.

Nomenclature in different fields varies. Instead of feedback control, *intelligence* is sometimes used, e.g., in intelligent software agents. From a systems point of view such a concept is another example of a feedback system. The level of 'intelligence' in a feedback system is related to how the measured information is processed and how sophisticated the control algorithms are. *Adaptive* and *learning* systems in control systems literature are well-defined and their theory profoundly developed.

The cases given above are examples of the so-called *single-input single output* (SISO) or *scalar control* systems. There are also many examples of *multiple-input multiple-output* (MIMO) or *multivariable* systems in wireless communication, such as *MIMO antennas* and *adaptive beamforming*.

This chapter reviews the basics of feedback control. The principle of feedback control is discussed, followed by the coverage of the simplest and most often used controllers.

Terminology

- **Feedback system**: a sensor measures the output of the system. The measurement is fed back to a device (or human), which determines if the desired goal has been achieved. If not, the difference (error) is used to form a control decision. The control signal guides the actions of the actuator, which makes the necessary changes in the system. Feedback systems can be static or dynamic.
- **Relay control**: relay control output has only two values: 0 or 1 in the asymmetric case; and −1 or +1 in the symmetric case. Can be implemented, e.g., by one bit or on-off relay.

- **Proportional-Integral-Derivative (PID) control**: the most common controller used practically everywhere where continuous control is needed. The proportional term reacts immediately to the error, the integral term remembers the error history and corrects slowly, and the derivative term predicts the future and makes fast corrections.
- **Adaptive control**: PID control parameters are fixed. The adaptive controller changes control parameter values based on the new measurements.
- **Single-input single-output (SISO) or scalar system**: the system has only one input and one output.
- **Multiple-input multiple-output (MIMO) or multivariable system**: the system has many inputs and many outputs. If the number of inputs and outputs is the same, the system is called a 'square' system. In a MISO system, there are multiple inputs and one output.
- **Phase-locked loop**: a feedback system which synchronizes the phase of a controlled oscillator to the phase of the reference signal. Used extensively in various communication systems.
- **State space model**: the state of the system $\mathbf{x}(t_0)$ at time t_0 is defined as the minimum amount of information needed for computing the new unique state $\mathbf{x}(t)$, $t \geq t_0$, when the input $\mathbf{u}(t)$, $t \geq t_0$ is known. The state space model consists of a first-order differential equation and an algebraic output equation both in vector form. It can also be discrete.
- **Kalman filter**: the Kalman filter is a recursive, model-based, computational algorithm that uses measurement data to calculate the minimum variance estimate of the state of the system in a least squares sense.
- **Fuzzy logic**: in classical set theory an element either belongs to a set or does not belong. The related logic is called binary or 'crisp' logic. In fuzzy set theory an element may belong only partially to a set. Fuzzy logic is a many-valued logic, which is described by a membership function assuming values between 0 and 1. In binary logic the membership function values are either 0 or 1.
- **Fuzzy control system**: a fuzzy control system is a feedback control system, in which the controller is formed using fuzzy logic and a rule base. One way this can be done is by interviewing a human operator and form the rule base based on his actions.

2.2 Feedback Control

According to *Webster's* dictionary a *system* is 'a set of arrangement of things so related or connected as to form a unity or organic whole'. The basic task of a feedback control system is to make the output follow the input signal as closely as possible in spite of noise and variations in the system dynamics. The system is also called a *plant* or *process.* The system, consisting of the controller and the plant to be controlled, is called a control system. The input signal has many names such as *reference signal, desired output signal, desired trajectory, target signal* and *set point.* The reference signal may be time varying such as in satellite tracking, or constant such as a set point in temperature control. In block diagram form the feedback system is displayed in Figure 2.2.

The basic components in a feedback control system are:

1. system to be controlled;
2. sensors in feedback path;

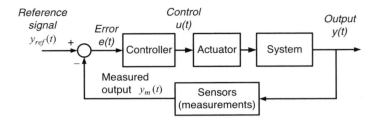

Figure 2.2   A block diagram of a feedback control system

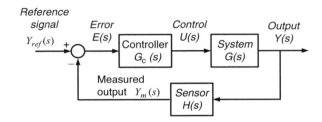

Figure 2.3   Block diagram presentation of a linear, time-invariant control system. s is the Laplace variable and is a complex number

3. controller;
4. actuator.

The following notation is used in Figure 2.2: $y_{ref}(t)$ is the *reference signal*; $u(t)$ is the *control signal*; $y(t)$ is the *output signal*; $y_m(t)$ is the *measured output signal*; and $e(t) = y_{ref}(t) - y_m(t)$ is the *error signal*. In general, these are vectors. Here t is *continuous time*, but an analogous theory can be presented for discrete-time systems. Continuous-time systems are mostly controlled with computers or microchips, in which case the controller is discrete time. Such systems are called *sampled-data* or *digital control* systems.

Most of the theory for feedback systems has been developed for linear, time-invariant systems, where linear algebra and Laplace transformation play a major role. Other basic machinery is similar to that of signal processing. In the linear case the system in Figure 2.2 is presented as a block diagram in Figure 2.3, where transfer functions are used. The variables are now in a Laplace domain.

In this chapter typical continuous-time controllers are first discussed and followed by digital controllers. Their characteristics are explained and behavior illustrated with simulated examples. Phase-locked loop is explained in detail. The chapter concludes with a state space representation of systems and a brief review of the Kalman filter theory.

2.3 Relay Control (ON–OFF Control)

The name of the relay controller comes from old mechanical relays and on–off switches. With digital technology, a relay controller is easily implemented in software, since only one bit is needed. Recall what was said about transmission power control. When the input is a continuous

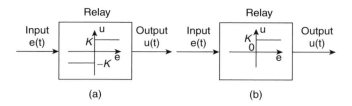

Figure 2.4 Relay characteristics. (a) Symmetric and (b) asymmetric case

signal, the output of a relay is either 0 or 1, or in a symmetric form, -1 or $+1$. Most of the time a gain K is associated with the relay, so instead of the output being -1 or $+1$, it is $-K$ or $+K$, where K is a positive constant. Figure 2.4 shows the characteristics of a relay controller: (a) in a symmetric and (b) in an asymmetric case. If the input to the relay is $e(t) = \sin t$ and $K = 1$, the output is a symmetric square wave with amplitude one. This is displayed in Figure 2.5.

Mathematically, the output u of a symmetric relay function is written as

$$u = \begin{cases} K, & \text{if input } e \geq 0 \\ -K, & \text{if input } e < 0 \end{cases} \tag{2.1}$$

Note that time t is not explicitly shown in Equation (2.1). This does not mean that u and e would be time-invariant as seen in the simulation of Figure 2.5, but rather that the relay characteristics are constant. Relays are often used as controllers, e.g. in temperature control.

Example 2.1 Temperature control in a room – relay control

Let the current room temperature be $15\,^{\circ}\text{C}$. It should be raised to $23\,^{\circ}\text{C}$. Adjusting the air-conditioner set point to $23\,^{\circ}\text{C}$ means that a step-function command is given to increase the

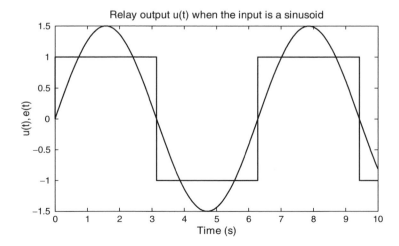

Figure 2.5 Output $u(t)$ is a square wave, when input $e(t) = 1.5 \sin t$ and relay gain $K = 1$

temperature by 8 °C. Assume that a wireless temperature sensor is used to measure the room temperature. The wireless transmission causes a constant time delay of $\tau = 0.2$ s. A constant time delay assumption is made to simplify the analytical treatment. Simulate the temperature response.

Solution

A block diagram of the temperature control system is shown in Figure 2.6. The current room temperature $T(t) = T_{previous}(t) + T_1(t)$ is the sum of the previous room temperature $T_{previous} = 15 \,°C$ and temperature change $T_1(t)$ caused by the step command at the input T_{ref}. The size of the step is $T_{ref} = (23 - 15)\,°C = 8\,°C$, the change required in the room temperature to reach the desired temperature $T_{desired} = 23\,°C$. The transfer function $G(s)$ between T_1 and the input u of the heating/cooling actuator together with the temperature dynamics of the room is usually by a first-order transfer function. Let the transfer function $G(s)$ be

$$G(s) = \frac{T_1(s)}{U(s)} = \frac{1}{s+1} \tag{2.2}$$

The transmission delay $\tau = 0.2$ s and relay gain $K = 8.5$. Assume an asymmetric relay, so that for positive input values the output of the relay is 8.5 and for negative zero (cf. Figure 2.4(b)). The corresponding Simulink® diagram is shown in Figure 2.7. Simulation results are shown in Figure 2.8.

The output response $T(t)$ achieves the desired 23 °C at 3.1 s, but then immediately drops to roughly 21.5 °C at time 3.2 s. At that point the relay controller kicks almost instantly into action and boosts the output back to the desired level and then again the temperature falls off to 21.5 °C. This is a typical, oscillatory behavior of a relay control system, especially if a delay appears in the loop.

Intuitively the oscillation makes sense, since the information needed for control decision is delayed, but it is also the quick step-like control decision which easily makes the feedback system oscillatory. One way to fix such a problem is to use a relay control with a dead zone, that is, if the error is small in amplitude nothing happens. For larger errors than the threshold, the control signal would be $-K$ or $+K$.

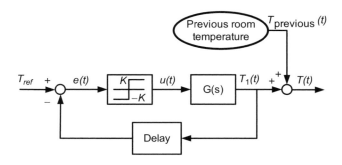

Figure 2.6   Temperature control system with a wireless temperature sensor

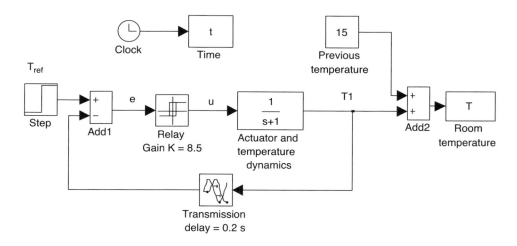

Figure 2.7 Simulink® configuration of the temperature control system with a relay controller

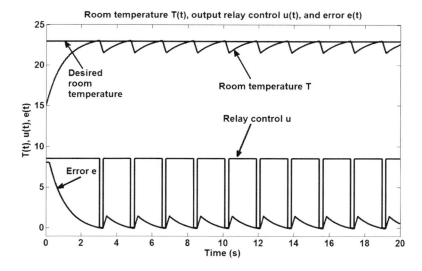

Figure 2.8 The output temperature $T(t)$, control signal $u(t)$, and the desired temperature $T_{desired} = 23\,^\circ\text{C}$

In transmission power control there have been suggestions that in 3G systems two bits should be used instead of one. This would allow the dead zone type of implementation of the power control algorithm.

2.4 Proportional-Integral-Derivative (PID) Control

The most common feedback controller in use is a PID controller. It can be seen in one form or another almost everywhere where feedback control is applied, including communication systems, cellular phones and personal computers. There are a few billion PID controllers in different embedded systems, industrial plants, cars, trains, and airplanes. The list of areas is quite awesome. The basic PID controller has three different control actions: proportional to the error (*P part*); proportional to the integral of the error (*I part*); and proportional to the derivative of the error (*D part*). These are used separately or in combination. In process control P and PI controllers are dominant. In motion control, as in hard disk drives or telecom power supplies, PD and PID controllers are in the majority.

At first glance the PID controller seems rather complicated, but when explained intuitively, the very essence of it becomes transparent. The P part is proportional to the current error. Therefore it reacts immediately to the sensed error without waiting. The *P controller* works on information of *present time*. The integral controller (*I controller*) integrates the *old history* of the error. It remembers, like an elephant, the old sins and corrects them eventually, but is much slower than the P controller. The derivative controller (*D controller*) tries to *predict the immediate future* and makes corrections based on the estimate of the error. It is faster than the P or I controller. It is not applied by itself, but is usually used in combination as a PD or PID controller. The reason becomes apparent if we consider taking the derivative of a noisy signal. In motion control systems, PD or PID controllers are absolutely necessary. Hard disk drives, satellite tracking or elevator control are typical applications of motion control systems.

The characteristics of a PID controller resemble a human being in the control loop. If an error is detected, a human tries to make an immediate correction. History from past experience is also remembered and corrective action taken. The ability to foresee the system behavior is necessary for the pilot both in flying an airplane or driving a car.

The conclusion is that a PID controller includes all the necessary ingredients needed in a feedback controller. It is possible to add more predictive capability or deductions from historical data, but the basic features of a PID controller appear in even complicated controllers – it uses the past, the present, and the future. This is very difficult to beat. All feedback controllers now and in the future must have such features. Intelligence, adaptive and learning features can be added, and this has been done, but the basic structure is very sound and beautiful. It has been applied for centuries in different forms. The speed control or governor in Watt's steam engine is one of the most famous controllers.

2.5 Proportional (P) Control

When an error is detected, the P controller takes immediate corrective action. The control signal $u(t)$ is directly proportional to the error $e(t)$

$$u(t) = K_P e(t) = K_P(y_{ref}(t) - y_m(t)) \tag{2.3}$$

where K_P is a positive constant, called the proportional gain. The transfer function of a P controller becomes a constant gain $G_P(s) = K_P$.

The basic behavior of a P controller can be understood when the relay control in Example 2.1 is replaced by a P controller.

Example 2.2 Temperature control in a room – P control

The configuration of the feedback system is the same as in Example 2.1 except that the relay controller is replaced with a P controller as shown in Figure 2.9. Compute the step response

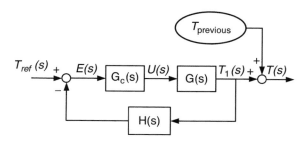

Figure 2.9   The block diagram of the temperature control system with a P controller, $G_c(s) = K_P$

when K_P varies from 0.7 to 5. Calculate also the steady-state value when $K_P = 0.7$. The input step and the delay are as in Example 2.1, $T_{ref} = 8\,°C$ and $\tau = 0.2$ s. The transfer function $G(s)$ is given in Equation (2.2), control transfer function is $G_C(s) = K_P$, and the delay transfer function $H(s) = e^{-0.2s}$.

Solution

The corresponding Simulink® configuration is shown in Figure 2.10. The simulation results are displayed in Figure 2.11, which also indicates the behavior of the response when K_P is increased.

Typically the P controller starts making corrective actions quickly, but cannot reach the desired temperature $T_{des} = T_{previous} + T_{ref} = (15 + 8)\,°C = 23\,°C$ in steady state as seen from the simulation results. For example, when $K_P = 0.7$, the steady-state value $T(\infty) = 18.3\,°C$ and the steady-state error in temperature $T_{des} - T(\infty) = 23\,°C - 18.3\,°C = 4.7\,°C$. This is confirmed by using the final-value theorem in an analytical calculation.

The steady-state error e_{ss} can be computed using the final-value theorem from Laplace transform theory. Since the input T_{ref} is a step function, its Laplace transform is $T_{ref}(s) = 8/s$.

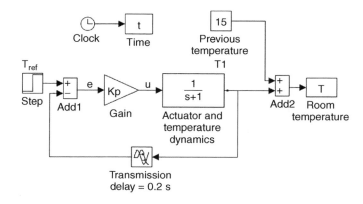

Figure 2.10   Simulink® configuration of the temperature control system of Example 2.1 with a P controller

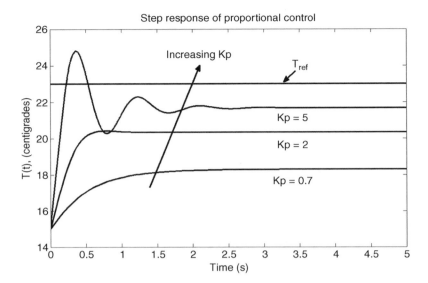

Figure 2.11 Output responses of temperature for increasing values of K_P. Note that the steady-state error remains finite

From Figure 2.9 after some algebraic manipulation

$$e_{ss} = \lim_{t \to \infty} e(t) = e(\infty) = \lim_{s \to 0} s \frac{1}{1 + G_c(s)G(s)H(s)} T_{ref}(s) = \lim_{s \to 0} s \frac{1}{1 + \frac{K_P e^{-0.2s}}{(s+1)}} \frac{8}{s} = \frac{8}{1 + K_P}$$

(2.4)

When $K_P = 0.7$, the steady state error $e_{ss} = 8/(1 + 0.7) = 4.7\,°C$ as observed from the simulation. Correspondingly, when $K_P = 5$, then $e_{ss} = 1.3\,°C$ and $T(\infty) = T_{des} - e_{ss} = 21.7\,°C$.

The steady-state error in the example can be made smaller by increasing the gain as seen in Figure 2.11, but the system becomes more and more oscillatory, even unstable, when the gain is increased. Instability is apparent, if $K_P > 8.5$. In this case the response grows without bound. In applications where the steady-state error is not critical and the delays are small, a P controller can be used. It provides a smaller control signal for small errors than the relay controller.

For feedback control systems stability becomes one of the prime requirements in the design. For linear, time-invariant (LTI) systems the interest lies in asymptotically stable systems. There are a number of definitions of stability, but for LTI systems they can be shown to be equivalent.

Definition 2.1 *Bounded-input bounded-output (bibo) stability*: a system is bounded-input bounded-output stable, if, for every bounded input, the output remains bounded for all time. An LTI system is BIBO stable, if the roots of the system's characteristic equation lie in the left half of the *s*-plane.

Since the P controller cannot produce a steady-state error equal to zero, a different controller is needed in the form of an integral controller.

2.6 Integral (I) Controller

The value of an integral controller is directly proportional to the history of the error. Since it uses the error history, it will respond more slowly to the current error than the P controller. In mathematical terms I control is expressed as

$$u(t) = K_I \int_0^t e(\alpha)d\alpha = K_I \int_0^t (y_{ref}(\alpha) - y_m(\alpha))d\alpha \tag{2.5}$$

The transfer function of an I controller becomes

$$G_I(s) = K_I/s \tag{2.6}$$

The behavior of an I controller can be observed by applying it to the temperature control problem as shown in Figure 2.12. The difference in Simulink® configuration compared with Figure 2.10 is that an integrator has been added with the integral gain K_I. This forms an integral controller.

Simulation results are presented in Figure 2.13. The effect of increasing the gain K_I is also demonstrated.

The integral controller removes the steady-state error for all values of K_I. Analytically this can be seen by computing as in Equation (2.4)

$$e_{ss} = \lim_{s \to 0} s \frac{1}{1 + G_c(s)G(s)H(s)} T_{ref}(s) = \lim_{s \to 0} s \frac{1}{1 + \frac{K_I e^{-0.2s}}{s(s+1)}} \frac{8}{s} = 0 \tag{2.7}$$

The closed-loop response of an integral controller is much slower than in the case of the P controller, where the steady state is reached in less than two seconds. Now it takes at least 10 seconds. The advantage here is that the steady-state error is forced to zero. This statement is not true for all kinds of inputs, but holds for step inputs.

Typically the response of an integral controller is slow and sluggish. If the integral gain is increased, the response becomes more and more oscillatory. Since P controller response is fast, but results in a non-zero steady-state error, and I controller response slow, but with zero

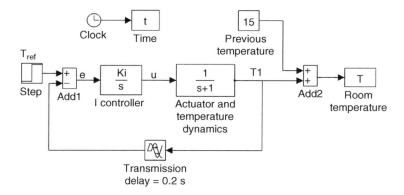

Figure 2.12 Simulink® configuration of the temperature control system with an I controller

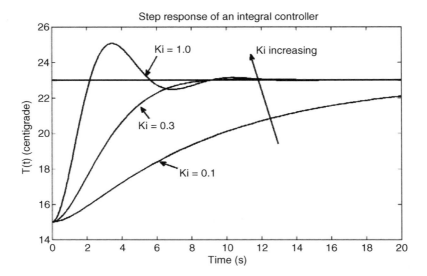

Figure 2.13   Output responses of temperature for increasing values of K_I. When K_I is increased, the response becomes more oscillatory. Note that the steady-state error is eventually removed

steady-state error, it makes sense to combine them to form a PI controller. In a PI controller the P part dominates in the beginning and the I part at the end.

2.7 Proportional-Integral (PI) Controller

The PI controller combines the strengths of the P and I controllers. It removes the basic problem of the plain P controller, since in the long run the I controller removes the steady-state error. The P part on the other hand fixes the problem of the slow, sluggish I controller, because it acts quickly.

Mathematically the PI controller is expressed as

$$u(t) = K_P e(t) + K_I \int_0^t e(\alpha)d\alpha \tag{2.8}$$

or

$$u(t) = K_P \left(e(t) + \frac{1}{T_I} \int_0^t e(\alpha)d\alpha \right) \tag{2.9}$$

where the integral time $T_I = K_P/K_I$. The corresponding transfer functions are

$$G_{PI}(s) = K_P + \frac{K_I}{s} = K_P \left(1 + \frac{1}{T_I s} \right) \tag{2.10}$$

This can be realized in two ways shown in Figure 2.14.

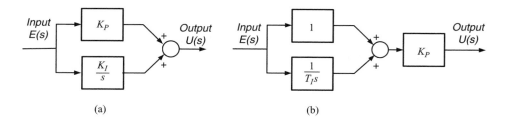

(a)   (b)

Figure 2.14   Block diagrams of the two transfer function forms of a PI controller

Applying a PI controller (with parameter values $K_P = 1$, $K_I = 1$) to the temperature control problem treated in Example 2.2 gives the closed-loop response shown in Figure 2.15.

The response behaves as expected for a PI controller – it is rapid and the steady-state error is zero. This is verified analytically as before

$$e_{ss} = \lim_{s \to 0} s \frac{1}{1 + G_c(s)G(s)H(s)} T_{ref}(s) = \lim_{s \to 0} s \frac{1}{1 + \frac{(K_P + \frac{K_I}{s})e^{-0.2s}}{(s+1)}} \frac{8}{s} = 0 \qquad (2.11)$$

The PI controller is commonly used in process control, where the response does not need to be as fast as for example in motor control. When a radar is tracking a satellite, it needs to predict the satellite's trajectory. Therefore, the controller needs derivative (D) action, an I controller only slows down the response.

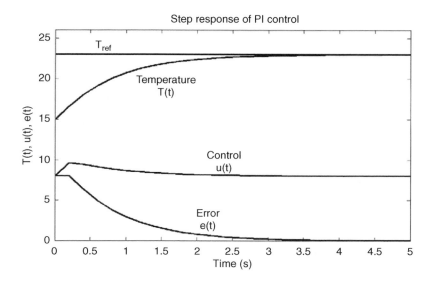

Figure 2.15   Output response of room temperature for a PI controller. The response is quite fast and the steady-state error is removed

2.8 Proportional Derivative (PD) Controller

The derivative controller is hardly ever used alone. Instead, it is combined with a P controller (or with both P and I controllers) and applied in situations where a very fast response is required. The PD controller is written as

$$u(t) = K_P e(t) + K_D \frac{de}{dt} \tag{2.12}$$

or

$$u(t) = K_P \left(e(t) + T_D \frac{de}{dt} \right) \tag{2.13}$$

where the derivative time $T_D = K_D/K_P$. The corresponding transfer functions are expressed as

$$G_{PD}(s) = K_P + K_D s = K_P(1 + T_D s) \tag{2.14}$$

Block diagrams of the two PD controller transfer functions in Equation (2.14) are shown in Figure 2.16.

When PD controllers are properly used, they increase the stability of a system. A well-known example of this is balancing a broom handle, which is set on your palm. This is also called the inverted pendulum problem. With a little bit of practice, the hand–eye coordination produces the desired result. The eyes will follow the angle of deflection, but this is not enough, also the angular speed (and direction) of the broom's top end has to be used. This is exactly what a PD controller does.

In motion control problems, also called servo problems, the system itself contains integral action, so only a PD controller is applied. On the other hand, further specifications often require that the whole PID controller is used.

Example 2.3   Satellite tracking

The satellite tracking radar consists of a motor drive system, a load (aerial dish) and the controller. The drive system includes a motor and its electronics. It produces the torque, which drives the load. The objective of the system is to track the satellite position moving in the sky. A simplified block diagram of the open-loop drive system with a load is shown in Figure 2.17.

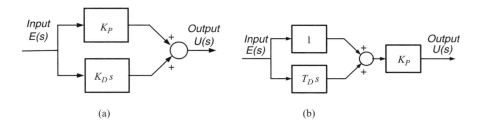

(a)   (b)

Figure 2.16   Block diagrams of the two transfer function forms of a PD controller

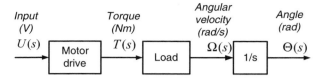

Figure 2.17 Block diagram of a typical motor drive–load system

The motor drive is modeled with a transfer function

$$G_m(s) = \frac{T(s)}{U(s)} = \frac{1}{s+1} \tag{2.15}$$

and the load with a transfer function

$$G_{load}(s) = \frac{\Omega(s)}{T(s)} = \frac{1}{s+1} \tag{2.16}$$

The output of the load is angular velocity $\omega(t)$. When it is integrated, we have the position angle $\theta(t)$.

A simplified model of a satellite tracking system is given in Figure 2.18. The closed-loop control system has a PD controller with gains $K_P = K_D = 3$.

The closed-loop step response is shown in Figure 2.19. Observe that the response is for the position, the angle $\theta(t)$, not for the angular velocity $\omega(t)$. $\square$

The step response of Figure 2.19 is a typical response for a second-order system. It has zero steady-state error. As indicated before, if the input is a step function, then the system together with its controller must include an integrator to produce a zero steady-state error. This is the case with a servo system. The other interesting issues are the transient properties of the system. These are characterized in the time domain e.g. with peak overshoot and rise time.

The response is fairly fast and the steady-state error for a unit step input is zero, when the gains are $K_P = K_D = 3$. The gain values are not chosen optimally, but rather with a little bit of experimentation. The issue of determining the controller parameters is called *tuning*. It will be discussed separately in its own section. The reason for the error being zero is that the system

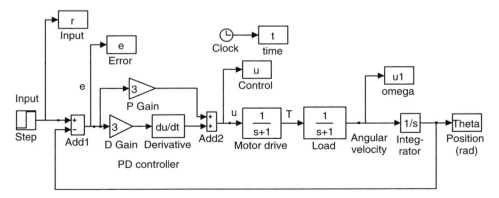

Figure 2.18 Simulink® diagram of a feedback control system for satellite tracking with a PD controller

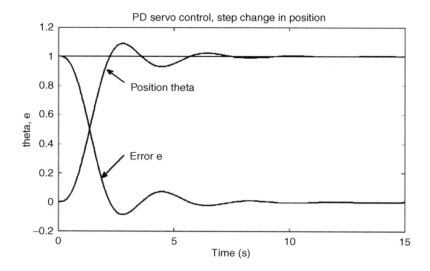

Figure 2.19   Unit step response of position $\theta(t)$ and the error $e(t)$ in the satellite tracking system

itself has an integrator. It is a model of the external input (step), and is therefore able to cope with it. Both have a Laplace transform $1/s$. This is called the internal model principle, which is a very general principle.

If a unit ramp input is applied to the system in Figure 2.18, the steady-state error is finite as seen in Figure 2.20. The Laplace transform of a unit ramp is $1/s^2$, but the system has only one

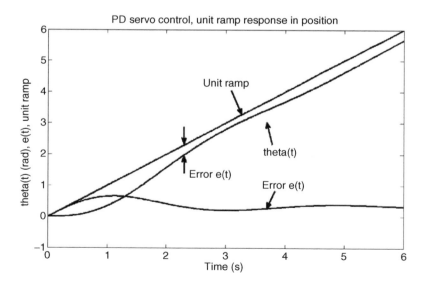

Figure 2.20   Unit ramp response of position $\theta(t)$ in the satellite tracking system results in a non-zero steady-state error unlike in the step response of Figure 2.19

integrator. Therefore it is not able to follow the ramp fast enough to produce zero steady-state error.

Note that in control design performance measures are usually set. Here we have only been interested in the steady-state performance and how 'fast' the response is. There can be much more detailed specifications related to for example overshoot, rise time, attenuation of disturbances at certain frequency range, etc.

It is evident that something can be gained by adding integral action to a PD controller. As mentioned before, the integral controller acts at low frequencies. Therefore, when a PD controller is applied, addition of the integral controller is especially useful in attenuating disturbances at lower frequencies.

2.9 Proportional-Integral-Derivative (PID) Controller

Until now we have discussed P, PI, and PD controllers. A PID controller contains all the best features of these. The D part is mostly left off in process control, say in temperature or flow control. Speed of the response is then not a big issue. When delays exist in the system, the prediction provided by the D part can be precarious because the measurements tend to be noisy. In motion control, on the other hand, the I controller can slow down the fast response required. Since the motor drive system already contains an integrator, it is not needed in the controller. These are generalized statements, which do not always hold. For example, in motion control it often happens that a PD controller alone cannot satisfy all the requirements. Adding an I controller will help to resolve the problem.

Mathematically the PID controller takes the form

$$u(t) = K_P e(t) + K_I \int_0^t e(\alpha)d\alpha + K_D \frac{de}{dt} \tag{2.17}$$

or

$$u(t) = K_P \left(e(t) + \frac{1}{T_I} \int_0^t e(\alpha)d\alpha + T_D \frac{de}{dt} \right) \tag{2.18}$$

The corresponding transfer functions are

$$G_{PID}(s) = K_P + \frac{K_I}{s} + K_D s \tag{2.19}$$

$$G_{PID}(s) = K_P(1 + \frac{1}{T_I s} + T_D s) \tag{2.20}$$

Block diagrams of PID controller transfer functions given in Equations (2.19) and (2.20) are shown in Figure 2.21.

Consider the satellite tracking system given in Figure 2.22, where the parameters have been changed for convenience. After a preliminary control design the control parameters of the PID controller in Equation (2.20) are chosen as $K_P = 2.5$, $K_D = 0.2$ and $K_I = 1$.

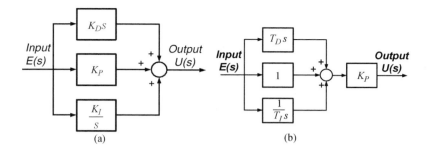

Figure 2.21 Block diagrams of the two transfer function forms of a PID controller

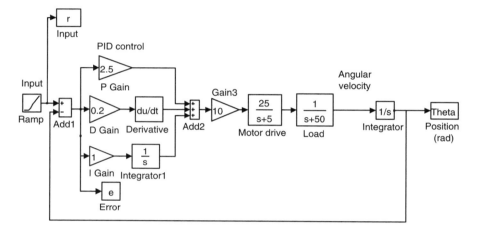

Figure 2.22 Feedback control system for satellite tracking with a PID controller

Using the same $K_P = 2.5$ and $K_D = 0.2$ for the PD controller, unit ramp responses are simulated for comparison. It is quite evident that the addition of an I controller is able to remove the steady-state error in the ramp response (Figures 2.23 and 2.24).

The unit ramp responses, the unit step responses do not differ significantly, especially when more optimal controller parameters are used. In general, a PID controller offers significantly more freedom in the overall control system.

2.10 Practical Issues

There are a number of important practical issues that must be taken into account when the PID controller is used. Some of these are described below.

An ideal derivative controller is not physically realizable, except for a certain frequency range. If analog realization is used, then the derivative part is implemented by a *lead compensator*. A simple transfer function of a lead network becomes

$$G_{LEAD}(s) = \frac{T_D s}{\frac{T_D}{N} s + 1} \tag{2.21}$$

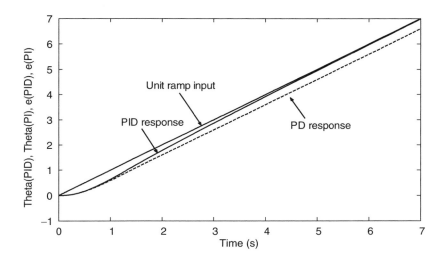

Figure 2.23 Unit ramp responses of the satellite tracking system. The PID controller removes the steady-state error. If a PD controller is used, the error remains finite

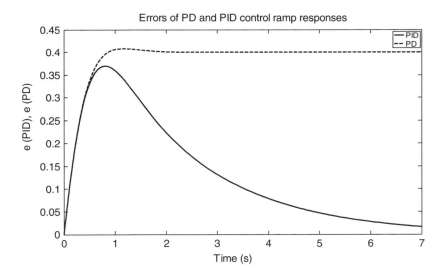

Figure 2.24 Errors in unit ramp responses for both PD and PID controllers. The PID controller eventually removes the steady-state error. If a PD controller is used, the error remains finite

where the constant N in practice assumes values between 5 and 20. A more general lead compensator would have a zero in the transfer function of Equation (2.21) in the left half plane ($T_D s \rightarrow T_D s + 1$) rather than at the origin, but the above presentation works in most cases. Similarly, an integral controller is realized with a *lag compensator*. Putting this together with a lead compensator, we have a *lead-lag compensator*, which is the analog realization of a PID controller.

When a step input is applied, this will generate an immediate, fast change in the error signal (error = input − measured output). A proportional controller will then take instantaneous action and therefore the control command to the actuator experiences a rapid change. For an actuator such a quick change is detrimental, wearing it down more quickly than a smoother control signal. This is called a 'proportional kick'. The derivative controller suffers a similar problem but even more adversely. Here the phenomenon is called a 'derivative kick'. There are a number of ways of avoiding the kick effect. The simplest one is to use the measured output instead of the error signal in computing the control signal. Then the kicks are not seen at all in the control signal.

The integral controller assumes that the error signal $e(t)$ in Equation (2.17) has no limit. In reality this is not the case, because the actuators will saturate if the control command is too large. The controller demands a greater corrective action to be performed by the actuator, which it cannot provide. The actuator is driven to its operation limit and it functions as in open loop. It cannot decrease the error as much as required by the controller, so the error remains large. The integral controller keeps integrating with this wrong premise resulting in a large value, that is, it is 'winding up'. This phenomenon is called *integrator windup*. If it is not taken care of, the control system easily oscillates and the transient due to this takes a long time to disappear. Basically, safeguards have to be developed to detect the time when the actuator saturation occurs at which point the integration should be stopped.

2.11 Tuning of PID Controllers

Since the PID controller (Equation (2.18)) is used overwhelmingly in feedback control, let us have a closer look what this implies in practice. The first observation is that the controller structure is fixed and there are three parameters to be determined: K_P, T_I, and T_D (or alternatively K_P, K_I, and K_D) in such a way that the required specifications are met. The PID controller is a parameterized controller and in many ways it is also standardized.

There are a number of ways that the PID control parameters can be determined. If the PID controller is part of an embedded device, the controller is designed beforehand, and the result is simulated and tested. The controller parameters are then fixed, because the realization is done with analog electronics. Changes even in microcontroller implementations can be hard to do. Fortunately, the PID controller is quite robust, so that changes in the environmental conditions, such as temperature or humidity, do not affect the control system performance adversely.

In other cases, as in process control, the PID control parameters are determined based on experiments performed either in open loop or closed loop. The experiments provide information about the plant to be controlled. Based on this information the PID control parameters are chosen. This procedure is called *tuning* the PID controller. Hundreds of different tuning rules have been developed for different cases, but only a handful of these are used in practice.

The most famous tuning rules are due to Ziegler and Nichols, who in early 1940s developed straightforward rules for PID controller tuning. They are still used in process control, but they tend to give too oscillatory closed-loop responses in other applications. It should be emphasized that tuning of PID controllers does not require an explicit model of the plants, the experiments performed are sufficient for control parameter determination.

In open-loop experiments a test signal, typically a unit step, is introduced at the input. The output response is measured and from it a few parameters are deduced. These are then fed to the chosen tuning rule, which provides the control parameter values. Further experimenting follows to fine tune the parameter values.

In closed-loop tuning experiments I and D controllers are first excluded completely (set the corresponding gains equal to zero) and the P controller gain is set to a small value. Apply a unit step at the input. Start increasing the P controller gain until the system begins to oscillate. At that point observe the oscillating frequency ω_{cr} and the critical gain K_{cr} at which the oscillation begins. Basically, we now have two important pieces of information needed for tuning the PID controller. If only the P controller is used, then the rule of thumb is to set the P controller gain to be $K_P = 0.5\ K_{cr}$. If a PI controller gain is used, then the P controller gain is set to a slightly smaller value $K_P = 0.45\ K_{cr}$ and $T_I = 0.8\ T_u$, where T_u is the ultimate period corresponding to the oscillating frequency. Analogous rule exist for PID controllers. Such rules only give rough values for PID controller parameters. If more precise tuning is needed, then further fine tuning is performed.

The above described tuning is often called one-shot tuning. It is done periodically and then the tuned parameters are kept fixed. In process control, the parameters can be left unchanged for many years. In varying circumstances a need for retuning continuously exists. Then adaptive or self-tuning is one avenue to follow. This means that the control parameters are continuously changed, if there are variations in the environment or the system parameters. In Chapter 6 a self-tuning controller will be discussed, which is able to adapt to changes in environment.

2.12 Digital Implementation of a PID Controller

The majority of PID controller implementations are done digitally with computers, micro-controllers, microchips or DSPs. Systems where the plant is continuous and the controller is digital are called *digital* or *sampled data control systems*. A typical digital control system is depicted in Figure 2.25. Compared with the analog control system of Figure 2.2, it consists of a computer (or a similar digital device), *digital-to-analog(D/A) converter, analog-to-digital (A/D) converter* and a clock. A clock is needed because the sampling is performed *synchronously*. In this brief discussion, asynchronous sampling, which is typical for example in

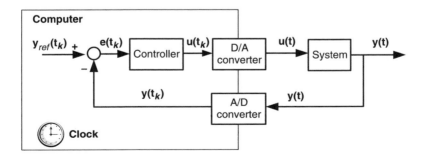

Figure 2.25 Display of a digital control system consisting of a computer, A/D and D/A converters and a clock

wireless sensor and actuator networks, is not discussed. Figure 2.25 indicates that the variables in the computer are handled only at sampled times, $t_k = kh$, where $k = 0,1, \ldots$ and h is the constant sampling time, $h = t_k - t_{k-1}$.

According to Shannon an analog signal can be reconstructed if it is sampled at least with Nyquist rate $2W$, where W is the highest frequency in the original baseband signal. In control such a sampling rate does not lead to sufficiently good results. What is often used is the dominant time constant of the system. Dividing dominant time constant of the system by a number in the range of 4–10 gives good results.

The analog PID controller is of the form

$$u(t) = K_P e(t) + K_I \int_0^t e(\alpha) d\alpha + K_D \frac{de}{dt} \tag{2.22}$$

Let $t = t_k$ in Equation (2.22). Hence

$$u(t_k) = K_p e(t_k) + K_I \int_o^{t_k} e(\tau) d\tau + K_D \frac{de(t_k)}{dt} \tag{2.23}$$

In the following let us use notation $u_k = u(t_k)$. Similar notation is used for all the variables. We will concentrate on approximating the continuous integral and derivative terms. Figure 2.26 shows the simplest way of approximating the integral in the interval $[t_{k-1}, t_k]$, $k = 1,2, \ldots$ The integral is the area under the curve $e(t)$ (if $e(t) \geq 0$ as in Figure 2.26). This area is approximated with a 'small' rectangle, which has an area $e(t_{k-1})h = e_{k-1}h$. If the sampling interval h is small, the error is quite small.

Since the integration starts from zero and runs to t, the integral term is approximated as follows

$$\int_0^{t_k} e(\alpha) d\alpha \approx \sum_{i=1}^{k} e_{i-1} h \tag{2.24}$$

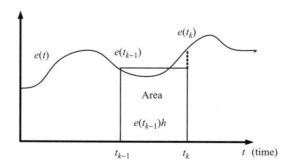

Figure 2.26   The area under curve $e(t)$ is approximated. When $e(t)$ is non-negative, area interpretation for integral is intuitively clear. The discretization procedure also works for negative values of $e(t)$

This is called *backward difference* approximation. *Forward difference* approximation would use rectangle area $e_k h$ resulting in

$$\int_0^{t_k} e(\alpha)d\alpha \approx \sum_{i=1}^{k} e_i h \tag{2.25}$$

Tustin's approximation or trapezoid rule is another way of giving an approximation for the rectangle area as $h(e_k + e_{k-1})/2$ and therefore

$$\int_0^{t_k} e(\alpha)d\alpha \approx h \sum_{i=1}^{k} \frac{(e_i + e_{i-1})}{2} \tag{2.26}$$

Denote the right-hand side of the equation with S_k. Then Equation (2.26) can be written in recursive form as

$$S_k = h \sum_{i=1}^{k} \frac{(e_i + e_{i-1})}{2} = S_{k-1} + \frac{h(e_k + e_{k-1})}{2} \tag{2.27}$$

The simplest backward approximation for the derivative term is

$$\frac{de(t_i)}{dt} \approx \frac{e_i - e_{i-1}}{t_i - t_{i-1}} = \frac{e_i - e_{i-1}}{h} \tag{2.28}$$

Combining the (backward difference) approximations in Equations (2.24) and (2.28) gives the discrete PID controller in the form

$$u_k = K_P e_k + K_I \sum_{i=0}^{k-1} e_i h + K_D \frac{e_k - e_{k-1}}{h} \tag{2.29}$$

Similar approximations can be written for the other approximations. To be mathematically exact, the 'equals' sign in Equation (2.29) should be replaced by $\approx$, but the equals sign is commonly used. The PID control algorithm of Equation (2.29) is called the *positional PID algorithm*.

In certain situations the control signal is driven by the difference of the error (that is, velocity). The corresponding PID algorithm is derived using u_k in Equation (2.29) and by forming $\Delta u = u_k - u_{k-1}$:

$$\Delta u = u_k - u_{k-1}$$

$$= K_P(e_k - e_{k-1}) + K_I \left(\sum_{j=0}^{k-1} e_j h - \sum_{j=0}^{k-2} e_j h \right) + K_D \left(\frac{e_k - e_{k-1}}{h} - \frac{e_{k-1} - e_{k-2}}{h} \right) \tag{2.30}$$

This simplifies to

$$\Delta u_k = K_P \Delta e_k + K_I e_{k-1} h + K_D \left(\frac{\Delta e_k}{h} - \frac{\Delta e_{k-1}}{h} \right) \tag{2.31}$$

or

$$u_k = u_{k-1} + K_P \Delta e_k + K_I e_{k-1} h + K_D \left(\frac{\Delta e_k}{h} - \frac{\Delta e_{k-1}}{h} \right) \tag{2.32}$$

This is called the *velocity form* of the PID algorithm. Again analogous formulae can be derived for the other approximations.

Summarizing the corresponding PI controllers:

Position form of PI

$$u(t_k) = K_p e(t_k) + K_I \sum_{j=0}^{k-1} e(t_j) h \tag{2.33}$$

$$u(t_k) = K_P \left(e(t_k) + \frac{1}{T_I} \sum_{j=0}^{k-1} e(t_j) h \right) \tag{2.34}$$

Velocity form of PI

$$u(t_k) = u(t_{k-1}) + K_p \Delta e(t_k) + K_I e(t_{k-1}) h \tag{2.35}$$

$$u(t_k) = u(t_{k-1}) + K_P \left(\Delta e(t_k) + \frac{1}{T_I} e(t_{k-1}) h \right) \tag{2.36}$$

2.13 Phase-Locked Loop

Phase-locked loop (PLL) is almost certainly the most extensively used demodulator for phase and frequency modulated signals. It is argued that there are uncounted billions of phase-locked loops in service.

The idea of PLL is analogous to tuning a guitar. The reference sound (frequency) can be taken from a piano or a tuning device. If the guitar string does not vibrate with the same frequency as the reference sound, the tension of the string is adjusted so that the beat frequency is impossible to hear. In this situation the guitar string is vibrating at the same frequency as the piano. The guitar is said to be in phase-lock with the piano.

In general, the phase-locked loop synchronizes the phase of a controlled oscillator to the phase of the reference signal. In order to do that a phase comparison must be carried out. The simplest comparator is a multiplier. The principle is illustrated in Figure 2.27.

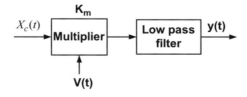

Figure 2.27   Multiplier acts as a phase comparator in the analog case

The incoming signal $X_c(t)$ and the signal from the oscillator $V(t)$ are given as

$$X_c(t) = A_c \sin(2\pi f_c t + \theta_{in}(t)) \tag{2.37}$$

$$V(t) = A_v \cos(2\pi f_c t + \theta(t)) \tag{2.38}$$

Here f_c is the carrier frequency. The signals are first multiplied together. The resulting signal is fed through a low-pass filter, which attenuates the high double-frequency carrier component. The remaining low-frequency output $y(t)$ is thus

$$y(t) = \frac{1}{2} K_m A_c A_v \sin(\theta(t) - \theta_{in}(t)) \tag{2.39}$$

where K_m is the multiplier gain. A more detailed derivation will follow shortly in the context of PLL.

Example 2.4   Multiplier as phase comparator

Let $X_c(t) = \sin(2\pi f_c t + \theta_{in}(t)) = \sin(100t + 1t)$ and $V(t) = A_v \cos(2\pi f_c t + \theta(t)) = \cos(100t + 3t)$ in Figure 2.27 or $\omega_c = 2\pi f_c = 100$ rad/s, $\theta_{in}(t) = 1t$ rad/s and $\theta(t) = 3t$ rad/s. The multiplier gain $K_m = 10$ and the first-order low pass filter $G(s) = \frac{1}{10s+1}$. Simulate the system up to $t = 50$ s and show the results.

Solution

The carrier frequency value f_c is chosen fairly low purposefully, so that the simulation results would be more transparent.

Analytically, the multiplier output is

$$\begin{aligned} y(t) &= K_m X_c(t) V(t) = 10\sin{(100t + 1t)}\cos{(100t + 3t)} \\ &= \frac{1}{2} 10(\sin(200t + 1t + 3t) + \sin(1t - 3t)) = 5(\sin(200t + 4t) + \sin(-2t)) \end{aligned} \tag{2.40}$$

The low pass filter dampens the first high frequency component and we are left with the second, so $y(t) = 5\sin(-2t)$. The Simulink® diagram of the system is shown in Figure 2.28 and the result in Figure 2.29.

The example demonstrates that a multiplier with a low-pass filter works as predicted.   □

The idea in PLL is to complete the feedback loop in Figure 2.27 by using a voltage-controlled oscillator (VCO). The oscillation frequency of a VCO is linearly dependent on the applied voltage. PLL systems can be implemented with analog or digital electronic components. Here the discussion is restricted to the analog system. A block diagram of the PLL is displayed in Figure 2.30.

The system consists of a phase detector, a loop filter, and a VCO. The aim of the PLL control loop is that the output phase $\theta_{out}(t)$ of $V(t)$ follows the phase $\theta_{in}(t)$ of the input $X_c(t)$, except for a fixed difference of 90°. Here the control issue is servo type, since the reference signal $\theta_{in}(t)$ is time-varying.

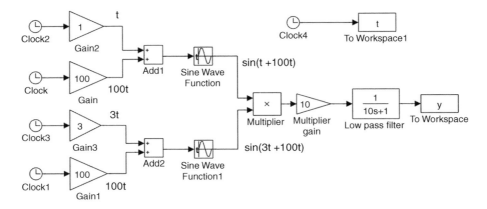

Figure 2.28 Simulink® configuration to demonstrate how a multiplier together with a low pass filter act as a phase detector

As discussed before the phase detector is a multiplier. Since both of the signals to be multiplied contain two frequency components – the high frequency carrier component and the low frequency phase component – the loop filter removes the high frequency component of the error signal leaving only the phase difference. The frequency deviation of the VCO is proportional to its input $u(t)$. Initially, when the control voltage $u(t)$ is zero, the frequency of the VCO is equal to the carrier frequency f_c, and the VCO output $V(t)$ has a 90° phase shift with respect to the input signal $X_c(t)$. Figure 2.30 represents a physical picture of a PLL and is

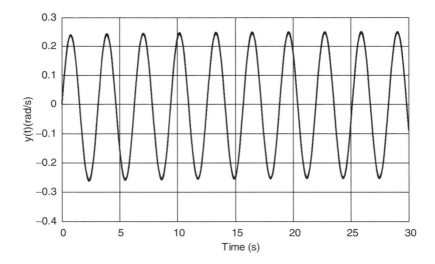

Figure 2.29 The resulting signal $y(t)$ of the phase detector in Figure 2.28. It is apparent that the carrier frequency component has been eliminated and that the phase of $y(t)$ is $\theta(t) - \theta_{in}(t) = (3 - 1)t = 2t$ rad/s ($\omega = 2$) as expected

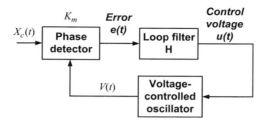

Figure 2.30 Phase-locked loop system

not easily interpreted in control theory terms. This is now mapped to the previously discussed feedback system framework.

The input signal in Figure 2.30 has the form

$$X_c(t) = A_c \sin(2\pi f_c t + \theta_{in}(t)) \tag{2.41}$$

Here A_c is the amplitude of the input signal and

$$\theta_{in}(t) = 2\pi K_1 \int_0^t V_m(\alpha) d\alpha \tag{2.42}$$

where K_1 is the gain (called frequency sensitivity of the frequency modulator). The input phase $\theta_{in}(t)$ is often a result of modulation and $V_m(t)$ the modulating voltage. Therefore, $\theta_{in}(t)$ should be recovered in order to have an estimate of $V_m(t)$.

The output voltage of the VCO can be written as

$$V(t) = A_v \cos(2\pi f_c t + \theta_{out}(t)) \tag{2.43}$$

Here A_v is the amplitude of the VCO output signal and

$$\theta_{out}(t) = 2\pi K_o \int_0^t u(\alpha) d\alpha \tag{2.44}$$

where K_o is the gain (also called frequency sensitivity of the VCO) and $u(t)$ the 'control' voltage. The phase detector, which is a multiplier, produces

$$e(t) = K_m V_c(t) V(t) = K_m A_c \sin(2\pi f_c t + \theta_{in}(t)) A_v \cos(2\pi f_c t + \theta_{out}(t)) \tag{2.45}$$

Here K_m is the gain of the multiplier. Using trigonometric formula Equation (2.45) can be written as

$$e(t) = \frac{1}{2} K_m A_c A_v \{\sin(\theta_{in}(t) - \theta_{out}(t)) + \sin(4\pi f_c t + \theta_{in}(t) + \theta_{out}(t))\} \tag{2.46}$$

When $e(t)$ is fed to the low pass loop filter, the latter, double carrier frequency term is removed and need not be considered further. Denote the phase error by $\theta_{error}(t) = \theta_{in}(t) - \theta_{out}(t)$.

$$\theta_{error}(t) = \theta_{in}(t) - \theta_{out}(t) = \theta_{in}(t) - 2\pi K_o \int_0^t u(\alpha) d\alpha \tag{2.47}$$

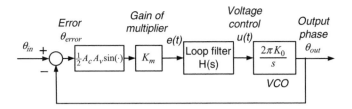

Figure 2.31   Nonlinear model of a phase-locked loop

This implies that the input to the loop filter is

$$e(t) = \frac{1}{2} K_m A_c A_v \sin(\theta_{in}(t) - \theta_{out}(t)) = \frac{1}{2} K_m A_c A_v \sin(\theta_{error}(t)) \qquad (2.48)$$

Figure 2.30 can now be redrawn in a simpler, phase-based form of Figure 2.31, so that we can make the phase lock issue more transparent in the control framework. The aim is to make the output phase $\theta_{out}(t)$ (or frequency) follow the input phase $\theta_{in}(t)$ (or frequency). The system is nonlinear, because of the sine function appearing in Equation (2.48).

Once the phase-locked loop operates properly, it may be assumed that phase error $\theta_{error} = \theta_{out} - \theta_{in} \approx 0$. Since $\sin \alpha \approx \alpha$ for small α, Equation (2.48) implies that

$$e(t) = \frac{1}{2} K_m A_c A_v (\theta_{out} - \theta_{in}) \approx 0 \qquad (2.49)$$

When $\theta_{error} = 0$ becomes zero, the system is said to be in phase lock.

A linear model of the phase-locked loop is displayed in Figure 2.32.

Let gain K_L be defined as

$$K_L = \frac{1}{2} K_0 K_m A_c A_v \qquad (2.50)$$

Note that we lose no generality by including amplitude A_v from the sinusoidal input in K_L. It is done here purely for notational convenience.

The analysis of the linearized system is based on basic control theory in Laplace domain.

Example 2.5   First-order phase-locked loop

Consider the system in Figure 2.32, where a unit step change in the input phase θ_{in} at time zero occurs. Assume further that the loop gain is K_L and that the loop filter is unity, $H(s) = 1$. This is called a first-order phase-locked loop. Compute the steady state error.

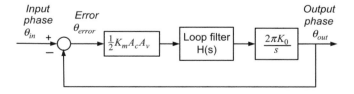

Figure 2.32   Linearized phase-locked loop. The output of VCO tracks the input phase

Solution

Computing the Laplace transform of the error from Figure 2.32 and substituting unit step $\Theta_{in}(s) = 1/s$ and the loop filter $H(s) = 1$ results in

$$\Theta_{error}(s) = \frac{1}{1 + \frac{2\pi K_L H(s)}{s}} \Theta_{in}(s) = \frac{s}{s + 2\pi K_L} \frac{1}{s} \qquad (2.51)$$

The final value theorem gives

$$\lim_{s \to 0} s\Theta_{error}(s) = \theta_{error}(\infty) = \lim_{s \to 0} s \frac{s}{s + 2\pi K_L} \frac{1}{s} = 0 \qquad (2.52)$$

This means that system will be in phase lock at steady state. The unit step response for the closed-loop system becomes

$$\Theta_{out}(s) = \frac{1}{1 + \frac{2\pi K_L}{s}} \Theta_{in}(s) = \frac{s}{s + 2\pi K_L} \frac{1}{s} = \frac{1}{s + 2\pi K_L} \qquad (2.53)$$

or in the time domain $\theta_{out}(t) = e^{-2\pi K_L t}$.   □

The problem with a first-order PLL is that the loop gain parameter K_L is responsible for both the loop bandwidth and hold-in frequency range of the loop. This is seen in the simulation shown in Figure 2.33, where gain $K_L = 10$ and the loop filter is unity, $H(s) = 1$. A sinusoidal input $\theta_{in}(t) = 3 \sin \omega t$, $\omega = 1, 5, 10$ rad/s, representing the phase of the incoming signal is applied (recall that the original incoming signal $X_c(t)$ is given by Equation (2.41) as $X_c(t) = A_c \sin(2\pi f_c t + \theta_{in}(t))$). The other difference compared with Example 2.5 is that the sinusoidal nonlinearity is taken into account.

The result is shown in Figure 2.34. First-order PLL tracks the low frequency ($\omega = 1$ rad/s) sinusoidal input. When $\omega = 5$ rad/s, tracking is poor and when $\omega = 10$ rad/s, the first-order PLL cannot cope with the situation anymore. Therefore first-order PLLs are rarely used in practice due to their several shortcomings.

For hold-in frequencies the loop remains phase locked. To avoid the problems that occur with first-order PLLs with high frequency signals, *second-order PLLs* are used instead. The loop

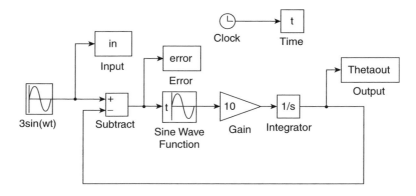

Figure 2.33 First-order PLL with a sinusoidal signal input and sinusoidal nonlinearity in the loop

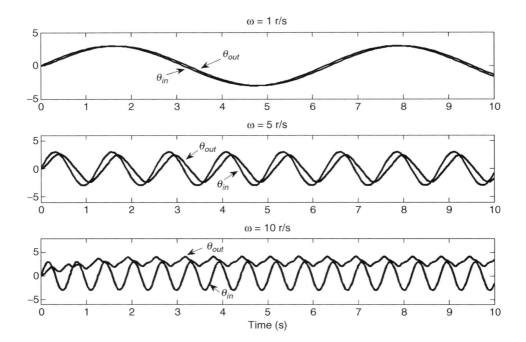

Figure 2.34 First-order PLL tracks the sinusoidal input, when $\omega = 1$ rad/s (above). The tracking result becomes poor when $\omega = 5$ rad/s (middle). When $\omega = 10$ rad/s (below), the tracking ability is completely lost

filter takes the form

$$H(s) = \frac{s + \alpha}{s} = 1 + \frac{\alpha}{s} \tag{2.54}$$

Comparing this with Equation (2.10), we can conclude that the loop filter is a special case of a PI controller. Other forms of loop filters are also frequently used.

Example 2.6   Nonlinear phase-locked loop

Let $\theta_{in}(t)$ in Figure 2.33 be $\theta_{in}(t) = 3 \sin \omega t$, $\omega = 10, 40, 41$, and 50 rad/s, ($\theta_{in}(t) = 3 \sin 10\, t$ created a problem for the first-order PLL). Design a second-order loop filter for the PLL, where the gain is $K_L = 20$. Simulate the responses for different $\theta_{in}(t)$ and compare them with those of Figure 2.34. Study the PLL with other inputs such as a step input and a pure sinusoidal input, when Gaussian noise appears.

Solution

Choose $\alpha = 500$ in Equation (2.54). The Simulink® configuration of the PLL is given in Figure 2.35.

The simulation results are displayed in Figure 2.36.

Comparing the response with that in Figure 2.34, the result shows that after a brief transient the second-order PLL is able to follow the input phase angle of $\omega = 10$ rad/s almost perfectly,

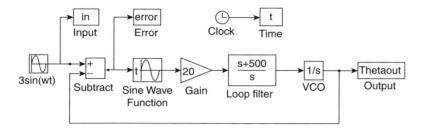

Figure 2.35 Simulink® configuration of a second-order PLL, when the sinusoidal nonlinearity is taken into account

when the first-order PLL could not. Once ω increases to 41 rad/s, the tracking is not successful anymore and fails completely at 50 rad/s. A second-order PLL functions much better than a first-order PLL. This exercise implies that a more sophisticated loop filter and a more careful design of PLL provides a wider hold-in frequency range.

If the input in Figure 2.35 is a unit step change, instead of a sinusoid, then the result is as shown in Figure 2.37.

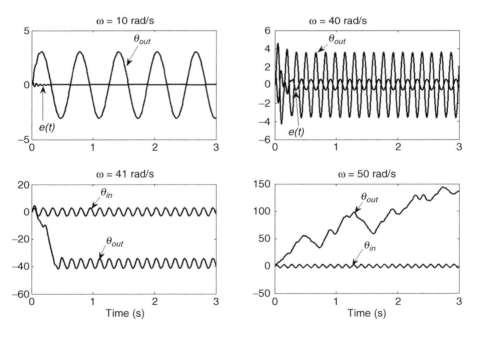

Figure 2.36 Responses of a second-order PLL. Tracking of input sinusoid of 10 rad/s is almost perfect. Tracking is still possible with decreased performance, when $\omega = 40$ rad/s (left and right figures above). When $\omega = 41$ rad/s the tracking fails and is completely lost when $\omega = 50$ rad/s (left and right figures below). Observe that in the figures below, $\theta_{in}(t)$ and $\theta_{out}(t)$ are drawn in the same figure instead of $\theta_{out}(t)$ and error $e(t)$ (figures above)

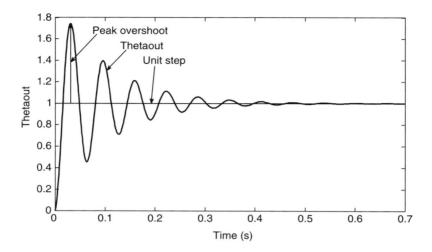

Figure 2.37   Unit step response in the input phase shows that second-order PLL quickly tracks the change

The response is a typical second-order system response. The peak overshoot shown in the Figure is 74% and could be made smaller with a more careful design. On the other hand, a proper design should take into account other factors, e.g. noise.

If the incoming signal $X_c(t)$ contains an additive, bandlimited Gaussian noise, the phase detector can be modeled as shown in Figure 2.38.

An example illustrates the effect of noise. Adding the effect of noise into the system of Figure 2.35, results in the Simulink® configuration of Figure 2.39.

Applying a sinusoid $\theta_{in}(t) = 3 \sin \omega t$, $\omega = 10$. The simulation result is shown in Figure 2.40. The effect of noise is not as significant as might be expected, but of course it depends on the noise level. Here the default block of band-limited white noise block is used. It has noise power of 0.1 and sample time 0.1. When ω is increased, tracking performance gradually decreases. Finally, when $\omega = 31$ rad/s, tracking fails The area of having noise in PLL is a much deeper subject area requiring much more attention than the short treatment given here.   □

From the above examples, it is clear that more sophisticated loop filters, like Butterworth filters, can satisfy even more stringent performance requirements. In any case, simulation provides a powerful tool in studying the characteristics of PLLs.

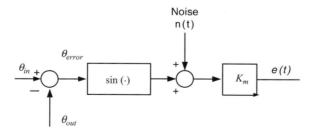

Figure 2.38   Equivalent noise model for a multiplier phase detector

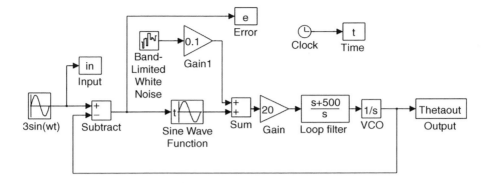

Figure 2.39   Equivalent noise model in a second-order PLL

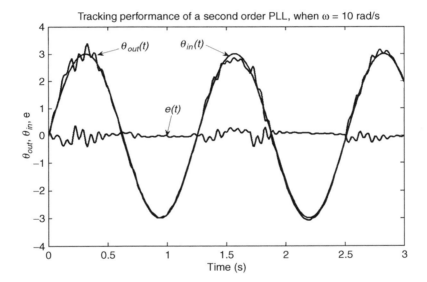

Figure 2.40   Output $\theta_{out}(t)$, input $\theta_{in}(t)$, and error $e(t)$ of a second-order PLL, when $\omega = 10\,\text{rad/s}$ and band-limited noise has been added to the feedback system

2.14 State-Space Representation

In this section state-space representation is discussed. In engineering literature it is common to denote the time derivative of a variable with a dot for convenience: $dx/dt = \dot{x}$.

Example 2.7   Linear state-space representation

Consider a typical second-order differential equation

$$\ddot{x}(t) + 2\dot{x}(t) + 3x(t) = u(t) \tag{2.55}$$

where $x(t)$ is position, $\dot{x}(t)$ is velocity, $\ddot{x}(t)$ is acceleration, and $u(t)$ is external input, force. Denote $x_1 = x$, $x_2 = \dot{x}$. The first observation is that $\dot{x}_1 = x_2$ and $\dot{x}_2 = \ddot{x}$. Solving Equation (2.55) for $\ddot{x}$

$$\ddot{x}(t) = -2\dot{x}(t) - 3x(t) + u(t) \tag{2.56}$$

Substituting and summarizing the above

$$\begin{aligned}\dot{x}_1(t) &= x_2(t) \\ \dot{x}_2(t) &= -3x_1(t) - 2x_2(t) + u(t)\end{aligned} \tag{2.57}$$

In vector-matrix notation this becomes (dropping the argument t)

$$\dot{\mathbf{x}} = \begin{bmatrix} \dot{x}_1 \\ \dot{x}_2 \end{bmatrix} = \begin{bmatrix} x_2 \\ -3x_1 - 2x_2 \end{bmatrix} + \begin{bmatrix} 0 \\ 1 \end{bmatrix} u(t) = \begin{bmatrix} 0 & 1 \\ -3 & -2 \end{bmatrix} \begin{bmatrix} x_1 \\ x_2 \end{bmatrix} + \begin{bmatrix} 0 \\ 1 \end{bmatrix} u = \mathbf{Ax} + \mathbf{bu} \tag{2.58}$$

where $\mathbf{x} = \begin{bmatrix} x_1 \\ x_2 \end{bmatrix}$, is the state vector, $\mathbf{u} = [u]$ the input vector, and $\dot{\mathbf{x}} = \begin{bmatrix} \dot{x}_1 \\ \dot{x}_2 \end{bmatrix}$ the time derivative of the state vector. Further,

$$\mathbf{A} = \begin{bmatrix} 0 & 1 \\ -3 & -2 \end{bmatrix}, \quad \mathbf{b} = \begin{bmatrix} 0 \\ 1 \end{bmatrix} \tag{2.59}$$

Matrix $\mathbf{A}$ is the system matrix and $\mathbf{b}$ the input matrix (in this case a vector).

Usually not the all the states can be measured. Assume that only $x = x_1$ can be measured. The output (or measurement) equation becomes

$$\mathbf{y} = \mathbf{c}^T \mathbf{x} = \begin{bmatrix} 1 & 0 \end{bmatrix} \begin{bmatrix} x_1 \\ x_2 \end{bmatrix} \tag{2.60}$$

Vector $\mathbf{y}$ is called the *output* (or *measurement*) *vector* and matrix (in this case vector) $\mathbf{c}$ the *output* (or *measurement*) *matrix*.

Summarizing the state space form of Equation (2.55)

$$\begin{aligned}\dot{\mathbf{x}} &= \mathbf{Ax} + \mathbf{bu} \\ \mathbf{y} &= \mathbf{c}^T \mathbf{x}\end{aligned} \tag{2.61}$$

Equation (2.61) is a time-invariant linear system in state-space form.   □

In state-space form there are only first-order derivatives on the left-hand side of the equation. There are no derivatives on the right-hand side of the equation. The example above is given for a linear, time-invariant system, but the statement also holds for time-variant linear and even for nonlinear systems.

Definition 2.2   The state of the system $\mathbf{x}(t_0)$ at time t_0 is defined as the minimum amount of information needed for computing the new unique state $\mathbf{x}(t)$, $t \geq t_0$, when the input $\mathbf{u}(t)$, $t \geq t_0$ is known.

Two observations should be made. State-space representation is not unique. When state-space representation is used, the initial state $\mathbf{x}(t_0)$ contains all the necessary information about the system – no system history is needed. State definition is quite general in the sense that it applies in the continuous as well as in the discrete time case.

There are many advantages provided by the state-space form of analyzing differential equations. Equation (2.55) is a simple, scalar second-order equation. For more complicated systems there can be hundreds or thousands of scalar equations. The state-space model simplifies enormously the overall analysis, design, optimization, and simulation of the system.

In general the *system equations* in scalar form can be written as

$$
\begin{aligned}
\dot{x}_1 &= f_1(x_1, \cdots, x_n, u_1, \cdots, u_m, t) \\
&\;\vdots \\
\dot{x}_n &= f_n(x_1, \cdots, x_n, u_1, \cdots, u_m, t)
\end{aligned}
\tag{2.62}
$$

and *output, measurement equations*

$$
\begin{aligned}
y_1 &= h_1(x_1, \cdots, x_n, u_1, \cdots, u_m, t) \\
&\;\vdots \\
y_r &= h_r(x_1, \cdots, x_n, u_1, \cdots, u_m, t)
\end{aligned}
\tag{2.63}
$$

In vector notation the state-space representation becomes
System equation

$$
\dot{\mathbf{x}} = \mathbf{f}(\mathbf{x}(t), \mathbf{u}(t), t), \quad \mathbf{x}(t_0) = \mathbf{x}_0
\tag{2.64}
$$

Output (Measurement) equation

$$
\mathbf{y}(t) = \mathbf{h}(\mathbf{x}(t), \mathbf{u}(t), t)
\tag{2.65}
$$

where the vectors are (dropping the arguments for simplicity)

$$
\dot{\mathbf{x}} = \begin{bmatrix} \dot{x}_1 \\ \vdots \\ \dot{x}_n \end{bmatrix}, \;
\mathbf{x} = \begin{bmatrix} x_1 \\ \vdots \\ x_n \end{bmatrix}, \;
\mathbf{u} = \begin{bmatrix} u_1 \\ \vdots \\ u_m \end{bmatrix}, \;
\mathbf{f} = \begin{bmatrix} f_1 \\ \vdots \\ f_n \end{bmatrix}, \;
\mathbf{y} = \begin{bmatrix} y_1 \\ \vdots \\ y_p \end{bmatrix}, \;
\mathbf{h} = \begin{bmatrix} h_1 \\ \vdots \\ h_p \end{bmatrix}
\tag{2.66}
$$

The initial condition for Equation (2.64) is $\mathbf{x}(t_0) = \mathbf{x}_0$. The vector presentation gives a clever, geometric way of looking at dynamical systems. Writing Equation (2.62) in vector form, the solution can be interpreted as a vector in Euclidean space $\mathbb{R}^n$. For the linear case many powerful results from linear algebra can be brought in to play such as eigenvalue theory and singular value decomposition to mention a few.

A linear time-invariant system becomes

$$
\begin{aligned}
\dot{\mathbf{x}} &= \mathbf{A}\mathbf{x} + \mathbf{B}\mathbf{u} \\
\mathbf{y} &= \mathbf{C}\mathbf{x} + \mathbf{D}\mathbf{u}
\end{aligned}
\tag{2.67}
$$

where the dimensions of the matrices match the vectors in Equation (2.66)

$$\mathbf{A} = \begin{bmatrix} a_{11} & \cdots & a_{1n} \\ \vdots & \ddots & \vdots \\ a_{n1} & \cdots & a_{nn} \end{bmatrix}, \mathbf{B} = \begin{bmatrix} b_{11} & \cdots & b_{1m} \\ \vdots & \ddots & \vdots \\ b_{n1} & \cdots & b_{nm} \end{bmatrix}, \mathbf{C} = \begin{bmatrix} c_{11} & \cdots & c_{1n} \\ \vdots & \ddots & \vdots \\ c_{p1} & \cdots & c_{pn} \end{bmatrix},$$

$$\mathbf{D} = \begin{bmatrix} d_{11} & \cdots & d_{1m} \\ \vdots & \ddots & \vdots \\ d_{p1} & \cdots & d_{pm} \end{bmatrix}$$

Matrix $\mathbf{A}$ is the *system matrix*, matrix $\mathbf{B}$ the *input matrix*, matrix $\mathbf{C}$ the *output* (or the measurement) *matrix* and matrix $\mathbf{D}$ the *feedforward matrix*. The feedforward matrix is often a null matrix.

Here all the matrices are constant. A good example of a time-variant system is a radio channel. The matrices are then time dependent. Such a model will be discussed in Chapter 3.

The solution of Equation (2.67) can be written in closed form as

$$\mathbf{x}(t) = e^{\mathbf{A}(t-t_0)}\mathbf{x}(t_0) + \int_{t_0}^{t} e^{\mathbf{A}(t-s)}\mathbf{B}u(s)ds \tag{2.68}$$

Analytically, the computations quickly become tedious. Computer tools are used for solving state-space equations.

Example 2.8   Simulation of continuous state-space models

Simulate the solution of Equation (2.55) when the input $u(t)$ is a unit step function and the initial conditions equal zero, $x(0) = 0$, $\dot{x}(0) = 0$. The state-space matrices are given in Equations (2.59) and (2.60).

Solution

Simulink® and MATLAB® (a registered trademark of The Math-Works, Inc.) provide efficient means for numerical solution of state-space equations. In the Simulink® block library there is a State-Space block (Figure 2.41).

Clicking the block open and feeding in the matrices in Equations (2.59) and (2.60) in MATLAB® format:

$$\mathbf{A} = [0\ 1;\ -3\ -2];\ \mathbf{B} = [0;\ 1];\ \mathbf{C} = [1\ 0];\ \mathbf{D} = 0$$

The input is a unit step function starting at time $= 1$ s (default value in Simulink®). The output is state x_1. The result is shown in Figure 2.42. From Equation (2.55) the steady-state solution is easily seen to be (since $\ddot{x}(\infty) = 0$; $\dot{x}(\infty) = 0$)

$$3x(\infty) = u(\infty) = 1 \Rightarrow x(\infty) = \frac{1}{3} \tag{2.69}$$

This can be verified in Figure 2.42.

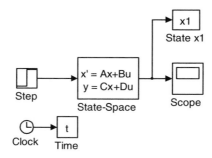

Figure 2.41   Simulink® configuration of a linear, time-invariant system in state-space form

Until now the state-space models have been continuous in time. Assume that the continuous signals are sampled. Denote the constant sampling interval by h and time $t_k = kh, k = 0,1,2, \ldots$. The sample interval is $h = t_k - t_{k-1}$.

Discrete state-space equations assume the form

$$
\begin{aligned}
\mathbf{x}_{k+1} &= \mathbf{E}\mathbf{x}_k + \mathbf{F}\mathbf{u}_k \\
\mathbf{y}_k &= \mathbf{H}\mathbf{x}_k + \mathbf{D}\mathbf{u}_k
\end{aligned}
\tag{2.70}
$$

Here the notation $\mathbf{x}_k = \mathbf{x}(kh) = \mathbf{x}(t_k)$ is used for all variables and also for time-variant matrices. The dimensions of the matrices are appropriate considering the vectors.

Discrete models exist by their own right. If the original model is in continuous form, MATLAB® Control Systems Toolbox provides a convenient way to obtain discrete models of continuous systems.

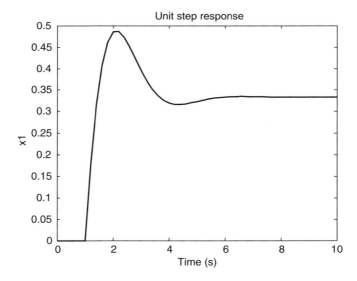

Figure 2.42   Unit step response of system (Equation (2.55))

Example 2.9   Simulation of discrete state-space models

Consider the example system given by Equations (2.58) and (2.60). The state-space system matrices are given as

$$A = [0\ 1;\ -3\ -2];\ B = [0;\ 1];\ C = [1\ 0];\ D = 0;$$

Let the sampling time be *Tsample* = .2 s. Two MATLAB® commands are needed. The first one defines the continuous system in state-space form and the second one transforms the continuous system into the corresponding discrete system, where the sampling time is known. The first command is *ss* and the second one *c2d*.

The MATLAB® commands are

```
A= [0 1; -3 -2]; B= [0;1]; C= [1 0]; D=0;
% Define the state-space system
csys=ss(A,B,C,D);
% Transform the continuous system into a discrete system, Tsample=
  0.2 s
dsys=c2d(csys,0.2)
```

Discretization of the system (Equations (2.58) and (2.60)) with sampling time *Tsample* = 0.2 s leads to a discrete-time state-space model:

$$\mathbf{E} = [0.9478\ \ 0.1616;\ \ -0.4847\ \ 0.6246];\ \mathbf{F} = [0.01741;\ \ 0.1616];\ \mathbf{H} = [1\ \ 0];\ \mathbf{D} = 0$$

or

$$\mathbf{E} = \begin{bmatrix} 0.9478 & 0.1616 \\ -0.4847 & 0.6246 \end{bmatrix};\ \mathbf{F} = \begin{bmatrix} 0.01741 \\ 0.1616 \end{bmatrix} \tag{2.71}$$

$$\mathbf{H} = [1\ \ 0];\ \mathbf{D} = 0$$

The system is simulated in analogous way to the continuous system before. The Simulink® configuration and the simulation result are displayed in Figure 2.43.

2.15 Kalman Filter

The Kalman filter is a recursive, model-based computational algorithm that uses measurement data to calculate the minimum variance estimate of the state of the system in a least squares sense. The filter employs the knowledge of the system and measurement dynamics, and noise statistics. The basic Kalman filter is linear and assumes white, Gaussian noise processes.

Before Kalman filtering was developed in the 1960s, filtering theory was mostly based on the frequency domain and applied to single-input single-output (SISO) systems. Space technology in the 1960s required accurate trajectory estimation in real time using computers. The Kalman filter was an ideal answer, since computations are done recursively in the time domain. State-space formulation is well suited for multi-input multi-output (MIMO) treatment. The Kalman filter in its various forms is used extensively in communications, navigation systems, robotics, computer vision, and data fusion. It is widely argued that it is the most useful result that systems theory has produced.

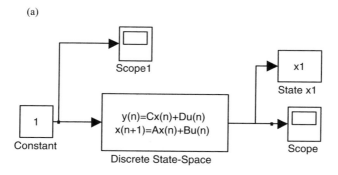

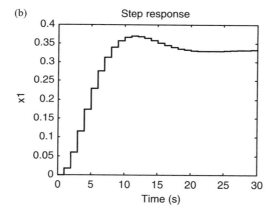

Figure 2.43 (a) Simulink® configuration of the discrete system (Equation (2.71)); (b) simulation result

Although the Kalman filter is called a filter, it is too narrow a name. The Kalman filter is used for estimating the state of the system. This includes filtering, prediction, and smoothing.

Example 2.10 Recursive computation of sample mean

Suppose we make N measurements of a constant a. Our measurement is corrupted with noise v_i, which is assumed to be a white noise sequence. The measurement equation is

$$z_i = a + v_i \tag{2.72}$$

Estimate of the sample mean becomes

$$m_N = \frac{1}{N} \sum_{i=1}^{N} z_i \tag{2.73}$$

If an additional measurement becomes available, the number of measurements is $N + 1$. The new estimate is given by

$$m_{N+1} = \frac{1}{N+1} \sum_{i=1}^{N+1} z_i \tag{2.74}$$

The computation is done over the whole batch instead of using the already computed m_N. It is often more convenient to derive a recursive form for the estimate subtracting (2.73) from (2.74).

$$m_{N+1} - m_N = \frac{1}{N+1} \sum_{i=1}^{N+1} z_i - \frac{1}{N} \sum_{i=1}^{N} z_i \tag{2.75}$$

After some manipulation this can be written in the form

$$m_{N+1} = \frac{N}{N+1} m_N + \frac{1}{N+1} z_{N+1} \tag{2.76}$$

or

$$m_{N+1} = m_N + \frac{1}{N+1} (z_{N+1} - m_N) \tag{2.77}$$

This is a simple state-space equation. In order to compute the new estimate, only the previous estimate m_N and the new measurement are needed. Observe the structure on the right-hand side more carefully. The first term is the state of the system, the old estimate. In the second term $z_{N+1} - m_N$ is called the measurement *residual*. It represents the estimated error. The multiplier in front of it is a weighting term. The second term makes a correction in our previous estimate. The amount of correction is related to our knowledge of the noise statistics. The general rule is that the less we know, the smaller the correction.

In verbal terms our algorithm can be written as

new estimate = old estimate + correction

In the sequel we will see that the same structure is repeated in a more general setting. This helps us to better understand the Kalman filter idea even when the detailed equations themselves seem quite complicated.

2.16 Linear Kalman Filter

Consider a linear, discrete-time system

$$\mathbf{x}(k+1) = \mathbf{E}(k)\mathbf{x}(k) + \mathbf{F}(k)\mathbf{u}(k) + \mathbf{v}(k) \tag{2.78}$$

with the measurement equation

$$\mathbf{z}(k) = \mathbf{H}(k)\mathbf{x}(k) + \mathbf{w}(k) \tag{2.79}$$

where $\mathbf{x}(k)$ is the state vector, $\mathbf{u}(k)$ the input vector, $\mathbf{z}(k)$ the measurement vector at time step k, $\mathbf{E}(k)$ the transition matrix, $\mathbf{F}(k)$ the input matrix, $\mathbf{H}(k)$ the measurement matrix, and $\mathbf{K}(k)$ the feedforward matrix all of appropriate dimensions. Vectors $\mathbf{v}(k)$ and $\mathbf{w}(k)$ are independent sequences of zero mean, white Gaussian noise with covariance matrices $\mathbf{Q}(k)$ and $\mathbf{R}(k)$, respectively:

$$E\{\mathbf{v}(k)\} = 0, \ E\{\mathbf{w}(k)\} = 0 \tag{2.80}$$

$$E\{\mathbf{v}(k)\mathbf{v}^T(k)\} = \mathbf{Q}(k)\delta_{kj}$$

$$E\{\mathbf{w}(k)\mathbf{w}^T(k)\} = \mathbf{R}(k)\delta_{kj} \qquad (2.81)$$

$$E\{\mathbf{w}(k)\mathbf{v}^T(k)\} = 0 \text{ for all } k \text{ and } j$$

where δ_{kj} is the Kronecker delta and E the expected value.

Introduce the notation $\hat{\mathbf{x}}(k+1|k)$ where $\hat{\mathbf{x}}$ is the estimate of $\mathbf{x}$, argument $k+1$ means the estimate at time step $k+1$ when the information up to time step k is available. From Equation (2.78) the predicted value of the estimate $\hat{\mathbf{x}}(k+1|k)$ can be calculated

$$\hat{\mathbf{x}}(k+1|k) = \mathbf{E}(k)\hat{\mathbf{x}}(k|k) + \mathbf{F}(k)\mathbf{u}(k) \qquad (2.82)$$

when the available measurement history is

$$\hat{\mathbf{x}}(k|k) = E\{\mathbf{x}(k)|Z^k\} \qquad (2.83)$$

where the measurement data is $Z^k = \{\mathbf{z}(1), \mathbf{z}(2), \ldots, \mathbf{z}(k)\}$.

Once the predicted value of the state is known, the predicted value of the next measurement becomes

$$\hat{\mathbf{z}}(k+1|k) = \mathbf{H}(k+1)\hat{\mathbf{x}}(k+1|k) \qquad (2.84)$$

When the new measurement $\mathbf{z}(k+1)$ becomes available, the residual or innovation can be computed

$$\mathbf{e}(k+1) = \mathbf{z}(k+1) - \hat{\mathbf{z}}(k+1|k) \qquad (2.85)$$

The new estimated state $\hat{\mathbf{x}}(k+1|k+1)$ is the sum of the predicted state at time step k and the residual weighted with a filter gain $\mathbf{W}(k+1)$

$$\hat{\mathbf{x}}(k+1|k+1) = \hat{\mathbf{x}}(k+1|k) + \mathbf{W}(k+1)\mathbf{e}(k+1) \qquad (2.86)$$

This is exactly the same form as in the simple Example 2.10 The new estimate is obtained from the old estimate by making a correction with the weighted residual (innovation). The key issue is the computation of the weighting matrix $\mathbf{W}(k+1)$. Here knowledge of the noise statistics is needed.

The gain $\mathbf{W}(k+1)$ is called the *Kalman gain matrix* and is computed from the *state prediction covariance* $\mathbf{P}(k+1|k)$, which is defined as

$$\mathbf{P}(k+1|k) = E\{[\mathbf{x}(k+1) - \hat{\mathbf{x}}(k+1|k)][\mathbf{x}(k+1) - \hat{\mathbf{x}}(k+1|k)]^T|Z^k\} \qquad (2.87)$$

and can be expressed as

$$\mathbf{P}(k+1|k) = \mathbf{F}(k)\mathbf{P}(k|k)\mathbf{F}^T(k) + \mathbf{Q}(k) \qquad (2.88)$$

where $\mathbf{x}(k+1) - \hat{\mathbf{x}}(k+1)|k)$ is the *a priori* estimate error. The *measurement prediction (innovation) covariance* $\mathbf{S}(k+1)$ is computed from

$$\mathbf{S}(k+1) = E\{[\mathbf{z}(k+1) - \hat{\mathbf{z}}(k+1|k)][\mathbf{z}(k+1) - \hat{\mathbf{z}}(k+1|k)]^T|Z^k\} \qquad (2.89)$$

The result becomes

$$\mathbf{S}(k+1) = \mathbf{H}(k+1)\mathbf{P}(k+1|k)\mathbf{H}^T(k+1) + \mathbf{R}(k+1) \tag{2.90}$$

The Kalman gain is calculated as follows

$$\mathbf{W}(k+1) = \mathbf{P}(k+1|k)\mathbf{H}^T(k+1)\mathbf{S}(k+1)^{-1} \tag{2.91}$$

The updated state covariance $\mathbf{P}(k+1|k+1)$ becomes

$$\mathbf{P}(k+1|k+1) = \mathbf{P}(k+1|k) - \mathbf{W}(k+1)\mathbf{S}(k+1)\mathbf{W}^T(k+1) \tag{2.92}$$

The Kalman gain matrix $\mathbf{W}$ minimizes the *a posteriori* error covariance in minimum mean square sense. Rewriting Equation (2.91) using Equation (2.89) gives

$$\mathbf{W}(k+1) = \frac{\mathbf{P}(k+1|k)\mathbf{H}^T(k+1)}{\mathbf{H}(k+1)\mathbf{P}(k+1|k)\mathbf{H}^T(k+1) + \mathbf{R}(k+1)} \tag{2.93}$$

This expression provides a useful qualitative interpretation of how the measurement uncertainty $\mathbf{R}$ effects gain $\mathbf{W}$ and how this reflects on the prediction $\hat{\mathbf{x}}(k+1|k+1)$ in Equation (2.86). For argument's sake, let all the matrices in Equation (2.86) be scalars. If the measurement error covariance $\mathbf{R}$ is 'small', the gain $\mathbf{W}$ weights the innovation term more in Equation (2.86). This implies that the new measurement $\mathbf{z}(k+1)$ includes better information than the predicted state $\hat{\mathbf{x}}(k+1|k)$ and the correction is quite significant. If, on the other hand, the *a priori* error covariance estimate $\mathbf{P}(k+1|k+1)$ is small, the gain $\mathbf{W}$ is also small making the correction term minute. Our predicted state is fairly reliable and need not be corrected very much.

If the noise covariance matrices $\mathbf{Q}$ and $\mathbf{R}$ remain constant, the estimation error covariance $\mathbf{P}(k|k)$ and the Kalman gain $\mathbf{W}(k)$ will reach steady-state quite quickly. In practice, constant values are commonly used for $\mathbf{P}$ and $\mathbf{W}$. This implies that they can be computed beforehand.

Consider the stationary Kalman filter. MATLAB® provides many powerful tools for making the complicated computations straightforward. For simulation purposes a more convenient form of the Kalman filter is derived as follows. Substitute first the innovation Equation (2.85) and measurement estimate Equation (2.84) expressions into Equation (2.86)

$$\hat{\mathbf{x}}(k+1|k+1) = \hat{\mathbf{x}}(k+1|k) + \mathbf{W}\mathbf{v}(k+1) = \hat{\mathbf{x}}(k+1|k) + \mathbf{W}[\mathbf{z}(k+1) - \mathbf{H}\hat{\mathbf{x}}(k+1|k)] \tag{2.94}$$

This is exactly analogous form for the recursive computation of the sample mean.

Next use this form for $\hat{\mathbf{x}}(k|k)$ (take one step back and replace $k+1 \to k$) in Equation (2.82) resulting in

$$\begin{aligned} \hat{\mathbf{x}}(k+1|k) &= \mathbf{E}\hat{\mathbf{x}}(k|k) + \mathbf{F}\mathbf{u}(k) \\ &= \mathbf{E}\{\hat{\mathbf{x}}(k|k-1) + \mathbf{W}[\mathbf{z}(k) - \mathbf{H}\hat{\mathbf{x}}(k|k-1)]\} + \mathbf{F}\mathbf{u}(k) \end{aligned} \tag{2.95}$$

which can be further written as

$$\hat{\mathbf{x}}(k+1|k) = \mathbf{E}(\mathbf{I} - \mathbf{W}\mathbf{H})\hat{\mathbf{x}}(k|k-1) + \mathbf{E}\mathbf{W}\mathbf{z}(k) + \mathbf{F}\mathbf{u}(k) \tag{2.96}$$

Similarly the predicted value of the measurement becomes

$$\hat{\mathbf{z}}(k+1|k) = \mathbf{H}(k+1)\hat{\mathbf{x}}(k+1|k) \tag{2.97}$$

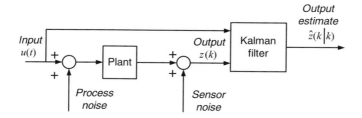

Figure 2.44   Principle of computing the output and the output estimate

In the stationary case the estimation error covariance **P** can be computed beforehand. The block diagram in Figure 2.44 indicates the way of simulating the output and the output estimate. Computation by hand becomes tedious and impossible when the dimension is larger than two. Simulation is then the most convenient way of computing the Kalman and the estimates. This is illustrated by Example 2.11.

Example 2.11   Simulation of continuous Kalman filter

Consider the continuous plant given in Equation (2.58) and the measurement equation (2.60). Form a discrete Kalman filter, configure it in Simulink® and simulate the result, when the noise covariances $Q = R = 1$.

Solution

The procedure is as follows:

1. Define the state-space system. It can be continuous or discrete. If continuous, transform it into a discrete system (sample time must be known).
2. Compute the Kalman gain.
3. Compute the Kalman estimator matrices.
4. Configure the overall system in Simulink® and compute the estimates.

The discretization of system Equations (2.58) and (2.60) with sampling time $Tsample = 0.2$ s leads to a discrete-time state space model

$\mathbf{E} = [0.9478\,0.1616; \quad -0.4847 \quad 0.6246]; \quad \mathbf{F} = [0.01741; \quad 0.1616]; \quad \mathbf{C} = [1 \quad 0]; \quad \mathbf{D} = 0;$

The Kalman gain is computed using the discrete Kalman filter command:

1. The state-space model is defined

```
% Give matrices A, B, C, D for state-space system
Ac=[0 1;-3 -2]; Bc=[0;1];
Cc=[1 0]; Dc=0;
sys_cont=ss(Ac,Bc,Cc,Dc)
```

The discrete equivalent is computed

```
sampletime=0.2;
% Determine the corresponding discrete system
sys_disc=c2d(sys_cont,sampletime,'zoh')
Ad=sys_disc.a;
Bd=sys_disc.b;
Cd=sys_disc.c;
Dd=sys_disc.d;
```

2. Compute the Kalman gain and filter matrices in (2.100) and (2.101)

```
% Give noise covariances
q=1; r=0.5;

%Compute Kalman gain
[KEST,L,P]=kalman(sys_disc,q,r)
Aaug=Ad*(eye(2)-L*Cd)
Baug=[Bd L]
Caug=Cd*(eye(2)-L*Cd)
Daug=[0 Cd*L]
```

3. Compute the Kalman estimator matrices (in an alternative way)

```
% Compute the Kalman estimator matrices
Aa=KEST.a;
Ba=[Bd KEST.b];
Ca=KEST.c(1,:);
Da=[0 KEST.d(1,1)];
```

4. Configure the overall system in Simulink® and compute the state estimates. Using Equations (2.94) and (2.97) and replacing $k + 1 \rightarrow k$ results in

$$
\begin{aligned}
\hat{\mathbf{x}}(k|k) &= \hat{\mathbf{x}}(k|k-1) + \mathbf{W}[\mathbf{z}(k) - \mathbf{H}\hat{\mathbf{x}}(k|k-1)] \\
&= (\mathbf{I} - \mathbf{WH})\hat{\mathbf{x}}(k|k-1) + \mathbf{Wz}(k)
\end{aligned}
\tag{2.98}
$$

Substituting this into Equations (2.82) gives

$$
\begin{aligned}
\hat{\mathbf{x}}(k+1|k) &= \mathbf{E}\hat{\mathbf{x}}(k|k) + \mathbf{F}(k)\mathbf{u}(k) \\
&= \mathbf{E}(\mathbf{I} - \mathbf{WH})\hat{\mathbf{x}}(k|k-1) + \mathbf{EWz}(k) + \mathbf{F}(k)\mathbf{u}(k)
\end{aligned}
\tag{2.99}
$$

or state estimation

$$
\hat{\mathbf{x}}(k+1|k) = \mathbf{E}(\mathbf{I} - \mathbf{WH})\hat{\mathbf{x}}(k|k-1) + [\mathbf{F} \quad \mathbf{EW}]\begin{bmatrix} \mathbf{u}(k) \\ \mathbf{z}(k) \end{bmatrix}
\tag{2.100}
$$

From Equations (2.97) and (2.98) output estimation

$$
\hat{\mathbf{z}}(k|k) = \mathbf{H}\hat{\mathbf{x}}(k|k) = \mathbf{H}(\mathbf{I} - \mathbf{WH})\hat{\mathbf{x}}(k|k-1) + \mathbf{HWz}(k)
\tag{2.101}
$$

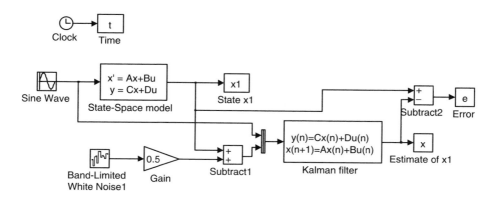

Figure 2.45 Simulink® configuration of Kalman filter example. The plant is continuous and the Kalman filter discrete

Equations (2.100) and (2.101) are in a form, which provides a convenient way of simulating the Kalman filter.

Once the Kalman filter gains have been computed, a Simulink® configuration of the estimation can be constructed using Equations (2.100) and (2.101) (see Figure 2.45).

The results of the simulation are shown in Figure 2.46. Although the noise level is quite significant, the Kalman filter provides quite good estimates.   □

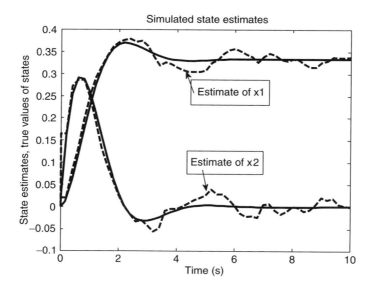

Figure 2.46 Simulated state estimates

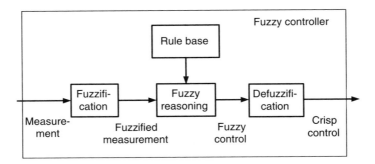

Figure 2.47   Principle of a fuzzy controller

2.17 Fuzzy Control

Fuzzy logic is rigorous and hard mathematics with a solid basis. It is not possible to review all fuzzy logic elements in any detail here. Fortunately, in engineering problems, we can survive with much less. The core of the fuzzy controller is the rule base, which uses logical expressions. This may be approached with much less sophisticated mathematics. Still the end result will be quite useful. For example, there is similarity between a PID controller and fuzzy logic controller. When the PID control structure is fixed, then the parameters of the controller need to be chosen. This might require only a few simple experiments. As we will see, a fuzzy controller is also a parameterized controller. Often human knowledge of the plant behavior is enough to determine the fuzzy controller parameters.

The basic structure of a fuzzy controller is displayed in Figure 2.47. Let us study each block in more detail.

2.17.1 Fuzzification

Consider a physical variable such as temperature T. We would like to describe the temperature in a room. First we must determine the range of temperatures we are interested in. Let us decide that the temperature range is [0 °C, 40 °C]. In everyday language we could use terms like 'cold temperature', 'warm temperature', and 'hot temperature'. This is a crude description of how people would describe the room temperature. Because of this fuzzy logic computing is described by Lotfi Zadeh, the inventor of fuzzy logic, as *computing with words*.

An engineer must now determine what these vague terms mean. We will first look into how we can quantify the uncertainty of a set 'cold temperature'. We use so-called *membership functions* to describe the degree of how cold we feel the temperature to be. We normalize the situation, so that membership functions can only take values in the interval [0 1]. Thus membership functions are defined as real-valued functions $\mu: R \rightarrow [0\ 1]$. This is the big difference between classical, 'crisp' logic, where the membership or characteristic function would assume a value of 0 or 1.

Many types of membership functions are commonly used, such as triangular, Gaussian, trapezoidal, and singletons. Typical to all of them is that they are parameterized and therefore

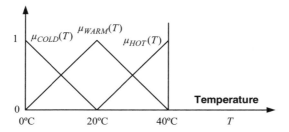

Figure 2.48 Room temperature has been divided into three fuzzy sets described by their corresponding membership functions. This figure describes the fuzzification block

easily handled by computer computations. Often the membership function has one single maximum, but not always, as in the case of trapezoids.

We can now return to our set 'cold temperature'. A fuzzy set is defined and completely characterized with its membership function. In fact, in computations only membership functions are used. Let us choose triangular membership functions for our example. Room temperature $0\,°C$ is certainly cold, so there is no uncertainty and membership function takes value 1. Since our range of interest does not go beyond $0\,°C$ we let the maximum of our membership function occur at that point. Once the temperature is above $0\,°C$, the value of the membership function becomes smaller until its value is zero. Assume that this occurs at $20\,°C$. Now we have a complete description of our fuzzy set 'cold temperature', and its corresponding membership function. We now repeat the same for fuzzy sets 'warm temperature' and 'hot temperature'. For 'warm temperature' we start at $0\,°C$ where the corresponding membership function takes value 0. At $20\,°C$, the membership function assumes its maximum value of 1 and at $40\,°C$ it is again zero. For 'hot temperature' the situation is the mirror image of 'cold temperature'. The overall description of the fuzzification block is shown in Figure 2.48, where all three membership functions have been drawn on the same figure.

How does the fuzzification process proceed if we have measured the room temperature to be $15\,°C$. This is shown in Figure 2.49.

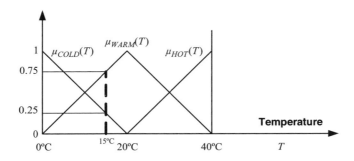

Figure 2.49 Measured temperature is $15\,°C$. Result of fuzzification process in which each membership function is evaluated

The membership functions are evaluated at $15\,^\circ C$: $\mu_{COLD}(15\,^\circ C) = 0.25$, $\mu_{WARM}(15\,^\circ C)$ 0.25 and $\mu_{HOT}(15\,^\circ C) = 0$. From this we could say that the temperature is not very cold, but it is not warm either. It is luke warm.

We are now ready to move to the next phase in our design: rules and fuzzy reasoning.

2.17.2 Rule Base

Rules for controlling room temperature are quite simple. We want the room temperature to be a comfortable $20\,^\circ C$. If it is below that, we increase the temperature setting and the heater turns on to heat the room. If it is too hot, then we want to decrease the temperature and turn the air conditioning on. Since we are dealing here with the action output, we need to define membership functions at the output. Our air conditioning (AC) actuator is considered to be able to both cool and heat.

The simplest rule base becomes:

1. If temperature T is '*cold*' then turn AC on '*heating*'.
2. If If temperature T is '*warm*' then AC is to do '*nothing*'.
3. If If temperature T is '*hot*' then turn AC on '*cooling*'.

The fuzzy membership functions for the output variables are defined as percentages from -50 to 50% or for an AC system this covers the whole scale of the temperature setting. Note that fuzzy set '*nothing*' does not mean zero. Figure 2.50 describes all three output membership functions.

2.17.3 Fuzzy Reasoning

How to evaluate the rules? There are a number of ways of doing this, here we use the *min–max* reasoning. This means that if the rules include logical AND, it corresponds to taking a minimum of the corresponding membership functions. Similarly, if there is a logical OR, it corresponds to a maximum of the membership functions. The implication THEN corresponds to taking a minimum.

When the input temperature is $15\,^\circ C$, then a graphical representation of the procedure for each rule is shown in Figure 2.51. The rules correspond to rows and evaluation proceeds from left to

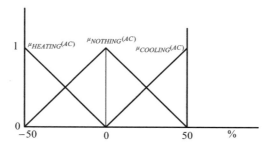

Figure 2.50 Output membership functions

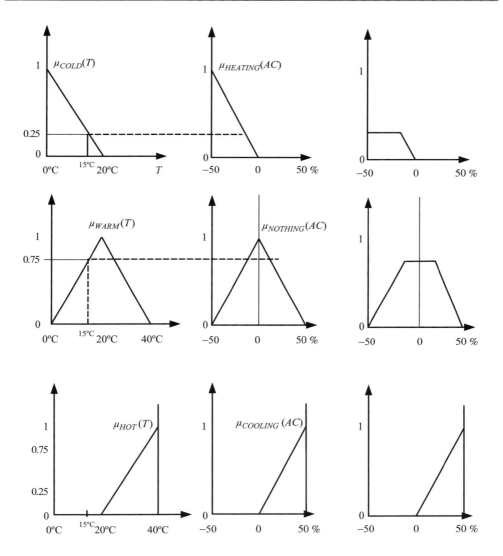

Figure 2.51 Fuzzy reasoning with min-max principle. Each row represents one rule. The end result of each rule is shown in the last column

right. At 15 °C, $\mu_{COLD}(15\,°C) = 0.25$. Computing graphically the minimum of value 0.25 and $\mu_{COOLING}(AC)$ results in the right-most figure. The other two rules are evaluated correspondingly.

The results of rules 1–3 are combined by using logical OR between the rules. Since OR implies maximum, this means that we take the maximum of the individual results shown on the last column. This still results in a fuzzy set (Figure 2.52). There are several methods of defuzzification. One of the most common is to compute *center of gravity* for the area under the curve. The final result is then about −5%, which makes sense. If our temperature is 15 °C and our desired temperature is 20 °C, then turn on the heat slightly.

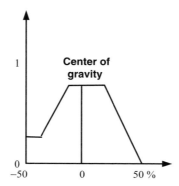

Figure 2.52 The figure shows the end result of combining of all rules with logical ORs. Defuzzificaton is performed with center of gravity operation resulting in about -2% or slight heating

The graphical computation is tedious and is done only to illustrate the process. Normally, computer software is used to the computations. Here is a quick introduction to the MATLAB® Logic Fuzzy Toolbox.

Once MATLAB® has been opened, type *fuzzy*. This will open a GUI (Figure 2.53) called the FIS Editor.

Figure 2.53 shows the fuzzy system with Input and Output blocks. In the middle there is an Untitled block, Mamdani type. This contains the rules and the inference part of the system. The other kind of reasoning is called Sugeno type, in which the output membership functions are singleton functions. We restrict our presentation to the Mamdani type of systems.

Use your mouse to click the Input block. This will open the Membership Function Editor. Here you can define the range of the variable, how many membership functions will be used and

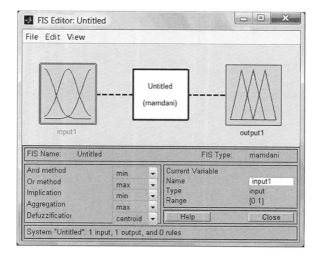

Figure 2.53 FIS Editor of Fuzzy Logic Toolbox

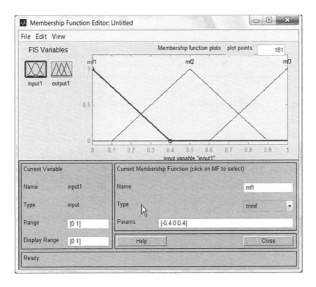

Figure 2.54 Membership Function Editor showing default triangular membership functions

their names and type. Default membership functions are triangular. We have already decided to have three triangular membership functions, *cold*, *warm*, and *hot* (Figure 2.54). The range is from 0 °C to 40 °C. At the output we also have three membership functions, *heating*, *nothing*, and *cooling* and their range goes from −50 to 50%.

To define the rules, choose *Edit* and then *Rules* (Figure 2.55).

The final result of the design is shown in Figure 2.56 giving the same result as in Figure 2.52.

□

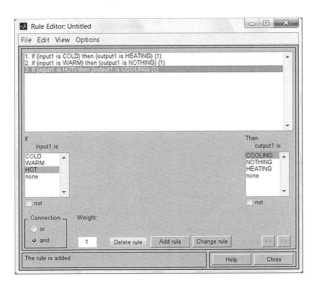

Figure 2.55 Rule Editor is easy and quick to use, because the basic rule forms are already given

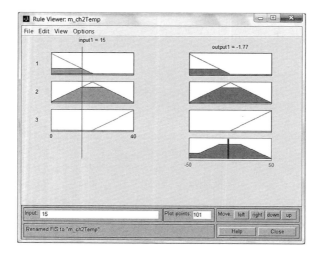

Figure 2.56 The Rule Viewer gives the same information as Figures 2.51 and 2.52. Resulting output is −1.77%, meaning slight heating

2.18 Summary

In this chapter the basics of control theory have been reviewed. There are many excellent textbooks in analog control theory, such as Dorf and Bishop (2008). Similarly many outstanding textbooks exist in digital control, such as Franklin, Powell, and Workman (2006). The most commonly used controller, the PID controller, has been discussed. The most complete coverage of this important controller is given by Åström and Hägglund (2006). In communications one widely used feedback system, the phase-locked loop, is given in-depth coverage. A comprehensive coverage of the topic is given by Egan (1998). State-space representation is studied, which is a topic covered in books on analog and digital control mentioned above. The Kalman filter fundamentals are also discussed. The book by Grewal and Andrews (2001) provides a nice overview. A quick review of fuzzy logic systems is given at the end of the chapter (Yen and Langari (1999)). Throughout the chapter MATLAB® and Simulink® have served as powerful simulation tools to demonstrate the concepts. A nice presentation on the topic is given by Gran (2007).

Exercises

2.1   A Simulink® configuration of a relay control system is depicted in Figure 2.57.
      Configure the system and simulate the unit step response. The relay controller causes an oscillatory response. In order to avoid oscillations, replace the relay with a relay with dead zone. A (symmetric) relay with dead zone is described as follows:

$$u(t) = \begin{cases} K, & \text{if } e \geq \delta \\ 0, & \text{if } -\delta < e < \delta \\ -K, & \text{if } e \leq -\delta \end{cases} \qquad (2.102)$$

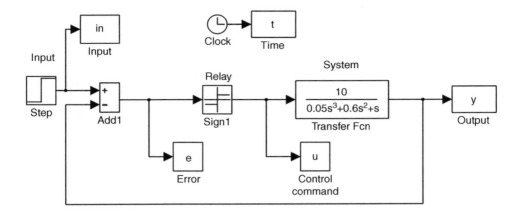

Figure 2.57   Control system with a relay

The characteristics are shown in Figure 2.58.

(a) If a sinusoidal input $e(t)$ of amplitude 1 and frequency 1 (rad/s) is applied to the relay with dead zone, compute and plot the output $u(t)$. Assume $\delta = 1, K = 1$.

(b) Generate the relationship in Equation (2.102) with the existing Simulink® blocks and repeat the task in (a).

(Hint: If you cannot solve this, look at Figures 6.9 and 6.10 for help.)

(c) Replace the relay controller with a relay with dead-zone controller. Set up the dead-zone parameters in such a way that the step response of the output temperature $T(t)$ does not oscillate. Simulate the step response with the same parameter data as in Example 2.1.

(Hint: You can concentrate on adjusting the width of the dead zone.)

2.2   Consider the temperature control of Example 2.2, where a P controller is used.

(a) Produce the same Simulink® configuration as in Figure 2.10. Apply the same parameters as in the example. Change the gain of the P controller K_P in order to reproduce the results of Figure 2.11.

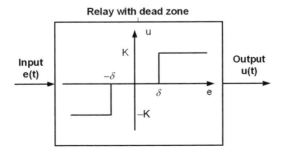

Figure 2.58   Relay with dead zone

(b) One way to tune PID controllers is to set the gains of I and D part equal to zero, so that we are left only with a P controller, exactly as in Figure 2.10. In closed-loop tuning K_P is increased from a small value, say $K_P = 0.7$, until the output starts to oscillate. Denote this value of K_P by K_{cr} and the period of oscillation by P_{cr}.

(c) Once the oscillation is achieved, information about K_{cr} producing the oscillation and the period of oscillation P_{cr} are used in various rules of thumb formulae (the most famous one being Ziegler–Nichols) for computing the approximate values for K_P, K_I and K_D in Equation (2.17) or K_P, T_I, and T_D in Equation (2.18). Consider a PI controller. One tuning rule suggests that $K_P = 0.45K_{cr}$, $T_I = P_{cr}/2$. Use the values obtained in (b) to compute values for K_P and T_I. Apply the PI controller (instead of K_P) of Equation (2.10) with these parameter values in the system in Figure 2.10. Simulate the step response.

2.3 Consider the first-order PLL of Example 2.5 in Figure 2.33. Let the amplitude of the sinusoidal input be $= 4$. Increase the radian frequency of the input starting from $\omega = 1$. At each new value study the tracking performance by plotting the output, the input and the error. Determine ω at which the system loses its ability to track the input. Is there anything that can be done to push this frequency higher? If so, test it by simulation.

2.4 Repeat the simulations carried out in Example 2.6.

2.5

(a) Form a state-space representation for

$$\dddot{x} + 4\ddot{x} + 3x + 2x = u$$

Set up a Simulink® configuration for the state-space representation. Simulate the unit step response for the system.

(b) If the system is given as

$$\dot{x}_1(t) = 3x_2(t) + 7u_1(t)$$
$$\dot{x}_2(t) = -6x_1(t) - 4x_2(t) + 2u_2(t)$$
$$y(t) = 2x_1(t) + 5x_2(t) + 0.5u_1(t) + 1.5u_2(t)$$

Identify the matrices $\mathbf{A}$, $\mathbf{B}$, $\mathbf{C}$, and $\mathbf{D}$ in Equation (2.67). Set up a Simulink® configuration. Determine $x_1(t)$, $x_2(t)$, and $y(t)$, when $u_1(t) = 1$ (unit step), $u_2(t) = 0$ by simulating the system. Assume zero initial conditions.

(c) The system is given as

$$\ddot{x}(t) + 4\dot{x}(t) + 6x(t) = 2u(t) + 3\dot{u}(t)$$

Determine a state-space representation of the system.

2.6 Consider a linear time-invariant system

$$\dot{\mathbf{x}} = \mathbf{A}\mathbf{x} + \mathbf{B}\mathbf{u}$$
$$\mathbf{y} = \mathbf{C}\mathbf{x} + \mathbf{D}\mathbf{u}$$

where

$$\mathbf{A} = \begin{bmatrix} 1.5 & 3 \\ -4 & -8 \end{bmatrix}, \mathbf{B} = \begin{bmatrix} -2 \\ 2.2 \end{bmatrix}$$

$$\mathbf{C} = \begin{bmatrix} 5 \\ 3 \end{bmatrix}, \mathbf{D} = [0]$$

Form a discrete Kalman filter for the system similar to that in Figure 2.45. Configure it in Simulink® and simulate the result. The noise covariances in Equation (2.81) are $Q = R = 1$.

2.7 Repeat the steps taken in Section 2.7 in formulating a fuzzy controller.

3

Channel Modeling

3.1 Introduction

The radio channel is the key component in wireless communication system. The signal is transmitted from the sending base station to the receiving mobile unit or vice versa. It is therefore essential to understand the complex behavior of the channel. Since the transmitted signal is electrical, Maxwell's equations govern its propagation in the environment. In the most general case the equations would be time-varying, three-dimensional partial differential equations. For practical purposes they are quite complicated. Therefore, other simpler models have been developed which are approximations of the Maxwell's equations. These models describe the spatio-temporal characteristics of the radio channel using electric field strength.

The chapter begins with static, spatial models, which give a good basis for understanding the attenuation of the transmitted signal or signal path loss. The basic propagation mechanisms are reflection, diffraction, and scattering. They are called *large-scale propagation* or *path-loss* models. These are followed by temporal, time-varying models, which are required to model multipath fast changes in the channel. They form *small-scale fading* models.

Terminology

- **Propagation channel:** physical medium that supports the wave propagation *between the antennas.*
- **Radio channel:** propagation channel and the transmitter and the receiver antennas viewed collectively.
- **Digital channel:** also incorporates the modulation and demodulation stages.
- **Flat fading:** maximum differential delay spread is *small* compared with the symbol duration of the transmitted signal.
- **Frequency selective fading:** maximum differential delay spread is *large* compared with the symbol duration of the transmitted signal.
- **Intersymbol interference:** the received signals of the successively transmitted signals will overlap in the time domain.
- **Narrowband signal:** the message bandwidth is much smaller than the carrier frequency.

Systems Engineering in Wireless Communications   Heikki Koivo and Mohammed Elmusrati
© 2009 John Wiley & Sons, Ltd

- **Isotropic antenna:** antenna that transmits equally in all directions. It has a spherical radiation pattern. In free space it radiates power equally in all directions. It is an ideal device, which cannot be realized physically.
- **Downlink or forward channel:** the radio channel from the base station (BS) to a mobile station (MS) or from a satellite to the ground station.
- **Uplink or reverse channel:** the radio channel from an MS to the BS.

3.2 Large-Scale Propagation Models

Ray-optical approximations are commonly adopted in modeling the propagation of electromagnetic waves in wireless communication, where the signals have frequencies over 800 MHz. This is a reasonable approximation, because the signal wavelengths are very small compared to the buildings, hills, and forests. Thus the electromagnetic signal is assumed to propagate through space in one of the following ways (Figure 3.1): *directly from the transmitter to the receiver (line-of-sight path); reflected from earth or buildings; diffracted from larger objects*; or *scattered from objects that have smaller dimensions like lamp posts*. The approach can be applied to both indoor and outdoor areas.

1. *Line-of-sight (LOS) path.* Communication is by optical line of sight, which is the shortest direct path between the transmitter and the receiver. This is a good approximation in space communication, where no intervening obstacles exist. It provides the simplest propagation model.
2. *Reflection.* On earth the transmitted signal will not have a free path, but will face obstacles. If the signal wavelength is smaller relative to the surface encountered, it is reflected from the surface. Outdoors the ground or buildings, indoors the walls cause reflections. Reflection is more important indoors than outdoors.
3. *Diffraction.* Diffracted fields are generated by secondary wavelets. According to Huygen's principle these wavelets combine to produce a new wavefront propagating away from the diffracting edge as cylindrical waves. At the edges of large objects the radio wave is diffracted, when the signal wavelength is smaller relative to the surface encountered.

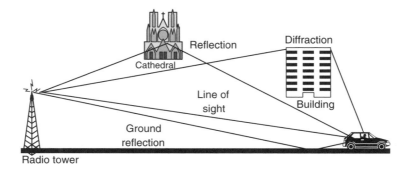

Figure 3.1 Different propagation mechanisms: *line-of-sight (LOS), reflection, and diffraction. Scattering* is not shown due to lack of space

This means that the electromagnetic wave 'bends' around the corner and can reach a receiver, which is not in line of sight. Analytical models for diffraction are not developed, but diffraction phenomenon is included in the empirical models implicitly.

4. *Scattering.* If the electromagnetic wave has a wavelength, which is comparable to the object encountered, it is scattered. An incoming signal is scattered into many weaker signals. The obstacles can be cars or trees outside, furniture inside. Because the power levels will be significantly reduced, scattering can mostly be ignored and will not be discussed further.

3.2.1 Line-of-Sight Path Model

A common situation in space communication is to have a communication satellite transmitting to receiving stations on earth or the other way around, as shown in Figure 3.2. Both transmitters can be considered point sources. The term *downlink* or *forward channel* is used to describe the transmission channel from the satellite to an earth station. The radio channel from the earth station to the satellite is called *uplink* or *reverse channel*. The same terminology is used for base stations and mobile units – base station corresponding to satellite and mobile unit to earth station.

Let the distance from the satellite to the receiver be d. This is a typical situation of an LOS communication in free space. It is governed by the *Friis free-space* equation

$$\bar{P}_r = c \frac{\bar{G}_t \bar{G}_r}{d^2} \bar{P}_t \tag{3.1}$$

Here $\bar{P}_t$ is the transmitted power, $\bar{P}_r$ the received power, $\bar{G}_t$ the transmitter gain, $\bar{G}_r$ the receiver gain, d the distance between the transmitter and the receiver, and c a constant that includes antenna geometry and wavelength. The model is in *linear scale*. To separate linear from log-scale (dB), a bar is used above the variable. A variable without the bar is given in dB, for example, P_t is in dB.

Equation (3.1) indicates that the received power $\bar{P}_r$ is inversely proportional to the square of the distance d. Heuristically speaking this makes sense: the further you are, the weaker

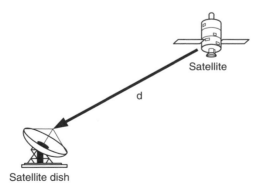

Figure 3.2 LOS signal path from the transmitting satellite to a receiving station in downlink direction

the signal. *Path loss* or *free-space loss* L_p is a useful concept describing the attenuation of the transmitted signal. The received power from Equation (3.1) can be written as

$$\bar{P}_r = \frac{\bar{G}_t \bar{G}_r}{\bar{L}_p} \bar{P}_t \qquad (3.2)$$

where the path loss

$$\bar{L}_p = \frac{d^2}{c} \qquad (3.3)$$

Equation (3.2) in dB becomes

$$P_r = P_t + G_t + G_r - L_p \qquad (3.4)$$

This represents a *balance equation*. The transmitted power P_t with the antenna gains G_t and G_r have to overcome the loss L_p, so that the received signal P_r is positive and can be observed.

In communications literature absolute values of power are often used instead of relative values, dB. The most common reference levels are 1 W and 1 mW. The corresponding decibel levels of power are decibel watt, dBW, and decibel milliwatt, dBm

$$Power_{dBW} = 10 \log_{10} \frac{Power_W}{1\ W} \qquad (3.5)$$

$$Power_{dBm} = 10 \log_{10} \frac{Power_{mW}}{1\ mW} \qquad (3.6)$$

3.2.2 Reflection Model

Friis equation (3.1) holds in free space, but on terrestrial environment the signal from the transmitter will reach the receiver via a multitude of paths. Taking the free-space model one step further leads to a *flat-earth* or a *two-ray* model (Figure 3.3). Two rays are observed, one signal via the LOS path and the other reflected from the ground, which is assumed to be flat and perfectly reflecting. The heights of the transmitting base station (BS) and receiving mobile station (MS) are taken into account in the two-ray model.

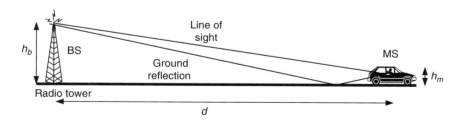

Figure 3.3 Flat-earth model for radio signal propagation: radio channel with two propagation paths

With a further assumption of narrowband signals it is possible to derive the following extension to Equation (3.1)

$$\bar{P}_r = h_b^2 h_m^2 \frac{\bar{G}_t \bar{G}_r}{d^4} \bar{P}_t \qquad (3.7)$$

Here h_b is the height of the BS, h_m the height of the MS, and d the distance between BS and MS. The heights h_b and h_m are assumed to be much smaller than the distance d. Observe that the path loss is inversely proportional to the fourth power of the distance d.

3.2.3 Statistical Path-Loss Models

In Friis equation (3.1) the received power was inversely proportional to the second power of distance. In the two-ray model (Equation (3.7)) it has grown to the fourth power meaning very fast attenuation of the signal. The flat-earth assumption holds reasonably well over distances of kilometers for wireless communication systems, when the base stations are tall.

Several other analytical models have been derived based on additional simplifying assumptions. In reality the signal will take many paths from the transmitter to the receiver and physical modeling becomes laborious. Fortunately, we can combine the analytical and empirical models. We rely on real measurements and statistical evidence to derive improved empirical relationships for signal propagation. Such models will take into account not only LOS and reflection, but also diffraction and scattering.

It can be seen in Equations (3.1) and (3.7) that the received power $\bar{P}_r$ is inversely proportional to the power of the distance, which varies. Such relationships are called *large-scale path-loss* models and take the form

$$\bar{P}_r = \left(\frac{\alpha}{d^n}\right) \bar{P}_t \qquad (3.8)$$

In free space model (Equation (3.1)) $n = 2$ and in the two-ray model (Equation (3.7)) $n = 4$. In practice, the measurements indicate that n varies, for example, in flat rural areas $n = 3$ and in dense urban areas $n = 4.5$. Indoors, the variation can be even greater, from $n = 2$ in corridors to $n = 6$ in obstructed buildings. Parameter α in our simple model is assumed to include the antenna gains, carrier frequency, and other environmental factors. It is usually determined from measurements.

As in Equations (3.2) and (3.3), the inverse of the term in front of $\bar{P}_t$ in Equation (3.8) represents the path loss L_p

$$\bar{L}_p = \frac{d^n}{\alpha} \qquad (3.9)$$

Let $\bar{L}_0 = d_0^n/\alpha$. Now Equation (3.9) can be written in dB form

$$L_p = L_0 + 10n \log\left(\frac{d}{d_0}\right), \; d \geq d_0. \qquad (3.10)$$

Here L_0 is the measured path loss (in dB) at the reference point d_0, and n depends on the propagation environment. For macrocells, d_0 is taken as 1 km, for smaller microcells 100 m. In practice, measurements have to be carried out over several wavelengths to determine the path-loss exponent n.

Using Equations (3.9) and (3.10), Equation (3.8) is rewritten in dB

$$P_r = P_t - L_p = P_t - L_0 - 10n \log\left(\frac{d}{d_0}\right). \tag{3.11}$$

Or more conveniently as

$$P_r = P_0 - 10n \log\left(\frac{d}{d_0}\right). \tag{3.12}$$

Where P_0 is the power received at the reference point d_0. Equation (3.12) can be used in designing cell coverage area.

One of the more complex empirical path-loss models for macrocellular areas was developed by Hata using data measured by Okumura. The model takes into account the carrier frequency, the heights of the base station and the receiver and the distance between them. The Hata–Okumura model is given by

$$L_p = 69.55 + 26.16 \log_{10} f_c - 13.82 \log_{10} h_b - a(h_m) + [44.9 - 6.55 \log h_b] \log_{10} d \tag{3.13}$$

where f_c is the carrier frequency in MHz ranging from 150 Mhz to 1500 MHz, h_b the height of the base station, and h_m the height of the receiver. Function $a(h_m)$ depends on the size of the city and frequency of operation:

For large cities

$$a(h_m) = 8.29(\log 1.54 h_m)^2 - 1.1 \, \text{dB} \quad \text{for} f_c \leq 200 \, \text{MHz} \tag{3.14}$$

$$a(h_m) = 3.2(\log 11.75 h_m)^2 - 4.97 \, \text{dB} \quad \text{for} f_c \geq 300 \, \text{MHz} \tag{3.15}$$

and for small- to medium-sized cities, $150 \geq f_c \geq 1500 \, \text{MHz}$

$$a(h_m) = (1.1 \log f_c - 0.7) h_m - (1.56 \log f_c - 0.8) \, \text{dB} \tag{3.16}$$

Radio signal propagation indoors can be modeled analogously to outdoor propagation by reflection, diffraction, and scattering. The path-loss models follow basically Equation (3.10), but because there are more attenuation factors inside the buildings extra terms are added to the right-hand side of the equation.

Empirical path-loss models for indoors can be written in the form

$$L_p = L_0 + 10n \log\left(\frac{d}{d_0}\right) + FAF \tag{3.17}$$

Here the exponent n is for the *same floor measurement* and *FAF* is the *floor attenuation factor*. A somewhat more complicated model is

$$L_p = L_0 + 10n \log\left(\frac{d}{d_0}\right) + \sum_{i=1}^{I}(FAF)_i + \sum_{j=1}^{J}(WAF)_j \tag{3.18}$$

where *WAF* is the *wall attenuation factor* and the running indices i and j refer to *floors* and *walls*, respectively, separating the transmitter from the receiver.

It is clear from above that empirical models play an important role not only in the path-loss modeling but also in other issues discussed later in the book. Therefore, we devote the next section to data-fitting techniques and illustrate how they can be efficiently used in path-loss modeling. The most common methods are briefly reviewed. They will be used in the upcoming chapters as well. Instead of analytical solutions, we take the user's point of view and use powerful MATLAB® (a registered trademark of The Math-Works, Inc.) commands.

3.2.4 Data-Fitting Methods in Path-Loss Modeling

Empirical models like the ones in the previous subsection are based on measurements. The basic task is to find the parameters in the model. The most straightforward approach to data fitting is the least squares approach. Consider first the linear case

$$y = ax + b \tag{3.19}$$

Here y is the output, x is the input, a and b are unknown, constant parameters. In order to determine the parameters N measurements are carried out and a data set $\{y_i, x_i\}$, $i = 1, ... N$ is obtained. Since measurements contain errors, the ith measurement is written as

$$y_i = ax_i + b + e_i \tag{3.20}$$

where e_i represents the measurement error.

In order to find the best parameters a and b, a natural cost function is the overall cumulative error

$$J(a, b) = \sum_{i=1}^{N} e_i^2 = \sum_{i=1}^{N} (y_i - ax_i - b)^2 \tag{3.21}$$

which is to be minimized with respect to a and b. It is straightforward to solve the analytical expressions for the optimal a and b. This is left as an exercise.

If the model is nonlinear and multidimensional, the least squares approach still applies. Let the model be described by

$$\mathbf{y} = \mathbf{f}(\mathbf{x}; \mathbf{a}) + \mathbf{e} \tag{3.22}$$

Here $\mathbf{y} = [y_1, \cdots, y_m]^T$ is an m-dimensional output vector, $\mathbf{y} \in \mathbf{R}^m$; $\mathbf{x} = [x_1, \cdots, x_n]^T$ is an n-dimensional input vector, $\mathbf{x} \in \mathbf{R}^n$; $\mathbf{a} = [a_1, \cdots, a_k]^T$ is an unknown, k-dimensional constant parameter vector, $\mathbf{a} \in \mathbf{R}^k$, and $\mathbf{e} = [e_1, \cdots, e_m]^T$ an m-dimensional error vector (including noise), $\mathbf{e} \in \mathbf{R}^m$. Function $\mathbf{f} \in \mathbf{R}^m$ is an m-dimensional vector, $\mathbf{f}(\mathbf{x}; \mathbf{a}) = [f_1(\mathbf{x}; \mathbf{a}), \cdots, f_m(\mathbf{x}; \mathbf{a})]^T$ of sufficient smoothness (can be differentiated). Again N measurements are performed resulting in a data set $\{\mathbf{y}_i, \mathbf{x}_i\}$, $i = 1, \cdots, N$.

The cost function to be minimized with respect to parameter vector $\mathbf{a}$ corresponding to Equation (3.19) becomes:

$$J(\mathbf{a}) = \sum_{i=1}^{N} ||\mathbf{e}_i||^2 = \sum_{i=1}^{N} ||\mathbf{y}_i - \mathbf{f}(\mathbf{x}_i, \mathbf{a})||^2 \tag{3.23}$$

The norm $||\cdot||$ is the Euclidean norm. Solution of this problem is usually analytically untractable and we need to resort to numerical computing. Many numerical optimization schemes can be

Table 3.1   Measurement data. The last column B is computed from the first two

Distance from transmitter d_i (m)	Received power P_i (dBm)	Note that $P_0 = P_i(d_o) = 0$ $B = 10^*\log(d_i/d_o)$
50	0	0
100	−15	3
400	−27	9
800	−34	12
1600	−45	15
5000	−86	20
$d_0 = 50\,\text{m}$		

applied to solve the minimization problem. These include gradient (steepest descent), Gauss–Newton, and Levenberg–Marquardt. They are all included in the MATLAB® Optimization Toolbox. The behavior of the algorithms can be studied by typing *optdemo*.

A suitable command for solving least squares problems in MATLAB® is *lsqnonlin*. It is applied to solve problems of the form *min{sum {FUN(a).^2}}*, where *a* and the values returned by *FUN* can be vectors or matrices. The command *a = lsqnonlin(FUN,a0)* starts at the vector *a0* and finds a minimum *a* to the sum of squares of the functions in *FUN*.

Example 3.1   Finding parameters of the path loss model with data fitting

Suppose that noise-corrupted measurements have been made to determine the exponent n in the path-loss model (Equation (3.10)). Power P_i is the received power at distance d_i. Given the data in Table 3.1 find n by the least squares fit. Reference distance is $d_0 = 50\,m$ and $P_0 = P_i(d_0) = P_i(50\,m) = 0\,\text{dBm}$.

Solution

Using Equation (3.12) the problem is formulated as finding the minimum of

$$J(n) = \sum_{i=1}^{N} e_i^2 = \sum_{i=1}^{N} [P_i - (P_0 - n \cdot 10\log(d_i/d_0))]^2 = \sum_{i=1}^{N} [P_i + nB]^2 \tag{3.24}$$

where $B = 10\log(d_i/d_0)$ to simplify notation.

Apply MATLAB® command *lsqnonlin*. Here function *FUN = 'P_i + n*B'*. Note that the answer will not be an integer.

```
n0=0; % Initial guess
% n is the unknown, which we iterate on; the data set from Table 2.1 is
  fed as vectors;
Pi=[0 -15 -27 -34 -42 -86]; B=[0 3 9 12 15 20];

% Apply least squares fit command to determine n
n=lsqnonlin(@(n) Pi+n*B,n0)
```

```
% Optimization terminated: first-order optimality less than OPTIONS.
TolFun and no negative/zero curvature detected in trust region
model.

n = 3.5460
% The best fit.
```

The received power becomes

$$P_r = P_0 - 10 \cdot 3.546 \log(d/d_0) = -10 \cdot 3.546 \log(d/50), \ d \geq 50 \, \text{m} \tag{3.25}$$

or the path-loss model from Equation (3.12)

$$L_p = -P_r = 10 \cdot 3.546 \log\left(\frac{d}{50}\right), \ d \geq 50 \, \text{m} \tag{3.26}$$

Plotting the measured received power P_i and the fitted P_r with MATLAB®

```
% Fig. 3.4 (a) Received power in dBm; Original data from Table 2.1
B=[0 3 9 12 15 20]; Pi=[0 -15 -27 -34 -42 -86];
% Computing from model (3.21)
d=[50 100 400 800 1600 5000]; Pr=-10*3.546*log10(d/50);
semilogx(d/50,Pr,'-k',d/50,Pi,'-.ok'); grid
```

The result is shown in Figure 3.4. The fit is not the best one. Therefore another approach is taken.   □

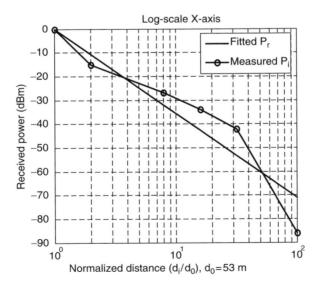

Figure 3.4   Least squares fit of received power P_r with $n = 3.546$; x-axis in log-scale

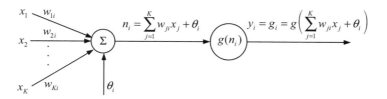

Figure 3.5   Single node i in an MLP network. Denote $g_i = g(n_i)$

If the data exhibit nonlinear and multivariable characteristics, neural networks can be used in path-loss prediction as alternative model structures for received power. Neural networks consist of a large class of different architectures. In path-loss models, the issue is approximating a static nonlinear mapping $f(\mathbf{x})$ with a neural network $f_{NN}(\mathbf{x})$, where $\mathbf{x} \in \mathbf{R}^K$. Here the K elements of $\mathbf{x} \in \mathbf{R}^K$ would typically be the same as in Equation (3.13): distance, heights of base and mobile stations, terrain profile parameters, and so on.

Here we concentrate on multilayer layer perceptron (MLP) networks.

An MLP network consists of an input layer, several hidden layers, and an output layer. Node i, also called a neuron, in an MLP network is shown in Figure 3.5. It includes a summer and a nonlinear activation function g.

The inputs x_k, $k = 1, \ldots, K$ to the neuron are multiplied by weights w_{ki} and summed up together with the constant bias term θ_i of the node i. The resulting n_i is the input to the activation function g. The activation function was originally chosen to be a relay function, but for mathematical convenience a hyberbolic tangent (*tanh*) or a sigmoid function are most commonly used. The hyberbolic tangent is defined as

$$\tanh(x) = \frac{1 - e^{-x}}{1 + e^{-x}} \tag{3.27}$$

The output of node i becomes

$$y_i = g(n_i) = g_i = g\left(\sum_{j=1}^{K} w_{ji} x_j + \theta_i \right) \tag{3.28}$$

By connecting several nodes in parallel and series, an MLP network is formed. A typical network is shown in Figure 3.6. It has K inputs x_k, two outputs y_1 and y_2, and all activation functions g are the same.

The output y_i, $i = 1, 2$, of the MLP network can be written as

$$y_i = g\left(\sum_{j=1}^{3} w_{ji}^2 g(n_j^1) + \theta_j^2 \right) = g\left(\sum_{j=1}^{3} w_{ji}^2 g\left(\sum_{k=1}^{K} w_{kj}^1 x_k + \theta_j^1 \right) + \theta_j^2 \right) \tag{3.29}$$

From Equation (3.29) we can conclude that an MLP network is a nonlinear parameterized map from input space $\mathbf{x} \in \mathbf{R}^K$ to output space $\mathbf{y} \in \mathbf{R}^m$ (here $m = 3$). The parameters are the weights w_{ji}^k and the biases θ_j^k. Superscript k refers to the layer, subscript j to the node and i to the input component. Activation functions g are usually assumed to be the same in each layer and known in advance. In the figure the same activation function g is used in all layers.

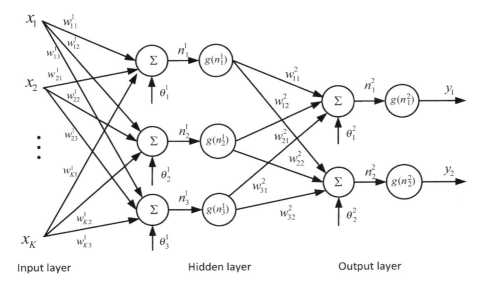

Figure 3.6 A multilayer perceptron network with one hidden layer. Here the same activation function g is used in both layers. The superscript of n, θ, or w refers to the layer, first or second

Given input–output data (x_i, y_i), $i = 1, \ldots, N$, finding the best MLP network, Equation (3.29) is formulated as a data-fitting problem as was done in Equation (3.22). The parameters to be determined are (w_{ji}^k, θ_j^k).

The procedure goes as follows. First the designer has to fix the structure of the MLP network architecture: the number of hidden layers and neurons (nodes) in each layer. The activation functions for each layer are also chosen at this stage, that is, they are assumed to be known. The unknown parameters to be estimated are the weights and biases, (w_{ji}^k, θ_j^k). These correspond to parameter vector **a** in Equation (3.22).

Many algorithms exist for determining the network parameters. In neural network literature the algorithms are called *learning* or *teaching* algorithms, in system identification they belong to *parameter estimation* algorithms. The most well-known are *back-propagation* and *Levenberg–Marquardt* algorithms. Back-propagation is a gradient-based algorithm, which has many variants. Levenberg–Marquardt is usually more efficient, but needs more computer memory. Here we will concentrate only on using the algorithms and omit the details.

Summarizing the teaching procedure for multilayer perceptron networks:

1. The structure of the network is first defined. In the network, activation functions for each layer are chosen and the network parameters, weights, and biases are initialized.
2. The parameters associated with the training algorithm such as error goal and maximum number of epochs (iterations) are defined.
3. The training algorithm is called.
4. After the neural network has been determined, the result is first tested by simulating the output of the neural network with the measured input data. This is compared with the measured outputs. Final validation must be carried out with independent data.

The MATLAB® commands used in the procedure are *newff, train,* and *sim.*

The MATLAB® command *newff* generates a new MLP neural network, which is called *net.*

$$net = newff\left(\underbrace{\text{x}}_{\substack{measured \\ input\ vector}}, \underbrace{\text{y}}_{\substack{measured \\ output\ vector}}, \underbrace{[S1\ S2...S(N-1)]}_{size\ of\ the\ ith\ layer}, \underbrace{\{TF1\ TF2...TFN\}}_{\substack{activation\ function \\ of\ ith\ layer}}, \underbrace{BTF}_{\substack{training \\ algorithm}}\right)$$

(3.30)

The inputs in Equation (3.30) are:

x = measured input vector
y = measured output vector
Si = number of neurons (size) in the *i*th layer, $i = 1,....,N-1;$ output layer size SN is determined from output vector y
N = number of layers
TFi = activation (or transfer) function of the *i*th layer, $i = 1,...,N;$ default = '*tansig*' (tanh) for all hidden layers and '*purelin*' (linear) for the output layer
BTF = network training function, default = '*trainlm*'

In Figure 3.6, $R = K$, $S1 = 3$, $S2 = 2$, $N1 = 2$ and $TFi = g$.

The default algorithm of command *newff* is Levenberg–Marquardt, *trainlm.* Default parameter values for the algorithms are assumed and are hidden from the user. They need not be adjusted in the first trials. Initial values of the parameters are automatically generated by the command. Observe that their generation is random and therefore the answer might be different if the algorithm is repeated.

After initializing the network, the network training is originated using the *train* command. The resulting MLP network is called *net*1.

$$net1 = train\left(\underbrace{net}_{\substack{initial \\ MLP}}, \underbrace{\text{x}}_{\substack{measured \\ input\ vector}}, \underbrace{\text{y}}_{\substack{measured \\ output\ vector}}\right)$$

(3.31)

The arguments are: *net,* the initial MLP network generated by *newff; x,* measured input vector of dimension K; and y measured output vector of dimension m.

To test how well the resulting MLP *net*1 approximates the data, the *sim* command is applied. The measured output is y. The output of the MLP network is simulated with the *sim* command and called *ytest.*

$$ytest = sim\left(\underbrace{net1}_{\substack{final \\ MLP}}, \underbrace{\text{x}}_{\substack{input \\ vector}}\right)$$

(3.32)

The measured output y can now be compared with the output of the MLP network *ytest* to see how good the result is by computing the error difference $e = y - ytest$ at each measured point. The final validation must be done with independent data.

Example 3.2

Consider the feedforward network in Figure 3.6 with two inputs, i.e. $K = 2$. The input vector is $\mathbf{x} = [x_1 \; x_2]$ and the output vector $\mathbf{y} = [y_1 \; y_2]$. Set up the MATLAB® command *newff* to generate a new MLP network with three neurons in the first layer and one neuron at the output layer.

Solution

The simplest command from Equation (3.30) becomes

$$net = newff \left(\underbrace{\mathbf{x}}_{input}, \; \underbrace{\mathbf{y}}_{output}, 3 \right) \tag{3.33}$$

The first argument gives the measured input vector $\mathbf{x}$, the second defines the measured output vector $\mathbf{y}$. The third argument defines the size of the hidden layer, which is three. Activation functions need not be defined, when the default *tansig* for the hidden layer and *purelin* (linear) in the second (output) layer are used. $\quad\square$

To illustrate the use of neural networks in path-loss modeling, we concentrate only on the simplest case, where $K = 1$, $\mathbf{x} = d$ (distance) a scalar, and $f(\mathbf{x})$ is the path-loss function $L_p(d)$. This is analogous to the problem treated in Example 3.1.

Example 3.3

Given the data set in Table 3.2, determine a neural network model (MLP) for the path loss L_p.

Solution

The input distance vector $\mathbf{d}_i$ is defined as $\mathbf{x}$ and the output vector $\mathbf{P}_i$ as $\mathbf{y}$. We follow the procedure suggested above using four neurons in the hidden layer and one at the output layer.

Table 3.2   Noisy measurement data for path-loss modelling

Distance from transmitter d_i (m)	Received power P_i (dBm)	Pi(d0)-10n log (di/do)	Pi(d0)-10n log (di/do)
50	0	0	0
100	−15	0.3	0.301029996
200	−19		
400	−27	0.9	0.903089987
600	−29		
800	−34	1.2	1.204119983
1000	−38		
1,600	−42	1.5	1.505149978
2,500	−56		
3,500	−71		
5,000	−86	2	2
d0 = 50 m			

```
% Input vector di
di =[50 100 200 400 600 800 1000 1600 2500 3500 5000];
% Output vector Pi
Pi =[0 -15 -19 -27 -29 -34 -38 -42 -56 -71 -86];

% Setting up the new MLPN ;
net=newff(di, Pi, 4)

% Train the network using the input-output data {di, Pi}; default
  training algorithm is trainlm
net1 = train(net, di, Pi);

%Simulate the output of the network when the input is di
  Ptest= sim(net1, di);

% Plot the result Ptest as a function of di and the measured data Pi as a
  function of di
plot(di,Ptest, '-k',di,Pi, '-.ok')
```

The result, which is now quite good, is displayed in Figure 3.7. It could be improved by further experimenting.   □

This straightforward example illustrates the use of neural networks in path-loss modeling. Neural networks provide an efficient means for finding powerful models for more complicated

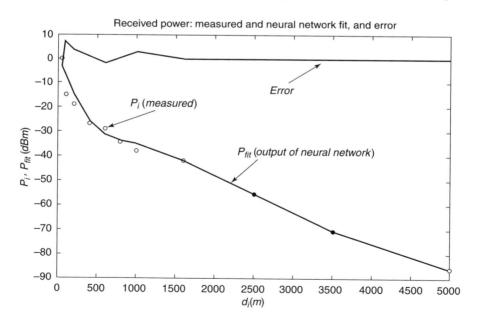

Figure 3.7   The result of a successful neural network fit

data sets. An important issue is that a sufficiently large data set exists for both data fitting and validation. In the example above this was not the case, because it was only meant to illustrate the learning process.

When neural networks are applied in data fitting, careful preprocessing of measurement signals must be done. In the case of path-loss modeling fast fading is removed by sampling the measurements over a certain distance and computing averages.

3.2.5 Shadow or Log-Normal Fading

The loss model in Equation (3.10) provides only the mean value of the signal strength. In reality, the signal is random. Measurements have shown that a good approximation can be achieved by adding normally distributed noise to Equation (3.10)

$$L_p = L_0 + 10n \log\left(\frac{d}{d_0}\right) + \xi \tag{3.34}$$

where ξ is a zero-mean normal random variable (in dB) with standard deviation σ (in dB). Therefore the path loss L_p is also a normal random variable ($\bar{L}_p$ is *log-normally* distributed).

Some clarifying words of the log-normal probability distribution are needed. If a random variable ξ is normally distributed, then $exp(\xi)$ has a log-normal distribution. Going the other way – the logarithm of a log-normal distributed variable is normally distributed. Log-normal distribution occurs when a variable is a product of many small independent factors, each of which multiplies the others. The *probability density function* (pdf) of a log-normal distribution is computed as follows.

Example 3.4   Computing probability density function of log-normal random variable

In Equation (3.34) ξ is a zero-mean normal random variable (in dB) with standard deviation σ (in dB). What is the probability density function of $\bar{\xi}$?

Solution

The probability density function of ξ is

$$f_\xi(\phi) = \frac{1}{\sigma\sqrt{2\pi}} \exp\left(-\frac{\phi^2}{2\sigma^2}\right) \tag{3.35}$$

Here σ is given in dB. Since ξ is in dB,

$$\xi = 10 \log\bar{\xi} \text{ or } \bar{\xi} = 10^{\frac{\xi}{10}} \tag{3.36}$$

We need to compute the pdf of $\bar{\xi}$. If Z and Y are random variables and $Z = g(Y)$, then according to probability theory the pdf of Z is obtained from

$$f_Z(z) = f_Y(g^{-1}(z)) \frac{dg^{-1}(z)}{dz} \tag{3.37}$$

Here $\bar{\xi} = Z$, $\xi = Y$ and $\bar{\xi} = g(\xi) = 10^{\frac{\xi}{10}}$, $\xi = g^{-1}(\bar{\xi}) = 10 \log\bar{\xi}$. Computing the derivative term in Equation (3.37) gives

$$\frac{d\xi}{d\bar{\xi}} = \frac{dg^{-1}(\bar{\xi})}{d\bar{\xi}} = \frac{1}{\bar{\xi}} \frac{10}{\ln 10}.$$

Using this and Equation (3.36) in Equation (3.37) results in

$$f_{\tilde{\xi}}(x) = \frac{10}{x(\ln 10)\sigma\sqrt{2\pi}} \exp(-(10\log x)^2/2\sigma^2)$$

(3.38)

This is the pdf of a log-normally distributed random variable.

3.3 Small-Scale Propagation Models and Statistical Multipath Channel Models

Large-scale propagation models are static, 'snapshot' models, where time is frozen. This means that time-varying signals can be considered as random variables (or constant, if no noise is present) at a given time instant *t*. In this way we can avoid treating them as stochastic processes. In this section time is introduced into models explicitly, but the situation remains the same – time is frozen. In Chapter 4, time series models are used to describe the time dependency of noisy signals. The reader should pay attention to this difference, because both situations are used in the literature. A noisy power signal *P(t)* at a fixed time *t* will represent a random variable and notation *P* should be used for it (dropping) *t*.

When a mobile unit is moving and transmitting a signal, the received signal can experience severe fading even in the space of a few meters. It is generally agreed that the small-scale fading is due to two phenomena: multipath fading and Doppler.

3.3.1 Multipath Fading

When the mobile is moving in an urban environment there is usually no LOS signal coming from the base station. Instead, several weaker signals are received, which have been reflected from buildings and other obstacles as seen in Figure 3.1. The transmitted signal produces a multitude of signals when bouncing off the ground and from other structures. The signals take different paths to reach the mobile unit. This is called the *multipath phenomenon*.

The signal paths have usually different lengths, so the signals are not received at the same time. They will have different amplitudes and phase shifts. Depending on their amplitudes and phase shifts they can either amplify or attenuate each other. Even at walking speed the signal fluctuations can be large, not to mention if the transceiver is in a moving car. The following simple example illustrates this.

Example 3.5

A narrowband sinusoidal signal is transmitted with carrier frequency $f_c = 1/2\pi$ Hz. The received signal $s(t)$ is a sum of the LOS signal $a_1(t)$ and a reflected signal $a_2(t)$

$$s(t) = a_1(t) + a_2(t) = A_1\cos(2\pi f_c t) + A_2\cos(2\pi f_c(t-\tau))$$

(3.39)

Let the amplitudes of the signals have numerical values $A_1 = 2$ and $A_2 = 3$ and let the time delay τ assume values in the interval $[0, 4\pi]$. Assume an ideal noiseless case. Study the effect of time delay τ on the amplitude of *s(t)*.

Solution

Substituting the numerical values into Equation (3.39) and using formula from trigonometry we rewrite $s(t)$ as

$$
\begin{aligned}
s(t) &= 2\cos t + 3\cos(t-\tau) \\
&= 2\cos t + 3\cos t \cos\tau + 3\sin t \sin\tau \\
&= \cos t(2+3\cos\tau) + 3\sin t \sin\tau
\end{aligned}
\tag{3.40}
$$

Or in a simpler form

$$
s(t) = A(\tau)\cos(t+\theta),
\tag{3.41}
$$

where the amplitude

$$
A(\tau) = \sqrt{13 + 12\cos\tau}
\tag{3.42}
$$

and the phase

$$
\theta(\tau) = \tan^{-1}(3\sin\tau/(2+3\cos\tau))
\tag{3.43}
$$

The amplitude $A(\tau)$ depends on delay τ. It achieves maximum $A_{max} = 5$ at $\tau = 0, 2\pi, 4\pi$ and minimum $A_{min} = 1$ at $\tau = \pi, 3\pi$. Using MATLAB$^\circledR$ A is plotted as a function of τ. The result is shown in Figure 3.8. The code is given as

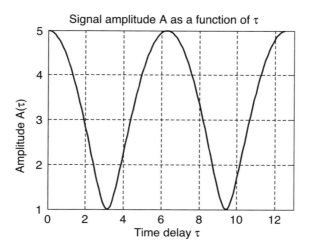

Figure 3.8 Amplitude A of the received signal $s(t)$ as a function of time delay τ. Observe how the multipath phenomenon can vary the amplitude of the signal significantly

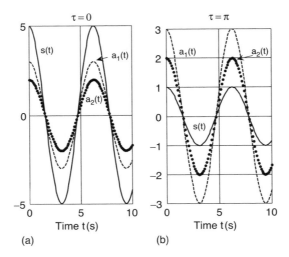

Figure 3.9 The effect of τ on amplitude of $s(t)$ (solid line): (a) signals $a_1(t)$ and $a_2(t)$ are constructive when $\tau = 0$; and (b) destructive, when $\tau = \pi$

```
%τ varies between 0 and 4π. Sampling interval is 0.1.
tau=0:0.1:4*pi;% create data
A=sqrt(13+12*cos(tau));% find values of A(t)
plot(tau,A)% plot A as a function of τ
```

The effect of strongly varying amplitude on the received signal is demonstrated in Figure 3.9, where the composite signal $s(t)$ in (a) has a much larger amplitude than the signal in (b). □

In reality the signals in the radio channel will not be deterministic, but random. Before discussing the general case of multipath signals we will consider characteristics of narrowband signals. Radio signals are typically narrowband signals, where the energy is centered around the carrier frequency f_c. A narrowband signal may be written in the form

$$s(t) = A(t) \cos\left(2\pi f_c t + \theta(t)\right) \tag{3.44}$$

where $A(t)$ is the *amplitude* or *envelope* of $s(t)$ and $\theta(t)$ the *phase* of $s(t)$. Both $A(t)$ and $\theta(t)$ are considered sample functions of low-pass stochastic processes with bandwidths much smaller than the carrier frequency. Expanding Equation (3.44) we obtain

$$s(t) = A(t) \cos\left(2\pi f_c t\right) \cos\theta(t) - A(t) \sin\left(2\pi f_c t\right) \sin\theta(t) \tag{3.45}$$

Defining

$$s_I(t) = A(t) \cos\theta(t) \quad s_Q(t) = A(t) \sin\theta(t) \tag{3.46}$$

Equation (3.45) can be expressed as

$$s(t) = s_I(t) \cos\left(2\pi f_c t\right) - s_Q(t) \sin\left(2\pi f_c t\right) \tag{3.47}$$

where $s_I(t)$ is called the *in-phase component* of $s(t)$ and $s_Q(t)$ the *quadrature component* of $s(t)$. Equation (3.47) is a *canonical representation* of a band-pass signal. Observe that Equation (3.44) may be also written in the form

$$s(t) = \Re\{u(t)e^{j2\pi f_c t}\} = \Re\{(s_I(t) + js_Q(t))e^{j2\pi f_c t}\} = \Re\{A(t)e^{j(2\pi f_c t + \theta(t))}\} \quad (3.48)$$

where $u(t)$ is called the *complex envelope* of the signal. From Equations (3.44) and (3.47) the amplitude $A(t)$ can be expressed as

$$A(t) = \sqrt{s_I^2(t) + s_Q^2(t)} \quad (3.49)$$

and the phase $\theta(t)$

$$\theta(t) = \tan^{-1}(s_Q(t)/s_I(t)) \quad (3.50)$$

For any fixed time t the in-phase component $s_I(t)$ and the quadrature component $s_Q(t)$ are random variables S_I and S_Q. They are assumed to be uncorrelated and thus independent and identically normally distributed variables with zero mean and same variance σ^2, or simply $S_I, S_Q \in N(0, \sigma^2)$. The joint density function of S_I and S_Q may be written as

$$f_{S_I S_Q}(s_I, s_Q) = \frac{1}{2\pi\sigma^2}\exp\left\{-\frac{1}{2}\frac{s_I^2 + s_Q^2}{\sigma^2}\right\} \quad (3.51)$$

The joint probability density function of the amplitude $A(t)$ and the phase $\theta(t)$ can be derived by first rewriting the mapping in Equations (3.49) and (3.50) in vector form (ignoring the dependence on t). Let $\mathbf{y} = [y_1 \quad y_2]^T = [A \quad \theta]^T$ and $\mathbf{x} = [x_1 \quad x_2]^T = [s_I \quad s_Q]^T$, then

$$\mathbf{y} = \begin{bmatrix} y_1 \\ y_2 \end{bmatrix} = \begin{bmatrix} A \\ \theta \end{bmatrix} = \begin{bmatrix} \sqrt{s_I^2 + s_Q^2} \\ \tan^{-1}(s_Q/s_I) \end{bmatrix} = \begin{bmatrix} g_1(s_I, s_Q) \\ g_2(s_I, s_Q) \end{bmatrix} = \begin{bmatrix} g_1(x_1, x_2) \\ g_2(x_1, x_2) \end{bmatrix} = \mathbf{g}(\mathbf{x}) \quad (3.52)$$

The inverse mapping (Equation (3.46)) is written similarly

$$\mathbf{x} = \begin{bmatrix} x_1 \\ x_2 \end{bmatrix} = \begin{bmatrix} s_I \\ s_Q \end{bmatrix} = \begin{bmatrix} A\cos\theta \\ A\sin\theta \end{bmatrix} = \begin{bmatrix} g_1^{-1}(A, \theta) \\ g_2^{-1}(A, \theta) \end{bmatrix} = \begin{bmatrix} g_1^{-1}(y_1, y_2) \\ g_2^{-1}(y_1, y_2) \end{bmatrix} = \mathbf{g}^{-1}(\mathbf{y}) \quad (3.53)$$

The pdf of $\mathbf{y}$ is then given as

$$f_y(\mathbf{y}) = f_x(\mathbf{g}^{-1}(\mathbf{y}))|J_I(\mathbf{y})| \quad (3.54)$$

where the Jacobian of the inverse transformation is

$$J_I(\mathbf{y}) = \det\left\{\frac{\partial \mathbf{g}^{-1}}{\partial \mathbf{y}}\right\} \quad (3.55)$$

Computing $J_I(\mathbf{y})$ from Equation (3.53) results in

$$J_I(\mathbf{y}) = \begin{vmatrix} \dfrac{\partial g_1^{-1}}{\partial y_1} & \dfrac{\partial g_1^{-1}}{\partial y_2} \\[2mm] \dfrac{\partial g_2^{-1}}{\partial y_1} & \dfrac{\partial g_2^{-1}}{\partial y_2} \end{vmatrix} = \begin{vmatrix} \cos\theta & -A\sin\theta \\ \sin\theta & A\cos\theta \end{vmatrix} = A \tag{3.56}$$

and thus the pdf of $\mathbf{y}$ which is the joint pdf of $(y_1, y_2) = (A, \theta)$ becomes

$$f_{\mathbf{y}}(A) = \begin{cases} \dfrac{A}{2\pi\sigma_A^2}\exp\left(-\dfrac{A^2}{\sigma_A^2}\right), & 0 \le A \\[3mm] 0, & A < 0 \end{cases} \tag{3.57}$$

From probability theory it follows that if S_I and S_Q are independent so are functions of individual variables, that is, A and θ are independent. From Equation (3.50) we can conclude that the range of θ is $[-\pi/2 \quad \pi/2]$. We are especially interested in the marginal distribution of A. This is obtained by integrating over θ, which is uniformly distributed over $[-\pi/2 \quad \pi/2]$

$$f_A(A) = \begin{cases} \displaystyle\int_{-\pi/2}^{\pi/2} f_y(\mathbf{y})dy_2 = \dfrac{A}{\sigma_A^2}\exp\left(-\dfrac{A^2}{\sigma_A^2}\right), & A \ge 0 \\[4mm] 0, & \text{elsewhere} \end{cases} \tag{3.58}$$

with $E\{A\} = \sigma_A\sqrt{\pi/2}$ and $E\{A^2\} = 2\sigma_A^2$.

This is the well-known *Rayleigh distribution function*. A good mental model of the distribution is to think of throwing darts at a dartboard. Then the probability of hitting the bull's eye is Rayleigh distributed, where A is the distance from the center.

The Rayleigh distribution of the envelope can be demonstrated with a Simulink® simulation. This is configured in Figure 3.10. The clock generates time t. Inputs from the source blocks $A1$ and $A2$ are normally distributed random numbers at any time instant t. The means and variances can be adjusted. Different seed numbers are chosen for random number generation. Default values are zero mean and unit variance. The rest of the configuration follows from Equations (3.49) and (3.50): The variables $A1$ and $A2$ are squared, summed, and a square root (*sqrt*) is taken of the sum so as to generate the time-dependent envelope function $A(t)$. The results are taken to workspace for plotting. MATLAB® command *plot(t,A1,t,A)* is then applied. Results are shown in Figure 3.11. The behavior of signal $A2$ is similar to that of $A1$ and is not plotted. Further evidence is seen in Figure 3.12, in which a histogram of variable A is presented. This is produced by MATLAB® command hist(A, 100), where 100 indicates the number of containers. The simulation is run up to 100,000 times, so that sufficient data is available.

The histograms in Figure 3.12 show that Simulink® configuration in Figure 3.10 produces appropriate distributions.

3.3.2 Rayleigh Fading Channel

We will now consider a more general case of *multipath fading* (multipath propagation was demonstrated in Figure 3.1).

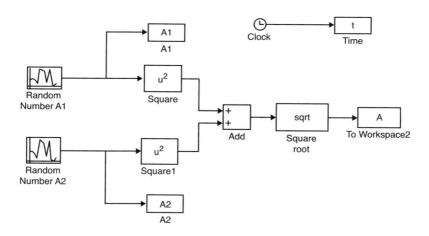

Figure 3.10 Simulink® configuration of Rayleigh distribution

To simplify matters, assume the transceivers are stationary and that there is no noise in the channel. Let the narrowband transmitted signal $x(t)$ be the real part of a complex baseband signal $\tilde{s}(t) = s_I(t) + js_Q(t)$ modulated with a carrier frequency f_c similarly to the signal in Equation (3.48)

$$x(t) = \Re e\{\tilde{s}(t)e^{j2\pi f_c t}\} \tag{3.59}$$

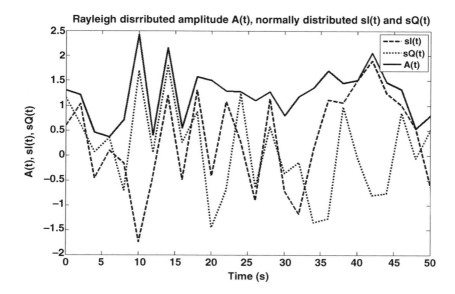

Figure 3.11 Normally distributed $s_I(t), s_Q(t) \in N(0, 1)$ can assume negative values. Rayleigh distributed amplitude A with $E\{A\} = \sqrt{\pi/2}$ and $E\{A^2\} = 2$. Observe especially the scales of the y-axis. Envelope $A(t)$ cannot have negative values

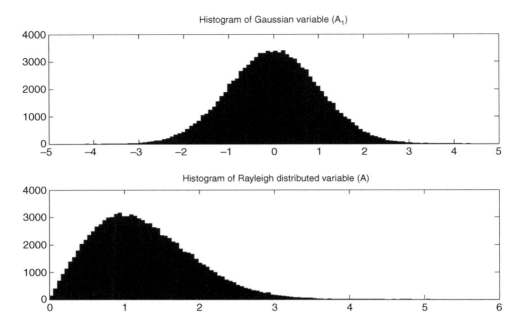

Figure 3.12 Histogram of Gaussian distributed A_1 above and Rayleigh distributed variable A below

The received signal $r(t)$ consists of the sum of attenuated transmitted signals $x(t)$ of Equation (3.59) with different delays $\tau_i(t)$, $i = 1, \cdots, N$ due to multipath propagation

$$
\begin{aligned}
r(t) &= \sum_{i=1}^{N} a_i(t) x(t - \tau_i(t)) = \Re\left\{ \sum_{i=1}^{N} a_i(t) \tilde{s}(t - \tau_i(t)) e^{j2\pi f_c(t - \tau_i(t))} \right\} \\
&= \Re\left\{ \left[\sum_{i=1}^{N} a_i(t) e^{-j2\pi f_c \tau_i(t)} \right] \tilde{s}(t - \tau_i(t)) e^{j2\pi f_c t} \right\}
\end{aligned}
\tag{3.60}
$$

where N is the number of paths (which can vary with time), $x(t)$ is the transmitted band-pass signal, $a_i(t)$ is the attenuation and $\tau_i(t)$ the corresponding propagation delay of channel i. As in the treatment of the previous section, all these may vary randomly and are described by stochastic processes. We assume them to be *stationary* and *ergodic*.

Equation (3.60) can further be written as

$$
r(t) = \Re\left\{ \tilde{r}(t) e^{j2\pi f_c t} \right\}
\tag{3.61}
$$

where the complex envelope of the received signal $\tilde{r}(t)$ is

$$
\tilde{r}(t) = \sum_{i=1}^{N} a_i(t) e^{-j2\pi f_c \tau_i(t)} \tilde{s}(t - \tau_i(t))
\tag{3.62}
$$

The complex gain $Z(t)$ of the channel in Equation (3.62) is

$$
Z(t) = \sum_{i=1}^{N} a_i(t) e^{-j2\pi f_c \tau_i(t)} = Z_{RE}(t) - j Z_{IM}(t)
\tag{3.63}
$$

The real and complex parts are defined as

$$Z_{RE}(t) = \sum_{i=1}^{N} a_i(t) \cos \theta_i \tag{3.64}$$

$$Z_{IM}(t) = \sum_{i=1}^{N} a_i(t) \sin \theta_i \tag{3.65}$$

where $\theta_i = 2\pi f_c \tau_i(t)$. Here $Z_{RE}(t)$ and $Z_{IM}(t)$ are orthogonal, that is, their product integrated over a period becomes zero. The *amplitude fading* and *carrier distortion* are given respectively as

$$A(t) = \sqrt{Z_{RE}(t)^2 + Z_{IM}(t)^2}, \tag{3.66}$$

$$\theta(t) = \tan^{-1}(Z_{IM}(t)/Z_{RE}(t)) \tag{3.67}$$

We assume that there is no LOS signal and further that for any fixed time t ('snapshot assumption') the amplitudes a_i are independent and identically distributed variables with zero mean and variance σ^2. The phases θ_i, $i = 1, \ldots, N$, are independent and identically uniformly distributed variables over the interval $[-\pi/2, \pi/2]$ due to isotropic transmission. Amplitudes a_i and phases θ_i, $i = 1, \ldots, N$, are considered independent, because the former depend on attenuation and the latter on delay.

Since the number of paths in radio transmission is large, we can invoke the *central limit theorem* from statistics. This tells us that for sufficiently large N the sum of independent and identically distributed random variables has approximately normal distribution. Since amplitudes a_i and phases θ_i, $i = 1, \ldots, N$, are considered independent, we can concentrate on the marginal distributions of a_i in Equation (3.64) and (3.65) (θ_i being uniformly distributed on $[-\pi/2, \pi/2]$ are integrated out). Thus $Z_{RE}(t)$ and $Z_{IM}(t)$ can be considered independent normally distributed random variables, $Z_{RE}(t), Z_{IM}(t) \in N(0, \frac{1}{2}\sum_{i=1}^{N} a_i^2)$, $i = 1, \ldots, N$. If this is the case, as in the previous section, the amplitude fading A follows a Rayleigh distribution

$$f_A(a) = \begin{cases} \dfrac{a}{\sigma_a^2} \exp\left(-\dfrac{a^2}{2\sigma_a^2}\right), & a \geq 0 \\ 0, & a < 0 \end{cases} \tag{3.68}$$

with $E\{A(t)\} = \sigma_a\sqrt{\pi/2}$ and $E\{A^2(t)\} = 2\sigma_a^2 = \sum_{i=1}^{N} a_i^2$.

The phase $\theta(t)$ in Equation (3.68) is uniformly distributed over $[-\pi/2, \pi/2]$. This is called a *Rayleigh fading channel*.

If the received signal also has an LOS component, then the channel becomes a *Rician fading channel*. The Rician probability density function for the amplitude A is

$$f_A(a) = \begin{cases} \dfrac{a}{\sigma_a^2} I_0\left(\dfrac{a a_0}{\sigma_a^2}\right) \exp\left(-\dfrac{a^2 + a_0^2}{2\sigma_a^2}\right), & a \geq 0 \\ 0, & a < 0 \end{cases} \tag{3.69}$$

Here a_0^2 is the power of the LOS component and I_0 is the zero-order *modified Bessel function* of the first kind. MATLAB® provides the command *BESSELI(NU,Z)* for computing a modified Bessel function of the first kind.

```
% Assume a₀ = 1; σ²_A = 1
% Modified 0th order Bessel 1 function I0(a) is computed on the
  interval (-6,6) at 100 points;

a=linspace(-6,6); % Data for argument a
y=BESSELI(0,a); % Computation of the I0(a)
plot(a,y) % Plotting result
```

At the origin $y = BESSELI(0,0) = 1$. The result is displayed in Figure 3.13.

To characterize the Rician channel a K-factor is defined

$$\bar{K} = \frac{\text{Power of deterministic LOS component}}{\text{Power of all multipath components}} = \frac{a_0^2}{2\sigma_A^2} \quad (3.70)$$

or in dB

$$K = 10 \log \frac{a_0^2}{2\sigma_A^2} \quad (3.71)$$

It is useful to rewrite Equation (3.70) as

$$f_A(a) = \underbrace{\frac{a}{\sigma_A^2} \exp\left(-\frac{a^2}{2\sigma_A^2}\right)}_{Rayleigh} \underbrace{\exp\left(-\frac{a_0^2}{2\sigma_A^2}\right) I_0\left(\frac{aa_0}{\sigma_A^2}\right)}_{Modifier} \quad (3.72)$$

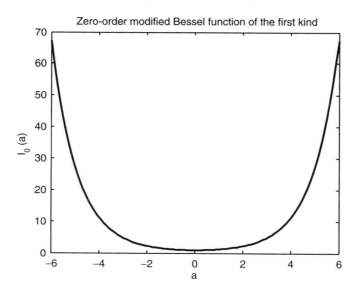

Figure 3.13   Zeroth order modified Bessel function of the first kind

and substituting the K factor from Equation (3.71) into Equation (3.72)

$$f_A(a) = \underbrace{\frac{a}{\sigma_A^2} \exp\left(-\frac{a^2}{2\sigma_A^2}\right)}_{Rayleigh} \underbrace{\exp(-\bar{K})I_0\left(2\bar{K}\frac{a}{a_0}\right)}_{Modifier} \qquad (3.73)$$

The K factor provides a measure for fading. For $\bar{K} = 0$ or $-\infty$ dB, there is no LOS component and we have Rayleigh fading. When $K \to \infty$ dB, there are no multipath components and we approach an *additive white Gaussian noise (AWGN)* channel.

It is instructive to plot the Rician pdf as K varies (in dB). This is straightforward with MATLAB® commands. Assume that $\sigma_A^2 = 1$, then for a given $\bar{K}$, a_0 can be computed. For example, if $\bar{K} = 1$, then from Equation (3.70) $a_0 = \sqrt{2}$.

Plot the Rayleigh pdf from Equation (3.73). Let $\bar{K}$ vary from 0 to 100.

```
a=linspace(0,10,100); % 100 points on the interval [0,10]K=[0 0.1
0.4 1 4 10 100] ; %
sq=sqrt(2*K);
for i=1:5
I = BESSELI(0,sqrt(2*K(i)).*a); % K=different values
hold on
mod0=exp(-K(i))*I; % compute the modifier function
ri=ry.*mod0; %
plot(a,ri)
end
XLabel('a')
YLabel('f_A (a) ')
hold off
```

The plots are shown in Figure 3.14, in which we have also collected Rician pdfs with different values of K. We can see how the Rician pdfs converge towards the Rayleigh pdf as $K \to -\infty$ dB. This figure verifies our previous statement that as the influence of the LOS component decreases with decreasing K, fading dominates. With increasing K, the channel model becomes an AWGN channel.

Although Rayleigh and Rician distributions give a good basic understanding of the channel behavior of fast fading, they do not cover all situations. Many other distribution models have been developed. The Nakagami distribution has been widely used to model multipath fading. The models provide first-order statistics over distances short enough for the mean level to be considered constant. The Jakes model is often used for a Rayleigh fading simulator to handle the time-correlation behavior due to the Doppler shift. It is based on a sum of sinusoids.

There are many sources that provide means of producing Rayleigh fading or something that is close to it. They all mimic quite well the characteristics of fast fading. For the purposes of this book any of them can be used. Here is a list of some of the software:

1. MATLAB® Communications Blockset provides ready-made blocks for both Rayleigh and Rician fading.

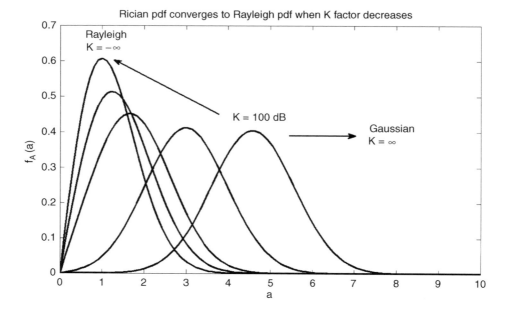

Figure 3.14 Convergence of Rician pdf to Rayleigh pdf, when *K* factor starts from 100 dB and approaches −∞ dB

2. MATLAB® Central has a few alternative programs for producing Rayleigh fading.
3. A way to model Rayleigh fading is to use MATLAB® command *exprand*.

We demonstrate the first two of these and show the results.

1. MATLAB® Communications Blockset. Here a *Multipath Rayleigh Fading* (and *Rician*) Simulink® block exists. Its basic use is straightforward. An example in the User's Guide helps you into it. Once you open a new model file in Simulink®, go to Signal Processing Blockset. There you'll find Signal Processing Sources. Opening it, choose *DSP constant block* and drag it to your file. It will serve as your input. Next go to Communications Blockset and choose Channels. There pick *Multipath Rayleigh Fading Channel* and drag the block to your model file. The final thing you need is a sink block, *To Workspace block*, which can be found from regular Simulink®. The configuration is shown in Figure 3.15. It looks innocent enough, but a number of block parameters need to be set up.
   − Open *DSP Constant block*: Set Constant value = j*ones(10000,1). In Output choice, pick Frame-based output. Set Frame period = 0.01.
   − Open *Multipath Rayleigh Fading Channel*. Set Delay vector = [0 2e-6 3e-6]. Set Gain vector = [0 −3 1]. In *To Workspace block* change Save Format from Structure to Array. (This is done in all simulation examples carried out in the book.)
   − The period of simulation is changed from 10 to 0.6. Once the simulation has been carried out go to MATLAB® command side. The simulation result is the faded signal's power and can be plotted versus the sample number by command: *P = P.'; plot(10*log10 (abs(P(:))))*.

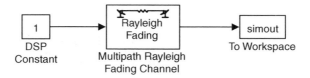

Figure 3.15 Configuration of Simulink® model of a Rayleigh fading signal using Multipath Rayleigh Fading block from Communications Blockset and DSP Constant from Signal Processing Toolbox

The result is shown in Figure 3.16. The reader is referred to the Communications Blockset User's Guide (2009) and to perform further simulations. Here we are mostly interested in the result, not further details.

2. MATLAB® Central has a few alternative programs for producing Rayleigh fading using Jakes model (Jakes, 1974) or its many modifications. A result of one is shown in Figure 3.17.

In this case carrier frequency $f_c = 900$ Mhz, speed is 5 km/h, and sample interval $= 10^{-4}$. The result is very similar to that of Figure 3.16.

Until now we have only considered spatial models. We have assumed the 'snapshot' approach, that is, time is frozen so that only spatial variation is considered. Rayleigh fading in Figure 3.16 (or Figure 3.17) is a representative at certain time instant t_1. In the figure the x-axis refers to distance. At a different time instant, t_2, we obtain a similar (not the same)

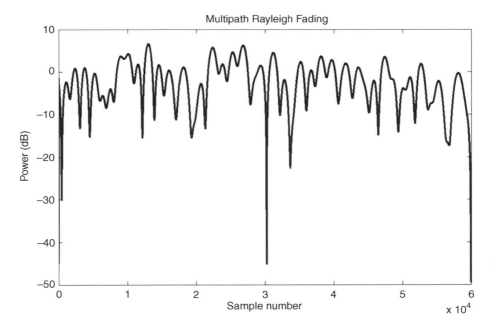

Figure 3.16 Result of Rayleigh fading signal simulation, when Figure 3.15 model is used

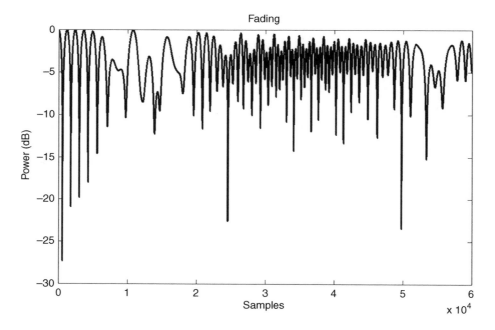

Figure 3.17   Result of Rayleigh fading, when MATLAB® code from MATLAB® Central is used

representative of Rayleigh fading. This means that in time-domain simulation we can use the above results.

We can summarize our findings from Equations (3.8), (3.36) and (3.68) using gains $\bar{G}$ and interpreting all variables in linear scale:

$$\bar{G}_1 = \alpha d^{-n}. \tag{3.74}$$

Large-scale path loss

$$\bar{G}_2 = 10^{\frac{\xi}{10}} \tag{3.75}$$

Shadow fading

$$\bar{G}_3 = \bar{P}_A = A^2. \tag{3.76}$$

Fast multipath fading

The overall attenuation of the signal power in linear scale is the product

$$\bar{G} = \bar{G}_1 \bar{G}_2 \bar{G}_3 = \alpha d^{-n} 10^{\frac{\xi}{10}} A^2. \tag{3.77}$$

In these equations d is the distance between the transmitter and the receiver, n is a real valued number ≥ 2 (n varies, in free space $n = 2$, in flat rural areas $n = 3$ and in dense urban areas $n = 4.5$), ξ is a log-normally distributed random variable with zero dB mean and a standard deviation of typically 4 to 12, the fast amplitude fading A is a random variable having Rayleigh, Rician, or Nakagami distribution.

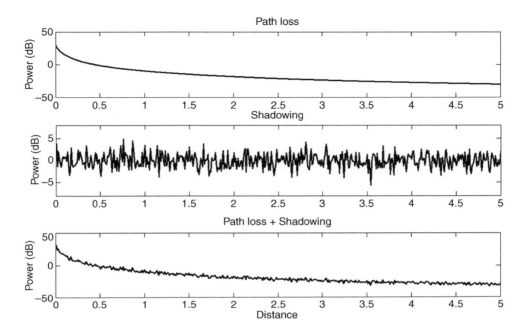

Figure 3.18 These simulations demonstrate the different amplitude attenuations in dB. Above path loss, in the middle shadowing, and below the sum of the two. Fast fading is shown in Figure 3.16

In logarithmic scale Equation (3.77) becomes

$$G = G_1 + G_2 + G_3 = \alpha_{dB} - n \log_{10}d + A^2. \tag{3.78}$$

It is worthwhile emphasizing here that this equation summarizes the 'snapshot' model for first-order statistics. The overall attenuation G in Equation (3.78) depends only on d at time t. The first term is constant and the second linear in $\log d$, the third term ξ is a normally distributed random variable, $\xi \in N(0, \sigma^2)$ and the fourth term A is Rayleigh distributed. Figure 3.18 illustrates the first two fadings in dB scale. Fast fading was depicted in Figure 3.16.

In Chapter 4 we will start studying the temporal aspects more closely.

3.4 Summary

In this chapter models of radio channels have been discussed. A complete model is provided by Maxwell's equations, but these are too detailed for use in radio communications. First, large-scale path-loss models were covered. Line-of-sight path, reflection, diffraction, and scattering were discussed. Since physical modeling easily becomes tedious, statistical path-loss models are used and were explained. The path-loss models describe the behavior of the mean value of the signal strength. Since radio signals are random, a good approximation is to add normally distributed noise (in dB) to the path-loss model. Therefore the path loss in linear scale is log-normally distributed. This is also called shadowing. Path-loss models are essential, basic tools in the design of cellular networks.

When a mobile unit is moving and transmitting, the received signal can fade very quickly in the space of a few meters. Compared to shadowing this happens very fast. Such a phenomenon is called small-scale fading and is due to multipath fading and the Doppler effect. Multipath fading was thoroughly discussed in the context of Rayleigh fading. In the following the Rayleigh fading model is used extensively. The model can be formulated in different ways. Software exists to provide tools for producing the channel data, such as, MATLAB® Communications Blockset, software in MATLAB® Central, or a simple model using exponential distribution.

Many textbooks in wireless communication cover the channel modeling topics of this chapter, such as Ahlin, Zander and Slimane (2006), Garg and Wilkes (1996), Haykin and Moher (2005), Mark and Zhuang (2003), Rappaport (2001) and Stüber (2000). Excellent textbooks exist in neural networks such as Haykin (2009). Here the Neural Network Toolbox (Demuth, Beale and Hagan (2009)) is applied. In addition to path-loss modeling, neural networks have gained popularity e.g. in cell design.

4

Channel Estimation and Prediction

4.1 Introduction

In this chapter we explore temporal, time-varying models, which are required to model multipath fast changes in the channel. These are called *small-scale fading* models. Both scalar and multivariable cases are discussed, first in continuous setting and then in discrete. In the latter part of the chapter linear time-invariant discrete models are emphasized leading to parameter estimation, prediction, and self-tuning prediction.

Terminology

- **Linear time-variant system:** a linear system in which parameters of the system vary with time.
- **Multiple-input multiple-output (MIMO) system:** has several inputs and several outputs. An example of such a system is multiple antennas at both transmitter and receiver. A **multiple-input single-output (MISO) system** has several inputs and one output; a **single input single-output (SIMO) system** has one input and many outputs.
- **Estimation:** either state estimation or parameter estimation is based on available measurement history. In this book the emphasis is on estimation theory of dynamical systems. The system can be linear or nonlinear, continuous or discrete.
- **Prediction:** forming a state estimate of a future state at current time based on measurement history. If current state is estimated, the estimator is called a filtered estimate. If earlier states are estimated, the estimate is called a smoothed estimate.

4.2 Linear Time-Variant (LTV) Channel Model

In our discussion of multipath fading we did not pay attention to the movement of transceivers. Let us return to Chapter 3, Section 3.3.2 on Rayleigh fading. The expression for the received signal $r(t)$ consists of a sum of attenuated transmitted signals $x(t)$ with different delays due to multipath fading and is given by Equation (3.60) and repeated here:

$$r(t) = \sum_{i=1}^{N} a_i(t)x(t - \tau_i(t)) \tag{4.1}$$

where the transmitted baseband signal $x(t)$ is delayed by a propagation delay $\tau_i(t)$ for each path. This is further written as

$$r(t) = \Re\left\{ \sum_{i=1}^{N} a_i(t)\tilde{s}(t - \tau_i(t))e^{j2\pi f_c(t - \tau_i(t))} \right\} \tag{4.2}$$

where $\tilde{s}(t)$ is the complex baseband signal.

When a mobile station is moving, a delay is seen not only in time t, but also in the carrier frequency f_c. Consider a car with a transceiver starting from the base station and driving away linearly from it with velocity v.

The radio signal experiences a phenomenon well known from basic physics, a *Doppler effect*. When an airplane flies over, its sound will have a different pitch depending if it is coming toward you or flying away. The same happens with radio waves. The Doppler shift effect changes f_c in Equation (4.1) to $f_c(1 - v/c)$ if the mobile is level with the transmitter

$$
\begin{aligned}
r(t) &= \Re\left\{ \sum_{i=1}^{N} a_i(t)\tilde{s}(t - \tau_i(t))e^{j2\pi(f_c - vf_c/c)(t - \tau_i(t))} \right\} \\
&= \Re\left\{ \sum_{i=1}^{N} a_i(t)\tilde{s}(t - \tau_i(t))e^{j2\pi(f_c + f_D)t}e^{-j2\pi(f_c + f_D)\tau_i(t)} \right\}
\end{aligned} \tag{4.3}
$$

where the Doppler frequency is defined as $f_D = -f_c v/c$. For a more general case, when the mobile unit moves away at an angle ϕ from the base station, as in Figure 4.1:

$$f_D = -f_c \frac{v\cos\phi}{c} \tag{4.4}$$

Equation (4.3) tells us that the received signal is not only time dependent but also frequency dependent. To simplify the notation introduce

$$\theta_i(t) = -2\pi(f_D t - (f_c + f_D)\tau_i(t)) \tag{4.5}$$

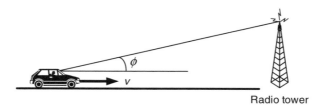

Radio tower

Figure 4.1   Mobile station moves away at an angle ϕ from the base station

Then Equation (4.3) becomes

$$r(t) = \Re\left\{ \sum_{i=1}^{N} a_i(t)\tilde{s}(t - \tau_i(t))e^{-j\theta_i(t)}e^{j2\pi f_c t} \right\}$$

$$= \Re\{\tilde{r}(t)e^{j2\pi f_c t}\} \tag{4.6}$$

where $\tilde{s}(t)$ is the complex envelope of the received signal and

$$\tilde{r}(t) = \sum_{i=1}^{N} a_i(t)\tilde{s}(t - \tau_i(t))e^{-j\theta_i(t)} \tag{4.7}$$

Since $a_i(t)$ depends on attenuation and $\theta_i(t)$ on delay and Doppler, a reasonable assumption is that the two stochastic processes are independent.

We assume that the channel is linear and noiseless. Equation (4.7) can be represented as an *impulse response model*

$$\tilde{r}(t) = \int_{-\infty}^{\infty} \tilde{h}(t, \tau)\tilde{s}(t - \tau)d\tau \tag{4.8}$$

Here the time-variant impulse response $\tilde{h}(t, \tau)$ is

$$\tilde{h}(t, \tau) = \sum_{i=1}^{N} a_i(t)\delta(\tau - \tau_i(t))e^{-j\theta_i(t)} \tag{4.9}$$

Impulse function $\delta(t)$ is defined by

$$\int_{-\infty}^{\infty} \delta(t - \tau)g(t)dt = g(\tau), \qquad \int_{-\infty}^{\infty} \delta(t)dt = 1 \tag{4.10}$$

Note: we invariably use *channel* or *impulse response* when referring to channel model.

In systems theory the impulse response formula, Equation (4.8), is often written in the equivalent form

$$y(t) = \int_{t_0}^{t} g(t, \tau)u(\tau)d\tau \tag{4.11}$$

where $y(t)$ is the output and $u(t)$ the input of the system. Special cases of Equation (4.9) are *time-selective channels* and *frequency-selective channels*:

4.2.1 Time-Selective Channel

In this case the path lengths are assumed to have the same delay and therefore the impulse response becomes

$$h(t, \tau) = \sum_{i=1}^{N} a_i(t)e^{-j\theta_i(t)}\delta(\tau) = \tilde{a}(t)\delta(\tau) \tag{4.12}$$

where

$$\tilde{a}(t) = \sum_{i=1}^{N} a_i(t) e^{-j\theta_i(t)} \tag{4.13}$$

4.2.2 Frequency-Selective Channel

If the receiver and the transmitter are stationary, the gains in Equation (4.9) are assumed constant and therefore

$$h(t,\tau) = \sum_{i=1}^{N} a_i e^{-j\theta_i} \delta(\tau - \tau_i(t)) = \sum_{i=1}^{N} \tilde{a}_i \delta(\tau - \tau_i(t)), \quad \tilde{a}_i = a_i e^{-j\theta_i} \tag{4.14}$$

Considering only the real parts in Equation (4.8) the impulse response model is written as

$$r(t) = \int_{-\infty}^{\infty} h(t,\tau) x(t-\tau) d\tau \tag{4.15}$$

where

$$h(t,\tau) = \sum_{i=1}^{N} a_i(t) \delta(\tau - \tau_i(t)) \tag{4.16}$$

If the channel is time-invariant, the impulse response becomes $h(t,\tau) = h(t - \tau)$. Therefore in the time-invariant case, if the input is a pulse applied at time $t = 0$ s, the response is exactly the same as if the pulse would be applied at time $t = 1$ s. The only difference is that in the latter case the response would be delayed by $t = 1$ s. The following example demonstrates that in the time-variant case this is not true.

Example 4.1   Effect of time varying impulse response

Let the impulse response of a noiseless time-variant channel be

$$h(t,\tau) = 20 \exp(-\tau) \cos(9t), \ \tau \geq 0, \tag{4.17}$$

(a) Determine the received signal $r(t)$, if the transmitted signal is

$$x_1(t) = \begin{cases} 1, |t| \leq 0.06 \text{ s}, \\ 0, |t| > 0.06 \text{ s}, \end{cases} \tag{4.18}$$

(b) Repeat part (a) if the transmitted signal is delayed by $T_1 = 0.26$ s or

$$x_2(t) = x_1(t - T_1). \tag{4.19}$$

Solution

(a) Analytical solution gives response

$$r_1(t) = \int_{-\infty}^{\infty} h(t,\tau) x_1(t-\tau) d\tau$$

$$= \int_{0}^{\infty} 20 e^{-\tau} \cos(9t) x_1(t-\tau) d\tau \tag{4.20}$$

At this point the integration variable is changed: $\tau = t - \tau'$ resulting in

$$r_1(t) = 20 \cos(9t) e^{-t} \int_{-\infty}^{t} e^{-\tau'} x_1(\tau') d\tau' = 20 \cos(9t) e^{-t} \int_{-0.06}^{t} e^{\tau'} d\tau' \qquad (4.21)$$

and finally in

$$= \begin{cases} 0, & t \leq -0.06s, \\ 20 \cos(9t)[1 - \exp(-t - 0.06)], & -0.06s < t < 0.06s, \\ 20 \cos(9t)[\exp(-t + 0.06) - \exp(-t - 0.06)], & t \geq 0.06s. \end{cases} \qquad (4.22)$$

(b) A similar computation is performed, when the input is given in Equation (4.19)

$$r_2(t) = \begin{cases} 0, & t \leq 0.2s, \\ 20 \cos(9t)[1 - \exp(-t + 0.2)], & -0.2s < t < 0.32s, \\ 20 \cos(9t)[\exp(-t - 0.32) - \exp(-t + 0.2)], & t \geq 0.32s. \end{cases}$$

$$(4.23)$$

The resulting responses for inputs (4.18) and (4.19) are shown in Figure 4.2. It can be clearly seen that the shape of response $r_2(t)$ is different compared with $r_1(t)$.

From the responses we can see typical characteristics of LTV systems. The received signals have been spread wider than the original pulse, which had a width of 0.06 s. In *linear*

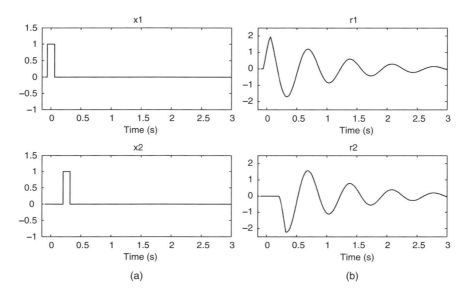

Figure 4.2 Responses of the system (Equation (4.17)) for inputs in Equation (4.18) above and in Equation (4.19) below

time-invariant (LTI) systems, if the input signal is delayed, so is the output signal and its shape remains the same as without delay. This is not true for LTV systems as seen in Figure 4.2

Simulation of LTV systems becomes complicated, except in special cases. We will return to the simulation of this example in Section 4.4.

4.3 Multivariable Case

The model in Equation (4.8) is easy to extend to the multivariable case, where there are M_T transmit antennas and M_R receiver antennas. A MIMO system is depicted in Figure 4.3.

The MIMO impulse response $\mathbf{H}(t, \tau)$ becomes

$$\mathbf{H}(t, \tau) = \begin{bmatrix} h_{11}(t, \tau) & h_{12}(t, \tau) & \cdots & h_{1M_T}(t, \tau) \\ h_{21}(t, \tau) & h_{22}(t, \tau) & \cdots & h_{2M_T}(t, \tau) \\ \vdots & \vdots & \ddots & \vdots \\ h_{M_R1}(t, \tau) & h_{M_R2}(t, \tau) & \cdots & h_{M_RM_T}(t, \tau) \end{bmatrix} \tag{4.24}$$

where $h_{ij}(t, \tau), i = 1, \ldots, M_R, j = 1, \ldots, M_T$ is the impulse response between the jth transmit antenna and the ith receiver antenna.

As a special case of the MIMO system we have the multiple-input single -output (MISO) system with M_T transmit antennas and $M_R = 1$ receiver antennas and $\mathbf{H}(t, \tau)$ reduces to a row $M_R \times 1$ vector

$$\mathbf{h}(t, \tau) = [\, h_{11}(t, \tau) \quad h_{12}(t, \tau) \quad \cdots \quad h_{1M_T}(t, \tau)\,] \tag{4.25}$$

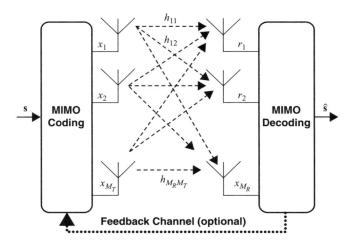

Figure 4.3 A MIMO system ($M_R \times M_T$), where the transmitted vector signal is $\mathbf{x}(t) = [x_1(t), x_2(t), \ldots, x_{M_T}(t)]^T$ and received vector signal $\mathbf{r}(t) = [r_1(t), r_2(t), \ldots, r_{M_R}(t)]^T$. If $M_R = 1$, then the system is a MISO system ($M_R \times M_1$)

or a single-input multiple-output (SIMO) system with $M_T = 1$ transmit antennas and M_R receiver antennas and $\mathbf{H}(t,\tau)$ reduces to a column $1 \times M_T$ vector

$$\mathbf{h}(t,\tau) = [h_{11}(t,\tau)h_{21}(t,\tau) \cdots h_{1M_R}(t,\tau)]^T \tag{4.26}$$

The received signal in the MIMO case is given as

$$\mathbf{r}(t) = \int_{-\infty}^{\infty} \mathbf{H}(t,\tau)\mathbf{x}(t-\tau)d\tau = \mathbf{H}(t,\tau)*\mathbf{x}(t) \tag{4.27}$$

where * denotes a convolution operation.

4.4 Simulation of LTV Systems

Simulation of linear time-variant systems (Equation (4.8)) is not straightforward. In time-invariant cases, a change to a *state-space* formulation will result in a much easier simulation task. In special cases, it is possible for *LTV* systems to resort to the same procedure.

LTV systems can be represented by the input–output description

$$\mathbf{y}(t) = \int_{t_0}^{t} \mathbf{G}(t,\tau)\mathbf{u}(\tau)d\tau \tag{4.28}$$

or in the state-space form

$$\begin{aligned}\dot{\mathbf{x}}(t) &= \mathbf{A}(t)\mathbf{x}(t) + \mathbf{B}(t)\mathbf{u}(t), \\ \mathbf{y}(t) &= \mathbf{C}(t)\mathbf{x}(t) + \mathbf{D}(t)\mathbf{u}(t). \end{aligned} \tag{4.29}$$

Given the impulse response matrix $G(t,\tau)$, determine the state-space representation, that is, matrices $\{\mathbf{A}(t), \mathbf{B}(t), \mathbf{C}(t), \mathbf{D}(t)\}$. This is called the *realization problem*.

The following theorem gives a useful realization result.

Theorem 4.1
Let the input–output description of a LTV system be

$$\mathbf{y}(t) = \int_{t_0}^{t} \mathbf{G}(t,\tau)\mathbf{u}(\tau)d\tau \tag{4.30}$$

where $\mathbf{u}(\tau)$ is the $p \times 1$ input vector, $\mathbf{y}(t)$ the output vector $q \times 1$, and $\mathbf{G}(t,\tau)$ $q \times p$. The impulse response matrix $\mathbf{G}(t,\tau)$ is realizable if and only if $\mathbf{G}(t,\tau)$ can be decomposed as

$$\mathbf{G}(t,\tau) = \mathbf{M}(t)\mathbf{N}(\tau) + \mathbf{D}(t)\delta(t-\tau), \ \forall t \geq \tau, \tag{4.31}$$

where $\mathbf{M}$, $\mathbf{N}$, and $\mathbf{D}$ are, respectively, $q \times n, n \times p$, and $q \times p$ matrices for some integer n.

A state-space realization becomes

$$\begin{aligned}\dot{\mathbf{x}}(t) &= \mathbf{N}(t)\mathbf{u}(t), \\ \mathbf{y}(t) &= \mathbf{M}(t)\mathbf{x}(t) + \mathbf{D}(t)\mathbf{u}(t). \end{aligned} \tag{4.32}$$

Example 4.2   How to simulate LTV systems?

Consider again Example 4.1. The impulse response of a noiseless LTV channel is given as

$$h(t, \tau) = 20 \exp(-\tau)\cos(9t), \ \tau \geq 0, \tag{4.33}$$

Note that we have causality ($h(t, \tau) = 0$ for $t < \tau$)

(a) If the transmitted signal is

$$x_1(t) = \begin{cases} 1, |t| \leq 0.06 \ \text{s}, \\ \\ 0, |t| > 0.06 \ \text{s}, \end{cases} \tag{4.34}$$

determine the received signal $r_1(t)$ using MATLAB®/Simulink®.
(b) Repeat part (a) when the transmitted signal is delayed, $x_2(t) = x_1(t - T_1)$, $T_1 = 0.26s$.

Solution

(a) Using Simulink®, we determine a realization for the given problem. First we change the integration variable exactly as in the analytical solution by setting $\tau' = t - \tau$:

$$y(t) = \int_{-\infty}^{t} 20e^{-(t-\tau')}\cos(9t)x_1(\tau)d\tau = \int_{-\infty}^{t} 20e^{-t}\cos(9t)e^{\tau'}x_1(\tau')d\tau' \tag{4.35}$$

The output is zero before we have any input, thus the infinite lower bound can be replaced with 0.06 s. The system is now of the form ($\tau' \to \tau$)

$$y(t) = \int_{-\infty}^{t} M(t)N(\tau)x_1(\tau)d\tau y(t) = \int_{0.06}^{t} \underbrace{20e^{-t}\cos(9t)}_{M(t)} \underbrace{e^{\tau}}_{N(\tau)} x_1(\tau)d\tau \tag{4.36}$$

Applying Theorem 4.1 gives the following realization

$$\begin{aligned} \dot{x} &= N(t)x_1(t) = \exp(t)x_1(t) \\ y &= M(t)x(t) = 20\exp(-t)\cos(9t)x(t) \end{aligned} \tag{4.37}$$

Simulation is now relatively easy as seen in Figure 4.4.

   For part (b), we need to add one delay block, which is shown in Figure 4.5. Simulation results are shown in Figure 4.6.

   As seen in Figure 4.6, the second output is not a delayed version of the first output, because the system is time varying.

4.5 Discrete-Time Models

The discrete time equivalent of Equation (4.8) is

$$r(t) = \sum_{i=0}^{\infty} h(t, i)x(t - i) + v(t), \ t = 0, 1, 2, \ldots, \tag{4.38}$$

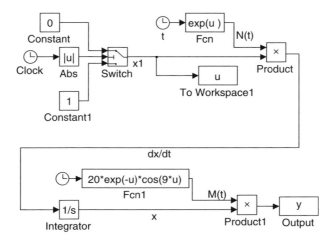

Figure 4.4 Simulink® configuration of the analytical solution

where $\{h(t,i), i = 1,\ldots,\infty, t = 0,1,2,\ldots\}$ is the impulse response. The transmitted signal is $x(t)$ and $v(t)$ is additive noise.

Over short time intervals it is customary to approximate the time-varying channel by a time-invariant impulse response. Another assumption is to assume that the channel can be described by a finite impulse response of length K, since the long paths are well attenuated falling below the background noise level. The received signal can then be written in discrete form

$$r(t) = \sum_{i=0}^{K-1} h(i)x(t-i) + v(t) \tag{4.39}$$

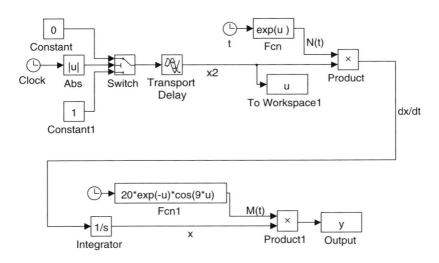

Figure 4.5 A delay block is added to simulate part (b) of the problem

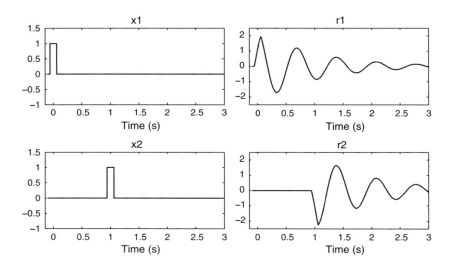

Figure 4.6 Simulation results: Part (a) above, part (b) below. On the left *inputs*, on the right *responses*

where K is chosen large enough to include all significant paths and $v(t)$ additive noise. The discrete channel is denoted by $h(i)$, where i is the index at baseband sampling period t_s.

In this book we choose to use the shift operator q in most cases instead of the z transform. The *forward shift operator* is defined as

$$qx(t) = x(t+1) \tag{4.40}$$

and the backward shift operator q^{-1} as

$$q^{-1}x(t) = x(t-1) \tag{4.41}$$

Hence we can write Equation (4.39) as

$$
\begin{aligned}
r(t) &= \sum_{i=0}^{K-1} h(i)x(t-i) + v(t) = \sum_{i=0}^{K-1} h(i)(q^{-i}x(t)) + v(t) \\
&= \left[\sum_{i=0}^{K-1} h(i)q^{-i}\right]x(t) + v(t)
\end{aligned}
\tag{4.42}
$$

Let the transfer function or transfer operator be defined as

$$H(q^{-1}) = \sum_{i=0}^{K-1} h(i)q^{-i} \tag{4.43}$$

Using this, Equation (4.42) becomes

$$r(t) = H(q^{-1})x(t) + v(t) \tag{4.44}$$

In the MIMO case Equation (4.39) takes the form

$$\mathbf{r}(t) = \sum_{i=0}^{K-1} \mathbf{H}(i)\mathbf{x}(t-i) + \mathbf{v}(t) \tag{4.45}$$

where the received vector signal is a column M_R − vector $\mathbf{r}(t) = [r_1(t), r_2(t), \ldots, r_{M_R}(t)]^T$, the transmitted signal is a column M_T − vector $\mathbf{x}(t) = [x_1(t), x_2(t), \ldots, x_{M_T}(t)]^T$, and the additive noise is a column M_R − vector $\mathbf{v}(t) = [v_1(t), v_2(t), \ldots, v_{M_R}(t)]^T$. The impulse response $M_R \times M_T$ matrix $\mathbf{H}(i)$ is

$$\mathbf{H}(i) = \begin{bmatrix} h_{11}(i) & h_{12}(i) & \cdots & h_{1M_T}(i) \\ h_{21}(i) & h_{22}(i) & \cdots & h_{2M_T}(i) \\ \vdots & \vdots & \ddots & \vdots \\ h_{M_R1}(i) & h_{M_R2}(i) & \cdots & h_{M_RM_T}(i) \end{bmatrix} \tag{4.46}$$

4.6 Discrete-Time Models with Noise

If a stationary process $v(t)$ in Equation (4.44) has a rational spectrum $\Phi_v(\omega)$, it can be represented as

$$v(t) = R(q^{-1})\xi(t) \tag{4.47}$$

where $\{\xi(t)\}$ is a sequence of independent, equally distributed random variables with zero mean $E\{\xi(t)\} = \mu_\xi = 0$ and with variance $E\{(\xi(i) - \mu_\xi)(\xi(j) - \mu_\xi)\} = E\{\xi(i)\xi(j)\} = \delta_{ij}\sigma^2$. It is known as *white noise*. The rational function $R(q)$ can be written as

$$R(q^{-1}) = \frac{C(q^{-1})}{A(q^{-1})} \tag{4.48}$$

where the polynomial operators

$$C(q^{-1}) = 1 + c_1 q^{-1} + \cdots c_{n_c} q^{-n_c} \tag{4.49}$$

$$A(q^{-1}) = 1 + a_1 q^{-1} + \cdots a_{n_a} q^{-n_a} \tag{4.50}$$

We say that the noise is 'colored', because Equation (4.47) implies that the noise model includes past values of noise. Colored noise is generated by passing white noise through a linear filter.

It is possible to write the transfer function in Equation (4.44) as

$$H(q^{-1}) = \frac{B(q^{-1})}{A(q^{-1})} \tag{4.51}$$

where

$$B(q^{-1}) = b_1 q^{-1} + \cdots b_{n_b} q^{-n_b} \tag{4.52}$$

Combining Equations (4.44), (4.47), (4.48) and (4.51) results in an *autoregressive moving average with external input* (*ARMAX*) model of a stochastic system.

$$
\begin{aligned}
r(t) &= H(q^{-1})x(t) + v(t) \\
&= H(q^{-1})x(t) + R(q^{-1})\xi(t) \\
&= \frac{B(q^{-1})}{A(q^{-1})}x(t) + \frac{C(q^{-1})}{A(q^{-1})}\xi(t)
\end{aligned}
\tag{4.53}
$$

which can also be written as

$$
A(q^{-1})r(t) = B(q^{-1})x(t) + C(q^{-1})\xi(t) \tag{4.54}
$$

Using Equations (4.49), (4.50) and (4.52) gives the model

$$
\begin{aligned}
r(t) + a_1 r(t-1) + \cdots + a_{n_a} r(t-n_a) &= b_1 x(t-1) + b_2 x(t-2) + \cdots \\
&\quad + b_{n_b} x(t-n_b) + \xi(t) + c_1 \xi(t-1) + \cdots + c_{n_c} \xi(t-n_c)
\end{aligned}
\tag{4.55}
$$

where $r(t)$, the output, is the received signal at time t, $x(t)$, the input, is the transmitted signal at time t, and $\xi(t)$ is white, Gaussian noise.

One of the main purposes of model (4.55) is to use it in prediction.

Example 4.3   One-step predictor

Consider the system

$$
r(t) + 0.1r(t-1) = 2x(t-1) + 0.15\xi(t), \quad t = 1, 2, 3, \ldots \tag{4.56}
$$

where $\xi(t)$, $t = 1, 2, 3, \ldots$ are independent stochastic variables with zero mean and variance one.

(a) Form a one-step prediction for the system when data of the received signal and the transmitted signal are known up until time k, $\{r(t), x(t); t = 0, 1, \ldots, k\}$. Compute the prediction error.

(b) Set up a Simulink® model of the original system (producing data) and the predictor. Simulate the system with $x(t) = 1$, $t = 0, 1, 2, \ldots$ and $r(0) = 1.5$.

Solution

(a) Write Equation (4.56) as

$$
r(t+1) = -0.15r(t) + 2x(t) + 0.1\xi(t+1) \tag{4.57}
$$

Since future noise $\xi(t+1)$ is unknown but of zero mean, the best predictor available is obtained by considering the mean of $r(t+1)$, that is, the conditional expectation of $r(t)$ denoted by $\hat{r}(t+1|t)$:

$$\hat{r}(t+1|t) = -0.15r(t) + 2x(t) \tag{4.58}$$

The $\hat{r}$ notation in $\hat{r}(t+1|t)$ means predictor. The first argument $t+1$ indicates how far we predict, in this case one step ahead, and the second argument t at which point of time the prediction is made. The resulting predictor minimizes the prediction error variance. In this example there is a gain in front of white noise. The prediction error becomes $\tilde{r}(t+1|t) = r(t) - \hat{r}(t+1|t) = 0.1\xi(t+1)$.

These issues are further discussed in Section 4.8.

(b) Figure 4.7 shows the Simulink® configuration of the original system together with the predictor. Equation (4.57) is used to generate the noisy system model. Forming the predictor is then straightforward from Equation (4.58).

The system is simulated. The output $r(t)$, one-step predictor $\hat{r}(t+1|t)$, and the error $\tilde{r}(t+1|t)$ are plotted in Figure 4.8. Since the prediction error depends completely on the noise term, it is immediately reflected on the prediction error. If the simple noise model is sufficient, prediction is accurate as in this example (Figure 4.8). If the error variance grows, then more sophisticated noise models and predictors have to be considered to obtain a more accurate prediction. □

If the noise model does not include colored but only white noise then Equation (4.55) becomes an *autoregressive with external input* (*ARX*) model

$$
\begin{aligned}
r(t) + a_1 r(t-1) &+ \cdots + a_{n_a} r(t-n_a) \\
&= b_1 x(t-1) + b_2 x(t-2) + \cdots + b_{n_b} x(t-n_b) + \xi(t)
\end{aligned} \tag{4.59}
$$

If no input appears ($x(t) = 0$) in Equation (4.55), then the model reduces to an *autoregressive moving average* (*ARMA*) model

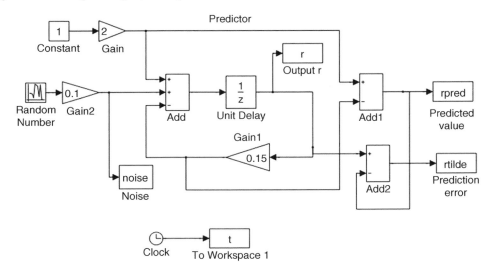

Figure 4.7 Simulink® configuration of the system and the one-step predictor of the output

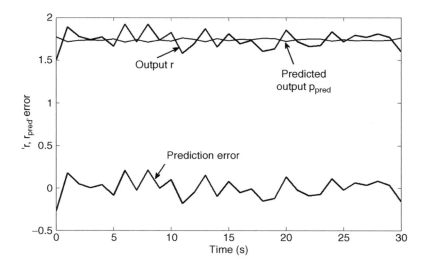

Figure 4.8 One-step predictor shown together with the output and prediction error

$$r(t) + a_1 r(t-1) + \cdots + a_{n_a} r(t-n_a)$$
$$= \xi(t) + c_1 \xi(t-1) + \cdots + c_{n_c} \xi(t-n_c) \tag{4.60}$$

Multivariable counterparts of the above models can be formulated in a straightforward manner starting from Equation (4.55). These are important, for example in *adaptive beamforming*. A MIMO extension of the *ARMAX* model becomes

$$\mathbf{r}(t) + \mathbf{A}_1 \mathbf{r}(t-1) + \cdots + \mathbf{A}_{n_a} \mathbf{r}(t-n_a) = \mathbf{B}_1 \mathbf{x}(t-1) + \mathbf{B}_2 \mathbf{x}(t-2) + \cdots$$
$$+ \mathbf{B}_{n_b} \mathbf{x}(t-n_b) + \xi(t) + \mathbf{C}_1 \xi(t-1) + \cdots + \mathbf{C}_{n_c} \xi(t-n_c) \tag{4.61}$$

Here the output vector is $\mathbf{r}(t) = [r_1(t), r_2(t), \ldots, r_{M_R}(t)]^T$, the input vector is $\mathbf{x}(t) = [x_1(t), x_2(t), \ldots, x_{M_T}(t)]^T$, and the white noise vector is $\xi(t) = [\xi_1(t), \xi_2(t), \ldots, \xi_{M_R}(t)]^T$. The $\mathbf{A}_i$ are $M_R \times M_R$, the $\mathbf{B}_i$ $M_R \times M_T$ and the $\mathbf{C}_i$ $M_R \times M_R$ matrices. Defining matrix operator polynomials

$$\mathbf{A}(q^{-1}) = \mathbf{I} + \mathbf{A}_1 q^{-1} + \cdots + \mathbf{A}_{n_a} q^{-n_a} \tag{4.62}$$

$$\mathbf{B}(q^{-1}) = \mathbf{B}_1 q^{-1} + \cdots + \mathbf{B}_{n_a} q^{-n_a} \tag{4.63}$$

$$\mathbf{C}(q^{-1}) = \mathbf{I} + \mathbf{C}_1 q^{-1} + \cdots + \mathbf{C}_{n_a} q^{-n_a} \tag{4.64}$$

Equation (4.61) can be written as

$$\mathbf{A}(q^{-1})\mathbf{r}(t) = \mathbf{B}(q^{-1})\mathbf{x}(t) + \mathbf{C}(q^{-1})\xi(t) \tag{4.65}$$

4.7 Least Squares Identification

Consider the ARX model of system (Equation (4.59)). It can be written as

$$r(t) = \boldsymbol{\theta}^T \boldsymbol{\phi}(t) + \xi(t) \tag{4.66}$$

where the parameter vector $\boldsymbol{\theta}$ is

$$\boldsymbol{\theta} = [a_1, ..., a_{n_a}, b_1, ..., b_{n_b}]^T \tag{4.67}$$

and the regression ()vector $\boldsymbol{\phi}$

$$\boldsymbol{\phi}(t) = [-r(t-1), ..., -r(t-n_{a)}, x(t-1), ..., x(t-n_b)]^T \tag{4.68}$$

The idea in the least squares method is to minimize the cost function of the accumulated error squares over the time $k = 1, \ldots, N$

$$V_N(\boldsymbol{\theta}) = \frac{1}{2} \sum_{k=1}^{N} \varepsilon_k^2 = \frac{1}{2} \sum_{t=1}^{N} (r(t) - \boldsymbol{\theta}^T \boldsymbol{\phi}(t))^2 \tag{4.69}$$

It is straightforward to find the value of $\boldsymbol{\theta}$ minimizing the cost function (Equation (4.69)):

$$\hat{\boldsymbol{\theta}}_N = \left[\sum_{t=1}^{N} (\boldsymbol{\phi}(t)\boldsymbol{\phi}(t)^T) \right]^{-1} \sum_{k=1}^{N} \boldsymbol{\phi}(t)r(t) \tag{4.70}$$

Introduce the following notation, matrix

$$R_N = \frac{1}{N} \sum_{t=1}^{N} (\boldsymbol{\phi}(t)\boldsymbol{\phi}(t)^T) \tag{4.71}$$

and vector

$$\mathbf{f}_N = \frac{1}{N} \sum_{t=1}^{N} \boldsymbol{\phi}(t)r(t) \tag{4.72}$$

With these notations Equation (4.70) becomes

$$\hat{\boldsymbol{\theta}}_N = R_N^{-1}\mathbf{f}_N \tag{4.73}$$

Well-known properties of the least squares estimate $\hat{\boldsymbol{\theta}}_N$ are that when the noise sequence $\xi(t)$ in Equation (4.66) is white (then $\boldsymbol{\phi}$ and $\xi(t)$ are independent) with variance σ^2, the parameter estimate converges with probability 1:

$$\hat{\boldsymbol{\theta}}_N \to \boldsymbol{\theta} \text{ as } N \to \infty. \tag{4.74}$$

and under mild conditions

$$R_N \to R = \lim_{N \to \infty} \frac{1}{N} \sum_{t=1}^{N} \left(\phi(t)\phi(t)^T \right) \tag{4.75}$$

The error variance has the asymptotic property

$$\text{var}[\hat{\boldsymbol{\theta}}_N - \boldsymbol{\theta}] \to \sigma^2 R^{-1} \qquad (4.76)$$

If the noise is not white, convergence $\hat{\boldsymbol{\theta}}_N$ occurs but a bias term is introduced to $\boldsymbol{\theta}$. Observe that if a one-step-ahead predictor is formed from Equation (4.66)

$$\hat{r}(t|t-1; \boldsymbol{\theta}) = \boldsymbol{\theta}^{\mathbf{T}} \boldsymbol{\phi}(t) \qquad (4.77)$$

and we consider the sum of prediction error squares

$$V_N(\boldsymbol{\theta}) = \frac{1}{N} \sum_{t=1}^{N} (r(t) - \hat{r}(t|t-1; \boldsymbol{\theta}))^2 \qquad (4.78)$$

the results are the same as above. In general, the predictor $\hat{r}(t|t-1; \boldsymbol{\theta})$ is a nonlinear function of $\boldsymbol{\theta}$ and numerical methods have to be used.

Example 4.4

Consider the case when the received power signal $\bar{r}(t)$ is given in dB, $\bar{r}(t) = 10^{r_{dB}/10}$ (cf. Equation (3.36)) and the corresponding channel model is

$$r_{\text{dB}}(t) = \boldsymbol{\theta}^T \boldsymbol{\phi}(t) + \xi(t) \qquad (4.79)$$

$$\boldsymbol{\phi}(t) = [-r_{dB}(t-1), ..., -r_{dB}(t-n), x_{dB}(t-1), ..., x_{dB}(t-n_b)]^T \qquad (4.80)$$

Derive the least squares estimate $\hat{\boldsymbol{\theta}}_N$.

Solution

From Equation (4.69)

$$V_N(\boldsymbol{\theta}) = \frac{1}{2} \sum_{t=1}^{N} \varepsilon_t^2 = \frac{1}{2} \sum_{t=1}^{N} (r_{dB}(t) - \hat{r}_{dB}(t|t-1; \boldsymbol{\theta}))^2$$

$$= \frac{1}{2} \sum_{t=1}^{N} (10 \log_{10} \bar{r}(t) - 10 \log_{10} \hat{r}(t|t-1; \boldsymbol{\theta}))^2 \qquad (4.81)$$

$$= \frac{100}{2} \sum_{t=1}^{N} \left(\log_{10} \frac{\hat{r}(t|t-1; \boldsymbol{\theta})}{\bar{r}(t)} \right)^2 = 50 \sum_{t=1}^{N} \left(\log_{10} \frac{\boldsymbol{\theta}^{\mathbf{T}} \boldsymbol{\phi}(t)}{\bar{r}(t)} \right)^2$$

The cost function is clearly nonlinear in $\boldsymbol{\theta}$. This means that a numerical procedure is needed to find the best parameter vector $\boldsymbol{\theta}$. On the other hand, since the received signal is a power signal, it has to be nonnegative. Therefore, the problem becomes a constrained optimization problem.

To simplify the problem develop the cost function in a Taylor series around $\hat{r}(t|t-1; \boldsymbol{\theta})/\bar{r}(t) - 1$ (since the ratio is ≈ 1). Then the cost can be approximated using only the first-order term x in the series $\log_{10}(1+x) \approx x - \frac{1}{2}(x-1)^2 + \cdots$:

$$V_N(\boldsymbol{\theta}) = 50\sum_{t=1}^{N}\left(\log_{10}\frac{\hat{r}(t|t-1;\boldsymbol{\theta})}{\bar{r}(t)}\right)^2 = 50\sum_{t=1}^{N}\left(\log_{10}\left(1+\underbrace{\frac{\hat{r}(t|t-1;\boldsymbol{\theta})}{\bar{r}(t)}-1}_{x}\right)\right)^2$$

$$\approx 50\sum_{t=1}^{N}\left(\underbrace{\frac{\hat{r}(t|t-1;\boldsymbol{\theta})}{\bar{r}(t)}-1}_{x}\right)^2 = 50\sum_{t=1}^{N}\left(\frac{\boldsymbol{\theta}^{\mathbf{T}}\boldsymbol{\phi}(t)}{\bar{r}(t)}-1\right)^2 = 50\sum_{t=1}^{N}\left(\boldsymbol{\theta}^{\mathbf{T}}\tilde{\boldsymbol{\phi}}(t)-1\right)^2$$

$$(4.82)$$

where $\tilde{\boldsymbol{\phi}}(t) = \boldsymbol{\phi}(t)/r(t)$. Denote $\tilde{f}_N = \frac{1}{N}\sum_{t=1}^{N}\tilde{\boldsymbol{\phi}}(t)\ r_{dB}(t)$ and $\tilde{R}_N = \frac{1}{N}\sum_{t=1}^{N}\left(\tilde{\boldsymbol{\phi}}(t)\tilde{\boldsymbol{\phi}}(t)^T\right)^2$. Then

$$\hat{\boldsymbol{\theta}}_N = \tilde{R}_N^{-1}\tilde{f}_N \qquad (4.83)$$

The result is equivalent to Equation (4.73).

The estimation algorithms of Equations (4.70) and (4.83) were batch types of algorithms. Often we are interested in recursive forms of estimation algorithms. These have been derived in standard textbooks on systems identification and only the result is given here.

Recursive least squares (RLS) algorithm

$$\hat{\boldsymbol{\theta}}(t) = \hat{\boldsymbol{\theta}}(t-1) + \mathbf{K}(t)[r(t) - \boldsymbol{\phi}^T(t)\hat{\boldsymbol{\theta}}(t-1)] \qquad (4.84)$$

$$\mathbf{K}(t) = \mathbf{P}(t)\boldsymbol{\phi}(t) = \frac{\mathbf{P}(t-1)\boldsymbol{\phi}(t)}{\lambda(t) + \boldsymbol{\phi}^T(t)\mathbf{P}(t-1)\boldsymbol{\phi}(t)} \qquad (4.85)$$

$$\mathbf{P}(t) = \frac{1}{\lambda(t)}\left[\mathbf{P}(t-1) - \frac{\mathbf{P}(t-1)\boldsymbol{\phi}(t)\boldsymbol{\phi}^T(t)\mathbf{P}(t-1)}{\lambda(t) + \boldsymbol{\phi}^T(t)\mathbf{P}(t-1)\boldsymbol{\phi}(t)}\right] \qquad (4.86)$$

Typically, a diagonal form $\mathbf{P}(0) = \delta\mathbf{I}$ is used for the initial value of the error covariance $\mathbf{P}(t)$. If the initial knowledge of the parameters is vague, large values for the scalar δ are chosen, $\delta = 100 - 1000$.

The forgetting factor $\lambda(t), 0 < \lambda(t) < 1$, can be chosen to be a constant, fairly close to one or variable forgetting factors may be applied. Constant values between 0.9 and 0.995 are typically chosen. The forgetting factor helps to cope with varying system dynamics, older values are 'forgotten' and newer values receive more weight.

4.8 Minimum Variance Prediction

Returning to the ARMA model (Equation (4.60))

$$\begin{aligned} r(t) + a_1 r(t-1) + \cdots + a_{n_a}r(t-n_a) \\ = \xi(t) + c_1\xi(t-1) + \cdots + c_{n_c}\xi(t-n_c) \end{aligned} \qquad (4.87)$$

This can be written as:

$$A(q^{-1})r(t) = C(q^{-1})\xi(t) \tag{4.88}$$

or

$$r(t) = \frac{C(q^{-1})}{A(q^{-1})}\xi(t) \tag{4.89}$$

where the polynomial operators $A(q^{-1})$ and $C(q^{-1})$ are defined as

$$A(q^{-1}) = 1 + a_1 q^{-1} + \cdots a_n q^{-n_a} \tag{4.90}$$

$$C(q^{-1}) = 1 + c_1 q^{-1} + \cdots c_n q^{-n_c} \tag{4.91}$$

and $\xi(t)$ is white noise as defined in Equation (4.47).

Denote the k-step ahead predictor for $r(t+k)$ at time t, based on the measurement history $Y_t = \{r(t), r(t-1), \ldots\}$, as $\hat{r}(t+k|t)$. The *minimum variance predictor* $\hat{r}(t+k|t)$ is determined so that the criterion

$$V = E\{\tilde{r}(t+k|t)^2\} \tag{4.92}$$

is minimized. Here $\tilde{r}(t+k|t)$ is the output prediction error

$$\tilde{r}(t+k|t) = r(t+k) - \hat{r}(t+k|t) \tag{4.93}$$

The solution of the predictor problem is then the conditional expectation of $r(t+k|t)$ given Y_t or $\hat{r}(t+k|t) = E\{r(t+k)|Y_t\}$.

The *Euclidean division algorithm* (Diophantine equation or Bezout identity) for polynomials gives

$$C(q^{-1}) = A(q^{-1})F(q^{-1}) + q^{-k}G(q^{-1}) \tag{4.94}$$

where

$$F(q^{-1}) = 1 + f_1 q^{-1} + \cdots f_{k-1} q^{-k+1} \tag{4.95}$$

$$G(q^{-1}) = g_0 + g_1 q^{-1} + \cdots g_{n-1} q^{-n_g+k} \tag{4.96}$$

When the system model Equation (4.89) is advanced by k time steps and Equation (4.94) is applied

$$r(t+k) = F(q^{-1})\xi(t+k) + \frac{G(q^{-1})}{A(q^{-1})}\xi(t) \tag{4.97}$$

Solving for $\xi(t)$ in Equation (4.89) leads to

$$r(t+k) = F(q^{-1})\xi(t+k) + \frac{G(q^{-1})}{C(q^{-1})}r(t) \tag{4.98}$$

Observe that

$$
\begin{aligned}
F(q^{-1})\xi(t+k) &= (1+f_1q^{-1}+\cdots f_{k-1}q^{-k+1})\xi(t+k) \\
&= \xi(t+k)+f_1\xi(t+k-1)+\cdots f_{k-1}\xi(t+1)
\end{aligned}
\tag{4.99}
$$

This implies that all the terms on the right-hand side are future noise, which are all independent of data $\mathbf{Y}_t$ (no information about future noise is available). The second term on the right-hand side of Equation (4.98) is a function of the measured data only, and is thus deterministic.

Substituting Equations (4.93) and (4.98) into Equation (4.92) gives, after some algebraic manipulation (using the independence of $\xi(t+k)$ and $r(t)$, and the fact that $E\{\xi(t+k)\} = 0$):

$$
V = E\left\{(F(q^{-1})\xi(t+k))^2\right\} + 2E\left\{F(q^{-1})\xi(t+k)\right\}\left(\frac{G(q^{-1})}{C(q^{-1})}r(t)\right) + \left(\frac{G(q^{-1})}{C(q^{-1})}r(t)\right)^2
$$

$$
-2E\left\{\left(F(q^{-1})\xi(t+k)+\frac{G(q^{-1})}{C(q^{-1})}r(t)\right)\right\}\hat{r}(t+k|t)+\hat{r}(t+k|t)^2
$$

$$
= E\left\{(F(q^{-1})\xi(t+k))^2\right\} + \left(\frac{G(q^{-1})}{C(q^{-1})}r(t)\right)^2 - 2\frac{G(q^{-1})}{C(q^{-1})}r(t)\hat{r}(t+k|t)+\hat{r}(t+k|t)^2
\tag{4.100}
$$

The optimal predictor is thus

$$
\hat{r}(t+k|t) = \frac{G(q^{-1})}{C(q^{-1})}r(t)
\tag{4.101}
$$

and the prediction error

$$
\tilde{r}(t+k|t) = r(t+k) - \hat{r}(t+k|t) = F(q^{-1})\xi(t+k)
\tag{4.102}
$$

Example 4.5   ARMA model for Rayleigh fading

(a) Generate Rayleigh fading data for 500 time steps. Apply the Jakes model for modeling fading. (It is available in MATLAB® Central in many forms and also in Communications Toolbox.) Here carrier frequency $f_c = 900$ MHz, the sampling time is 2 ms, and mobile speed $= 5$ km/h (pedestrian walk). For shadowing, a Gaussian distributed random signal with zero mean and variance 0.1 is applied. Determine an ARMA model for the data generated. Consider different structures of the ARMA model.

Solution

The resulting simulation is shown in Figure 4.9.

Estimation of parameters in the ARMA model (Equation (4.60))

$$
r(t) + a_1 r(t-1) + \cdots + a_{n_a} r(t-n_a) = \xi(t) + c_1 \xi(t-1) + \cdots + c_{n_c}\xi(t-n_c)
\tag{4.103}
$$

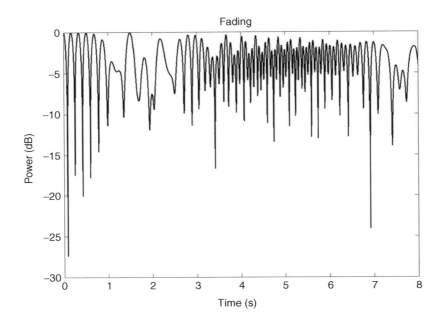

Figure 4.9 Result of Jakes model for Rayleigh fading and shadowing, when MS speed is 5 km/h

are carried out with commands from the System Identification Toolbox:

```
[ident % Initializes Identification Toolbox
%Preprocess data by making it zero mean
ze = detrend(fading_dB);
%Create data object for identification
% Remark: dat=iddata(y,u,ts) has input U. ARMA model has no input,
use[].
data = iddata(ze,[],1);
% Set na=4, nc=4 in model (4.103)
m=armax(data,[44])
```

The resulting model is

$$
\begin{aligned}
A(q) &= 1 - 1.37q^{-1} - 0.5103q^{-2} + 1.293q^{-3} + 0.4013q^{-4} \\
C(q) &= 1 + 0.1223q^{-1} - 0.6133q^{-2}0.1523q^{-3} - 0.04353q^{-4}
\end{aligned}
\tag{4.104}
$$

The value of the model is best tested by using it as a predictor. The MATLAB® commands for computing the p-step predictor are:

```
% predict psteps ahead
psteps = 1;
datahat = predict(m, data, psteps);
y = datahat.y;
```

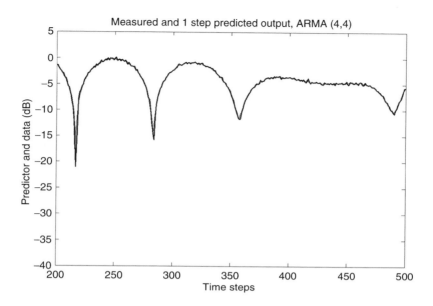

Figure 4.10 Jakes fading channel is modeled with ARMA(4,4) model. One-step ahead predictor together with the measured signal are displayed. They are indistinguishable

```
% plot
figure
plot([ze, y])
legend('measured',['prediction',num2str(psteps),'steps ahead']);
```

Figure 4.10 shows the one-step prediction together with the original data.

One-step prediction is quite good. This is the reason that a close-up from 200 to 500 time steps is shown. Otherwise it is not possible to see much difference at all. When MS speed is faster, prediction becomes more demanding. Other ARMA model structures with different n_a and n_c may also be tried.

When the speed of the MS is changed, the Jakes model gives a different data set. Estimation of parameters in this case leads to another ARMA model, and so on. Fixed parameter predictors could be set up for different speed classes, but a more useful way is to consider adaptive predictors. These are discussed in the next Section.

Instead of MATLAB® Command Window, it is possible to use the GUI (graphical user interface) of Identification Toolbox shown in Figure 4.11.

Follow the steps:

1. In the MATLAB® Command Window type *ident*. This opens the GUI window. All the GUI boxes are empty in the beginning.
2. Open *Import data* slot. Choose *Time domain data*. This opens the *Import Data* window. In *Workspace variable* area, use [] for input (since the ARMA model has no input) and *fading_dB* for output. Finally, press *Import*.
3. This fills the box in the upper left-hand corner with *mydata*.

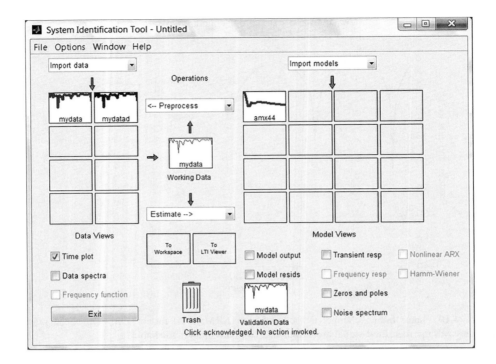

Figure 4.11 GUI of System Identification Toolbox, where data has been imported (*mydata*), preprocessed (*mydatad*), and an ARMA(4,4) model parameters estimated

4. Next go to *Preprocess slot* and choose *Remove means*. This fills the box next to *mydata* with data called *mydatad*. Move this with your cursor to the *Working Data* space situated in the center.

5. Choose *Estimate* slot and then *Linear* parametric models. Once this is open, change *AR* in structure choice to *ARMA(n_a, n_c)*. Default values are $n_a = 4$ and $n_c = 4$. Finally, press *Estimate*. This will fill the first box of the first row on the right-hand side. The result is called *armax44-*.

6. *Tick Model output*. The result is produced in *Model output: y1* window showing the original data *fading_dB* together with a 5-step-ahead prediction result (5 is by default). In the menu open options and *Set prediction horizon* choose 1. The result is the same as that in Figure 4.8.

4.9 Self-Tuning Predictor

In wireless communications the channel variations are fast and vary continuously. Nonlinearities are also common. Therefore constant parameter time-series models and predictors are inadequate. Since we continuously measure signals, e.g. power levels, it is possible to make predictors *adaptive* or *self-tuning*. In the following, we will use the term 'self-tuning'. Basically, once a new measurement is received, the parameters can be estimated using that measurement and the old data.

In this section we concentrate on *self-tuning predictors*. The self-tuning feature can be realized in two ways: *explicit* or *implicit*. In the explicit case, the parameters of the polynomials $A(q^{-1})$ and $C(q^{-1})$ are first estimated explicitly using Equations (4.84)–(4.86). Then the polynomials $F(q^{-1})$ and $G(q^{-1})$ are determined from Equation (4.94). The predictor is computed from Equation (4.101):

$$\hat{r}(t+k|t) = \frac{G(q^{-1})}{C(q^{-1})} r(t) \qquad (4.105)$$

In the implicit case, the predictor Equation (4.105) is written in the form

$$\hat{r}(t+k|t) = (1 - C(q^{-1}))\hat{r}(t+k|t) + G(q^{-1})r(t) \qquad (4.106)$$

This can be written as

$$\begin{aligned} \hat{r}(t+k|t) = {} & -c_1\hat{r}(t+k-1|t-1) - \cdots - c_{n_c}\hat{r}(t+k-n_c|t-n_c) \\ & + g_0 r(t) + g_1 r(t-1) + \cdots + g_{n_g} r(t-n_g) \end{aligned} \qquad (4.107)$$

The actual structure of the predictor is not known, so n_c and n_g have to be investigated beforehand. The value of g_0 is set to zero, to guarantee the domination of the value $r(t)$. The corresponding regression vector $\boldsymbol{\phi}(t)$ and estimated parameter vector $\hat{\boldsymbol{\theta}}(t)$ are

$$\boldsymbol{\phi}(t) = [-\hat{r}(t+k-1|t-1), ..., -\hat{r}(t+k-n_c|t-n_c), r(t), ..., r(t-n_g)]^T \qquad (4.108)$$

$$\hat{\boldsymbol{\theta}}(t) = [\hat{c}_1(t), ..., \hat{c}_{n_c}(t), \hat{g}_0(t), ..., \hat{g}_{n_g}(t)]^T \qquad (4.109)$$

The extended recursive least squares algorithm is given as in Equations (4.84)–(4.86):

$$\hat{\boldsymbol{\theta}}(t) = \hat{\boldsymbol{\theta}}(t-1) + \mathbf{K}(t)[r(t) - \boldsymbol{\phi}^T(t)\hat{\boldsymbol{\theta}}(t-1)] \qquad (4.110)$$

$$\mathbf{K}(t) = \mathbf{P}(t)\boldsymbol{\phi}(t) = \frac{\mathbf{P}(t-1)\boldsymbol{\phi}(t)}{\lambda(t) + \boldsymbol{\phi}^T(t)\mathbf{P}(t-1)\boldsymbol{\phi}(t)} \qquad (4.111)$$

$$\mathbf{P}(t) = \frac{1}{\lambda(t)}\left[\mathbf{P}(t-1) - \frac{\mathbf{P}(t-1)\boldsymbol{\phi}(t)\boldsymbol{\phi}^T(t)\mathbf{P}(t-1)}{\lambda(t) + \boldsymbol{\phi}^T(t)\mathbf{P}(t-1)\boldsymbol{\phi}(t)}\right] \qquad (4.112)$$

As indicated before, the initial value of the error covariance is taken as a diagonal matrix $\mathbf{P}(0) = \delta$, where the scalar $\delta = 100 - 1000$.

The forgetting factor λ is usually given a constant value between 0.9 and 0.995. One popular variable forgetting factor form is

$$\lambda(t) = \alpha\lambda(t-1) + (1-\alpha), \quad 0 < \alpha < 1 \qquad (4.113)$$

Here both α and $\lambda(0)$ are constant values between 0.95 and 0.99.

Implicit self-tuning predictor algorithm

1. Read the new received power measurement.

2. Update the new data vector $\boldsymbol{\phi}(t)$.
3. Calculate the new parameter estimate vector $\hat{\boldsymbol{\theta}}(t)$ from Equations (4.110)–(4.112).
4. Use the estimated parameters to compute the k-step ahead predictor from Equation (4.107).
5. Set $t \rightarrow t+1$ and go to Step 1.

4.10 System Identification with Neural Networks

Neural networks can provide a powerful tool if the system to be modeled is nonlinear. Their use has also been extended to system identification. The key ideas are briefly summarized here. The nonlinear system model is

$$r(t) = g(\boldsymbol{\phi}(t), \boldsymbol{\theta}) + \xi(t) \qquad (4.114)$$

or in the predictor form

$$\hat{r}(t+k|t) = g(\boldsymbol{\phi}(t), \boldsymbol{\theta}) \qquad (4.115)$$

Here $\boldsymbol{\phi}(t)$ is, as before, the regression vector and $\boldsymbol{\theta}$ is the parameter vector, which contains the adjustable weight parameters in the neural network. The neural networks used in these sorts of models vary. Typically, the MLP network, which was discussed in Section 3.2.4 (Equation (3.29)), are used. Radial basis networks are also popular. The predictor model can take any number of forms, e.g. NARX, NARMAX, NARMA, where N refers to the fact that neural networks are applied.

We restrict ourselves to consider only the NARMA model here. The latest System Identification Toolbox has a possibility of using nonlinear models including neural networks, but unfortunately it does not support the ARMA model (ARMAX is ok). Good software for treating dynamical systems is public domain NNSYSID software, which can be found on the MATLAB® Central web page. Rayleigh fading of Example 4.5 provides an avenue to become familiar with this powerful modeling approach.

Example 4.6   Neural network ARMA (NARX) model for Rayleigh fading

Generate Rayleigh fading data (e.g. as in Example 4.5). Consider an MLP network as in Figure 3.6, which is reproduced and slightly modified here. We have a scalar system, so there is only one output $\hat{r}(t+1|t)$, our one-step-ahead predictor. The input elements have been changed to NARMA regressor elements:

$$\boldsymbol{\phi}(t) = [-\hat{r}(t|t-1), -\hat{r}(t-1|t-2), r(t), r(t-1)]^T, (n_c, n_g) = (2, 2)$$

The structure can be changed, for example to $(n_c, n_g) = (1, 2)$. More sophisticated structures are usually not very good if the phenomenon is fast. In addition to the regressor form, the MLP network structure and activation functions must be fixed. NNSYSID supports a two-layer neural network.

Once the Rayleigh fading data is ready, the following command

```
[W1,W2,Chat,critvec,iteration,lambda]  =  nnarmax1(ND,NN,[],[],
[],settrain,y,[]);
```

will set up the neural network predictor. It proceeds by first defining the two-layer network structure. In the first hidden layer there are the *tanh* activation functions and in the output layer the linear activation functions. The number of 'H's in the command *ND* = *['HHHHHH';'L-----']* indicates how many parallel nodes there are in the hidden layer. At the output, *['L-----']*, the letter L indicates linear activation functions. The number of minus signs together with one L has to equal the number of H's. *ND* = [] can also be input. In this case the network size is determined automatically. Regression vector size is defined in the ARMA case as *NN* = [*na nc*], where *na* is the number of past outputs used for determining the prediction and *nc* is the number of past residuals.

```
clear;

tot = jakes; %load jakes (fading) data

ND = ['HHHHHH';'L- - - - -'];
% This defines the 2-layer network
% Activation functions: L = linear, H = tanh,
% first row defines the hidden layer, number of H's defines how many
  parallel nodes
% and second row defines output layer
NN = [2 1]; %[number of past outputs, number of residuals] for
  training

y = tot(2,:); %feed the fading data for nnarmax1

[W1,W2,Chat,critvec,iteration,lambda]  = nnarmax1(ND,NN,[],[],
[],settrain,y,[]);
% network identification

plot(critvec) % plot error convergence
title('Error convergence');
xlabel('Iteration');
ylabel('Criterion');

[yhat,NSSE] = nnvalid('nnarmax1',ND,NN,W1,W2,Chat,y,[]); %plot
  validation data
```

The predict or structure is shown in Figure 4.12. Figure 4.13 shows how the estimation of the network parameters has progressed. Convergence is quite fast. The prediction result is shown in Figure 4.14. Again the result is excellent.

4.11 Summary

In this chapter we have discussed the temporal aspects of radio channels. Both continuous and discrete models and the multivariable nature of the models have been treated. The chapter

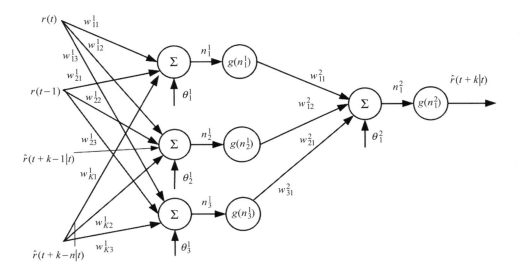

Figure 4.12 MLP neural network structure (2,2) for NARMA identification. Observe that the output is the predicted value of $r(t), \hat{r}(t+k|t)$, where k is the prediction horizon

started by explaining the Doppler effect and how modeling of that leads to LTV systems. Linear systems theory (Chen, 1999) provides ways of characterizing such systems using the impulse response. The approach was extended to cover MIMO systems, which become important when smart antennas are studied in Chapter 9. The early sections in this chapter used continuous time models, and the rest of the chapter treated discrete time systems. A quick review of

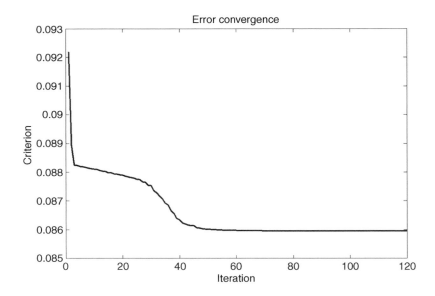

Figure 4.13 The convergence of parameter estimation (learning) was very good

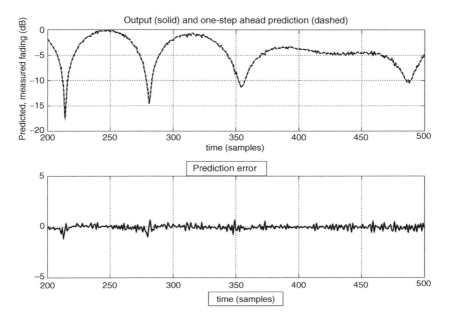

Figure 4.14   One-step ahead prediction result above. The prediction error is shown below. The result is excellent

discrete time models with noise was given, together with basic estimation and prediction theory. Their use was illustrated in modeling a fading channel (further details can be found in Ekman (2002)). Both a linear ARMA model and a neural network based NARMA model were used.

An elegant theory (Ljung, 1999) and many powerful tools (Ljung, 2009; Norgaard, 2000) have been developed for system identification. Modeling and simulation have become essential in studying or designing cellular radio systems witnessed in the books by Goldsmith (2005), Haykin and Moher (2005), Jeruchim, Balaban and Shanmugan (2000), Paulraj, Nabar and Gore (2003), Mark and Zhuang (2003), and Tse and Viswanath (2005) among others. Also system identification approaches have gained a lot of popularity in wireless communications. Still there seems to be room for educating the communications community about the advantages of this approach. It is hoped that this introduction will help to find the beginning of this path.

The Jakes model (Jakes, 1974) or similar Rayleigh channel models (Jeruchim, Balaban and Shanmugan 2000, Sklar (1997) have been extensively used in approximating fading channels. Jeruchim, Balaban and Shanmugan (2000) discuss the simulation of linear time-invariant systems extensively.

The results of this chapter are used extensively in later chapters.

Exercises

4.1   Let the impulse response of a noiseless channel be $h(t, \tau) = 3(t - \tau)e^{2(t-\tau)}$. Determine the state-space form (Equation (4.32)) of the system.
(Hint: Use Theorem 1 and Example 4.1.)

4.2   Let the impulse response of a noiseless channel be $h(t, \tau) = 4e^{-\tau}\cos(6t)$. Suppose the incoming signal is

(a) $u_1(t) = \begin{cases} 2, |t| \leq 0.3 \\ 0, |t| > 0.3 \end{cases}$. Compute the output $y(t)$ analytically. Set up a Simulink® model

      and simulate the output $y(t)$.

(b) Repeat the same tasks as in (a), when $u_2(t) = u_1(t - 0.7)$. Compare the results with those of part (a).

4.3   Consider the system

$$r(t) + 0.4r(t-1) + 0.8r(t-2) = x(t-1) + 0.1\xi(t)$$

where $\xi(t)$ are independent Gaussian variables with zero mean and variance one.

(a) Find the one-step predictor $\hat{r}(t+1|t)$ for the system output $r(t)$ when data of the received signal and the transmitted signal are known up until time k, $\{r(t), x(t); t = 0, 1, \ldots, k\}$. Compute the prediction error.

(b) Set up a Simulink® model of the original system (producing data) and the predictor. Simulate the system with $x(t) = 1, t = 0, 1, 2, \ldots$ and $r(0) = 1.5$ and plot the results.

4.4   Consider the system of Example 4.2 with colored noise:

$$r(t) + 0.1r(t-1) = 2x(t-1) + \xi(t) + 0.1\xi(t-1)$$

where $\xi(t)$ are independent Gaussian variables with zero mean and variance one.

(a) Find the one-step predictor $\hat{r}(t+1|t)$ for the system output $r(t)$ when data of the received signal and the transmitted signal are known up until time k, $\{r(t), x(t); t = 0, 1, \ldots, k\}$. Compute the prediction error.

(b) Set up a Simulink® model of the original system (producing data) and the predictor. Simulate the system with $x(t) = 1, t = 0, 1, 2, \ldots$ and $r(0) = 1.5$ and plot the results.

4.5   Consider identifying the Rayleigh channel as in Example 4.5. Use Jakes data for the Rayleigh channel. Apply system identification.

      Use both of the ARX and NARMA models. Study how well some other structures such as (2,1), (1,2) work.

4.6   Consider a Rician channel. Study if a linear ARMA or ARX can be used to model it.

5

Power Control, Part I: Linear Algebra Perspective

5.1 Introduction

There are several different ways to describe power control. However, in some cases, it becomes easier to analyze certain problems of power control within a certain perspective. For example, there is always a delay between the channel measurement and the transmit power update. This delay may cause acute performance degradation and even system instability. To analyze the loop delay and how to reduce its impacts, it is more convenient to analyze the problem as a closed-loop control system in the frequency domain. On the other hand, to specify the maximum channel capacity and the optimum power allocation, it becomes more suitable to analyze the problem using linear algebra tools. Therefore the power control problem will be introduced in two chapters. In this first part we will introduce and define the problem of power control and how to handle it using general and simple linear algebra tools. In the next chapter we will analyze the problem as a closed-loop control problem. We will cover the power control issues theoretically and practically without going into difficult mathematical proofs or very technical problems.

The main task of conventional power control algorithms is to keep the transmit power value of the transmitter at the minimum power required to achieve the target quality of service (QoS) in the communication link. The QoS of a communication system is a list of requirements to be fulfilled in order to achieve successful and accepted communication links. Some of these terms include the bit error rate (BER), data rate, packet delay and outage probability. The end users are not usually concerned with these technical details. However, they will consider the quality of the call like sound quality, echo, data download speed and cost. If they are not satisfied with the QoS, they may migrate to another network (churn behavior). In this chapter we will consider only the SINR as an indication of the QoS. Actually, most of the QoS parameters can be directly or indirectly mapped to the SINR. For example, the mapping between the BER and the SINR depends on the modulation type, channel condition, and interference structure. The mapping of fixed SINR to BER is well known and can be found in classical digital communication books such as in [1] and [2]. For more real situations where the SINR is a random process, we should average the BER over the probability density function (pdf) of the SINR [3].

Power control is essential for non-orthogonal multiuser communication environments such as CDMA systems. It can mitigate the near-far problem, increase the system capacity, reduce the interference, improve the QoS, increase the battery life of the mobile terminal, and decrease the biological effects of electromagnetic radiation (if any!). It is also required in MIMO communication systems which exploit the low correlation between multipaths to increase the system throughput. It determines the optimum power value to be transmitted over every antenna terminal. Power control should be adaptive for dynamic communication systems, e.g., for terminal mobility. In this case the optimum allocated power is changing with time.

The process of power control can be described as follows. Let the transmitter start transmission with an initial power value – the initial power should be high enough to be received and decoded correctly by the receiver. Other signals from co-channel and cross-channel interferences will be received as well. The received signal quality is compared to some target value. The signal quality can be measured by its SINR at the receiver input. If the received signal quality is less than the target, the power control algorithm will increase the transmitted power in the next time slot by some value depending on how far the received signal quality is from the target. On the other hand, if the received signal quality is higher than the target, reverse action is taken by the power control algorithm. This process is called distributed power control because the system needs to know only the signal quality of its communication link. However, if the power control is central and it is able to know the signal quality of all terminals as well as the channel gains of all links, it is possible to calculate the optimum power vector of all transmitters simultaneously in one time slot. This type of centralized power control has little practical value; however, it is important to start the power control analysis with it. The distributed power control algorithms should converge to the solution of the centralized one. We will show the mathematical details of the centralized as well as the distributed power control algorithms. But before that, it is important to mention the stochastic nature of the power control. The channel gain, the interference, and noise are all random processes in nature. Furthermore the channel gain and the interference are usually non-stationary random processes. This leads us to consider one assumption that the power control update duration time is less than the channel coherence time. In other words, when the power control algorithm receives the signal quality and then sends its command to update the transmit power, all these actions occur within a time that the channel has not changed significantly. In this case the channel is called quasi-static. We will assume that the channel, interference, and noise remain fixed during power control adaptation time. This assumption is sometimes called a 'snapshot' assumption [4].

Some notations and definitions are given first. Let the *transmit power control* vector be a Q-dimensional column vector $\mathbf{P} = [P_1, P_2, \ldots, P_Q]'$, where P_i is the transmit power of terminal i. The SINR of terminal i is denoted by Γ_i.

Mathematically the power control problem can be formulated as follows (other formulations are possible as well):

Find the power control vector $\mathbf{P}$ that minimizes the cost function

$$J(\mathbf{P}) = \mathbf{1}'\mathbf{P} = \sum_{i=1}^{Q} P_i \tag{5.1}$$

subject to

$$\Gamma_i = \frac{P_i G_{ii}}{\displaystyle\sum_{\substack{j=1 \\ j\neq i}}^{Q} P_j \theta_{ij} v_j G_{ij} + N_i} \geq \Gamma_{\min,i}, \qquad i = 1,\dots,Q, \tag{5.2}$$

and

$$P_{\min} \leq P_i \leq P_{\max}, \ \forall i = 1,\cdots,Q, \tag{5.3}$$

where

$$\mathbf{1}' = [1,\dots,1]$$

Q = number of terminals (or links)

G_{ij} = channel gain between transmitter j and receiver i, as shown in Figure 5.1

$0 \leq \theta_{ij} \leq 1$ = orthogonality factor between the signals of terminals i and j

$0 < v_j \leq 1$ = the voice activity factor

N_i = average power of the additive noise at receiver i

P_{max} = maximum power, which can be handled by the transmitter

P_{min} = minimum power to assume the terminal is active (not switched off)

$\Gamma_{min,i}$ = minimum predefined SINR for terminal i. If Γ_{min} is used without subscripts i, it means that all terminals have the same minimum required SINR

During a voice call we do not talk all the time, for some call duration we listen and also there are some periods where both parties are silent. This natural behavior is exploited by using discontinuous transmission (DTX) which cuts the transmission power of the traffic channel during this silent period [8]. This has direct influence in CDMA systems by reducing the average interference and hence increasing the network capacity. Mathematically, this is represented in Equation (5.2) by the voice activity factor. However, in most of our analysis in this book we consider $v_j = 1$. This can be considered as a worst case analysis. It is clear that

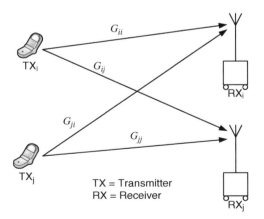

Figure 5.1   Channel gains notation

when the orthogonality factor $\theta_{ij} = 0$, then there is no effect of the interference from other users on the signal quality at the receiver. However, this case cannot be achieved in practical co-channel transmission systems such as CDMA. It is possible with perfectly synchronized TDMA or FDMA with perfect filters. When $\theta_{ij} \cong 0$, the power control becomes trivial and it is enough to update its value at a very small rate (e.g. 2 Hz) to mitigate the distance and shadow losses. There is no risk of using the maximum transmission power in orthogonal communication systems because there will be no interference for other terminals. However, it is better to use as small power as possible to prolong the battery life of the transmitter and also to reduce the cross-channel interference. For this reason power control is in use with GSM systems but with very low updating rate ($\sim$1 Hz) to compensate for the shadowing and distance-based attenuation and to reduce the co-channel interference within the reuse frequency clusters.

In this chapter we will concentrate on the nontrivial case, i.e., when $\theta_{ij} \neq 0$. If the SINR of user I is $\Gamma_i < \Gamma_{min}$, and the transmit power $P_i = P_{max}$, it means that the link is not able to achieve its minimum requirements, i.e., it is a congested link. One possible solution in this case is to drop one or more connections to relax the link (more analysis for congested networks is given in Chapter 7). Another important factor is the target SINR for terminal $i (\Gamma_i^T \geq \Gamma_{min,i})$. It should be noted that the superscript T means *Target*. The difference between the target SINR and the minimum predefined SINR is called the *SINR margin*. Actually, the target SINR is the value which has practical use; the minimum predefined SINR has only theoretical means. In UMTS systems, the target SINR value is determined by the outer-loop power control to achieve certain QoS in the cell as will be described later. The target (or the minimum predefined) SINR may be different from user to user because it depends on the type of service requested by the user. However, unless otherwise indicated, we will consider fixed target SINR.

Our power control analysis in this chapter is valid for an arbitrary wireless communication system. However, we concentrate on systems where the power control is essential such as cellular systems with non-orthogonal multiple access method (e.g. CDMA). In these systems we have one or more access points (e.g. base stations) and many terminals (e.g. mobile handsets). The transmit power control for terminals is called the uplink power control, and the transmit power control of the access point is called the downlink power control. Power control is important in both directions. For example, for the non-orthogonal system if power control is not utilized for uplink, the excellent link terminal may block all other terminals in the network. This problem is known as the near-far problem. For downlink, the power control reduces the interferences to other cells. The mathematical representation of power control for downlink is slightly different than the case of uplink as given in Equations (5.1)–(5.3). In the case of downlink, the total power of the access point is limited by some P_{max}. The transmit power to each terminal should be part of this total power. The constraints are also different. The mathematical representation of the downlink power control will be given later. Figure 5.2 shows the concepts of the access point and terminals.

To include the actual bandwidth of the digital transmission we use the bit energy to the noise spectral density ratio (E_b/N_0) to quantify the signal quality. There is a direct relation between the SINR and the E_b/N_0 as explained in Chapter 1. This relation depends on the modulation method, for example in binary direct sequence CDMA this relation is given by

$$E_b/N_0 = PG \times SINR \qquad (5.4)$$

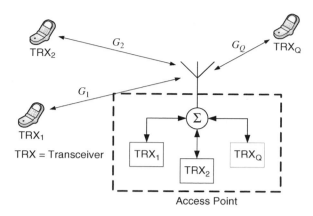

Figure 5.2   Single access point with Q terminals

where PG is the processing gain which can be approximated as the ratio of the chip rate to the data rate. The details of such a basic relation can be found in many textbooks of digital communications such as in [1].

Example 5.1

Formulate the power control problem for the downlink (DL) CDMA cellular systems.

Solution

The power control formulation is similar to the uplink (UL) but with some important differences. Let's define the DL power from the access point to user i as $\hat{P}_i$ and the DL channel gain between base station j and terminal i as $\hat{G}_{ij}$. The power control problem can be formulated such as follows:

Find the power control vector $\hat{\mathbf{P}} = [\hat{P}_1, \hat{P}_2, \cdots, \hat{P}_Q]$ that minimizes the cost function

$$J(\hat{\mathbf{P}}) = \mathbf{1}'\hat{\mathbf{P}} = \sum_{i=1}^{Q} \hat{P}_i \tag{5.5}$$

subject to

$$\hat{\Gamma}_i = \frac{\hat{P}_i G_{ii}}{\sum_{\substack{j=1 \\ j \neq i}}^{Q} \hat{P}_j \hat{\theta}_{ij} \hat{v}_j \hat{G}_{ij} + \hat{N}_i} \geq \hat{\Gamma}_{\min,i}, \quad i = 1, \ldots, Q, \tag{5.6}$$

and

$$\sum_{i=1}^{Q} \hat{P}_i \leq \hat{P}_{\max}, \quad \hat{P}_i \geq 0 \quad \forall i = 1, \cdots, Q$$

where $\hat{P}_{\max}$ is the maximum transmitting power of the base station. The DL power control is not as effective as the UL one. The reason is that for the UL power control decreasing the power of the near transmitters and increasing the power of the far ones can achieve some fairness between terminals and hence increase the capacity in terms of the number of served terminals. On the other hand, since the total power is limited in the DL, then the base station decreases the transmit power for the near terminals and increases the power for the far terminals. Therefore, increasing the power for the far terminals will increase the interference level for the near terminals. Furthermore, increasing the transmitted power for the near terminals will decrease the system capacity in terms of number of served users (why?) and also will increase the interference for the other cells. This is one reason why power control is omitted in high-speed downlink packet access (HSDPA) systems.

5.2 Centralized Power Control

If the information of the link gains and the noise levels are available for all users, then the centralized power control algorithm can be applied to solve the power control problem given in Equations (5.1)–(5.3) perfectly [4]. Although the central power control has very limited practical applications, studying the centralized power control can offer the following:

- Understanding the concepts of power control.
- Evaluating the feasibility of power control solutions in certain situations.
- Evaluating the maximum achievable capacity of communication networks.
- Can be used as a reference when comparing different distributed power control algorithms.

The maximum achievable system capacity can be evaluated in the case of noiseless systems. For the noiseless case, i.e., $N_i = 0$, Equation (5.2) becomes (for all $i = 1, \ldots, Q$) (using a paper and pencil, try to validate every step yourself)

$$\Gamma_i = \frac{P_i G_{ii}}{\displaystyle\sum_{\substack{j=1 \\ j \neq i}}^{Q} P_j G_{ij}} \geq \Gamma_{\min} \Rightarrow P_i \geq \Gamma_{\min} \sum_{\substack{j=1 \\ j \neq i}}^{Q} P_j \frac{G_{ij}}{G_{ii}} \tag{5.7}$$

Equation (5.7) can be written in a matrix form such as:

$$\mathbf{P} \geq \Gamma_{\min} \mathbf{HP} \tag{5.8}$$

where $\mathbf{H}$ is a nonnegative matrix with the following elements

$$(\mathbf{H})_{ij} = \begin{cases} 0 & i = j \\ \dfrac{G_{ij}}{G_{ii}} > 0 & i \neq j \end{cases} \tag{5.9}$$

The problem is how to find the power vector $\mathbf{P} > 0$ such that Equation (5.8) is satisfied. Equation (5.8) can be rewritten as

$$\left[\frac{1}{\Gamma_{\min}} \mathbf{I} - \mathbf{H} \right] \mathbf{P} = 0 \tag{5.10}$$

The inequality is dropped in Equation (5.10), since the equality sign holds for the minimum power vector. Now the optimum power vector is the solution of Equation (5.10). One direct solution of the linear system of equations in Equation (5.10) is that all transmit powers are zeros! This means that we must switch off all mobiles. Of course, no one is interested in such a solution. This is called a trivial solution. It is known from linear algebra that a nontrivial solution of Equation (5.10) exists if and only if $[\frac{1}{\Gamma_{min}}\mathbf{I} - \mathbf{H}]$ is a singular matrix. It is seen from Equation (5.10) that this happens if $\frac{1}{\Gamma_{min}}$ is an eigenvalue of $\mathbf{H}$, and the optimum power vector $\mathbf{P}$ is the corresponding eigenvector. The power vector $\mathbf{P}$ should be positive. The Perron–Frobenius theorem [5] says that for a nonnegative and irreducible $Q \times Q$ matrix $\mathbf{H}$, there exists a positive vector $\mathbf{P}$ associated with the maximum eigenvalue (which is also real and positive)

$$\lambda^* = \rho(\mathbf{H}) = \max_i |\lambda_i|, \quad i = 1, \ldots, Q, \qquad (5.11)$$

where λ_i is the ith eigenvalue of the matrix $\mathbf{H}$, and $\rho(\mathbf{H})$ is called the *spectral radius* of matrix $\mathbf{H}$. Hence, the maximum achievable SINR can be expressed as

$$\gamma^* = \frac{1}{\lambda^*} = \frac{1}{\rho(\mathbf{H})} \qquad (5.12)$$

The optimum power vector is the eigenvector associated with the spectral radius. It is important to mention that in the noiseless case the eigenvector, which represents the optimum power, is not unique. For example if the eigenvector is $\begin{bmatrix} c_1 \\ \vdots \\ c_Q \end{bmatrix}$, then $\alpha \begin{bmatrix} c_1 \\ \vdots \\ c_Q \end{bmatrix}$ is also an eigenvector for any real $\alpha \neq 0$. This means that in the noiseless case the absolute transmit power value of each terminal is not important. The importance is in the relation between them. For example, the second terminal sends at double the power of the first one and so on. It is straightforward to observe that the ith element in the optimum power vector is related to its channel gain such as $P_i \propto 1/G_{ii}$. We'll let the reader explain this relation! When additive noise is assumed (it always exists in reality), then the absolute value of the power is also important and must be determined.

Example 5.2

Four terminals are in connection with a single access point. The channel gain follows the relation $G_i = A_i/d_i^4$, where A_i is the shadowing factor of terminal i and d_i is the distance between terminal i and the access point (see Chapter 1). Find the normalized channel matrix $\mathbf{H}$, the maximum achievable SINR, and the optimum power vector. The distance and the shadowing factor of every terminal are given in Table 5.1.

Table 5.1 MS parameters of Example 5.2

	Terminal 1	Terminal 2	Terminal 3	Terminal 4
Distance (m)	60	50	70	120
Shadowing	0.1	1.0	0.5	0.4

Solution

From Table 5.1 we may calculate the channel gain vector as $\mathbf{G} = [0.008 \quad 0.160 \quad 0.021 \quad 0.002] \times 10^{-6}$. From Equation (5.9) we can compute the normalized channel matrix $\mathbf{H}$ as:

$$\mathbf{H} = \begin{bmatrix} 0 & 20.736 & 2.699 & 0.250 \\ 0.048 & 0 & 0.130 & 0.012 \\ 0.371 & 7.683 & 0 & 0.093 \\ 4.000 & 82.944 & 10.796 & 0 \end{bmatrix}$$

We can calculate the spectral radius of $\mathbf{H}$ as the maximum absolute eigenvalue. Using the *eig* command from MATLAB® we find that $\rho(\mathbf{H}) = 3.0$. The maximum achievable SINR is calculated as in Equation (5.12), $\gamma^* = 0.333$. The optimum power vector is the eigenvector associated with the maximum spectral radius which can be also calculated using same MATLAB® command (*eig*), and the result is: $\mathbf{P}^* = \alpha[0.242 \quad 0.012 \quad 0.090 \quad 0.967]'$, where α is any real number greater than zero. If we normalize the optimum power vector such as $\mathbf{P}^* = [0.25 \quad 0.012 \quad 0.093 \quad 1]'$, then the fourth terminal should transmit with the highest power value, and this is logical since it is the furthest terminal from the access point. The first terminal should transmit at power which is one-quarter of the transmit power of the last terminal. Can you explain why?

Example 5.3

Repeat the previous example with two access points. The distances and shadowing factors are given in Table 5.2. Both access points use the same frequency band. Assume perfect assignment so that every terminal communicates through the access point which has better channel quality. No soft handover is allowed. Repeat the calculations if the reverse is assumed.

Solution

Let's first calculate the channel gain for all terminals with both access points. This will help to decide with which access point every terminal should communicate. The channel gain values are: $G_{11} = 7.72 \times 10^{-9}$, $G_{12} = 1.60 \times 10^{-7}$, $G_{13} = 2.08 \times 10^{-8}$, and $G_{14} = 1.93 \times 10^{-9}$. For the second access point, the channel gain values are: $G_{21} = 1.23 \times 10^{-6}$, $G_{22} = 3.05 \times 10^{-9}$, $G_{23} = 4.82 \times 10^{-10}$, and $G_{24} = 6.4 \times 10^{-8}$. Now it is clear that we should assign the following terminals to the first access point: {2,3}, and {1,4} for the second access point. As done in the previous example we can calculate the normalized channel

Table 5.2 MS parameters of Example 5.3

	Terminal 1	Terminal 2	Terminal 3	Terminal 4
Distance from first access point (m)	60	50	70	120
Shadowing with first access point	0.1	1.0	0.5	0.4
Distance from second access point (m)	30	90	120	50
Shadowing with second access point	1.0	0.2	0.1	0.4

matrix as:

$$\mathbf{H} = \begin{bmatrix} 0 & 0.0025 & 3.9 \times 10^{-4} & 0.0518 \\ 0.048 & 0 & 0.130 & 0.012 \\ 0.371 & 7.683 & 0 & 0.093 \\ 19.29 & 0.0476 & 0.0075 & 0 \end{bmatrix}$$

The maximum achievable SINR is $\gamma^* = 0.963$ and the normalized optimum power vector is $\mathbf{P}^* = [0.018 \quad 0.130 \quad 1 \quad 0.349]'$. It is clear that there is a considerable improvement in the maximum achievable SINR compared to the single access-point case given in the previous example. It has been increased about three times! If we reverse the assignment such that the terminal is connected to the access point which has the worse channel gain, the channel matrix becomes:

$$\mathbf{H} = \begin{bmatrix} 0 & 0.0210 & 0.0030 & 0.0003 \\ 0.405 & 0 & 0.0002 & 0.0210 \\ 2.560 & 0.0063 & 0 & 0.1327 \\ 0.004 & 0.0829 & 0.0108 & 0 \end{bmatrix} \times 10^3$$

In this case the maximum achievable SINR is $\gamma^* = 0.0073$, which is a very small value, about 130 times less than the maximum achievable SINR in the optimum assignment. This shows the importance of the optimum assignment of the terminals to access points in a multi access-point environment. Remember that the maximum achievable SINR is the maximum SINR which all terminals can achieve in the noiseless case.

Exercise 5.1 Prove that for a single access point the maximum achievable SINR does not depend on the channel gain values of terminals and is given by $\frac{1}{Q-1}$, where Q is the number of terminals.

Exercise 5.2 Rewrite Equations (5.8) and (5.10) if every terminal has a different minimum predefined SINR (i.e., $\Gamma_{\min,i}$).

Exercise 5.3 Construct the gain matrix $\mathbf{H}$ in the DL single-cell case.

Example 5.4

When the target QoS parameters for all terminals are identical, the main objective of the UL power control is to make the received power from all terminals at their assigned base station be the same. For a single-cell scenario calculate the maximum number of supported terminals if the required data rate is 64 kb/s, the target $E_b/N_0 = 8dB$, and the chip rate is $f_c = 3.84\,Mchip/s$, for two scenarios

$$1. \quad \upsilon_j\theta_j = 1 \quad \forall j = 1, \cdots, Q$$
$$2. \quad \upsilon_j\theta_j = 0.6 \quad \forall j = 1, \cdots, Q$$

Ignore the additive noise effects.

Solution

From Equation (5.2) and (5.4), and in the case of a single cell and equal received power, it is straightforward to state that

$$\left(\frac{E_b}{N_0}\right)_j = \frac{f_c}{R_j}\frac{1}{(Q-1)\theta v} \tag{5.13}$$

Hence

$$Q = \frac{f_c}{R_j}\frac{1}{(E_b/N_0)\theta v} + 1 \tag{5.14}$$

In this first case when $v_j\theta_j = 1$, which represents the worst-case scenario, the number of supported terminals is $Q = \frac{3.84\times10^3}{64}\frac{1}{10^{0.8}} + 1 \cong 10$ terminals. When a good multiuser detection receiver is used, and also applying the DTX, the system capacity in terms of the number of terminals becomes: $Q = \frac{3.84\times10^3}{64}\frac{1}{0.6\times10^{0.8}} + 1 \cong 16$. This means that the system capacity has been increased by 60%. This shows the importance of exploiting all available possibilities to improve the system capacity.

Example 5.5

In the previous example, assume that you have used a more sophisticated coding method which gives a 1 dB gain. This means that you will achieve the same performance at $E_b/N_0 = 7dB$. Calculate the number of supported terminals in this case with ignoring other constraints such as the base-station total power. Discuss the influence of this gain over the cellular network.

Solution

By applying the same procedure as in the previous example we obtain

$$Q = \frac{3.84 \times 10^3}{64}\frac{1}{0.6 \times 10^{0.7}} + 1 \cong 21$$

which is very considerable increase for only a 1 dB improvement. If one base station can support about 15 terminals, by using this sophisticated coding we can support about 20 users, which means we save about 25% of the required base stations (a very rough estimate). If the cost of one base station is about $1 million, you can see the importance of this one 1 dB coding gain!

Now let's consider the additive white noise at the receivers, from Equation (5.2) the transmit power of terminal i should be (with $\theta_{ij} = 1$ and $v_j = 1$)

$$P_i \geq \Gamma_{\min}\left(\sum_{\substack{j=1 \\ j\neq i}}^{Q} P_j\frac{G_{ij}}{G_{ii}} + \frac{N_i}{G_{ii}}\right) \tag{5.15}$$

To be more specific in all coming analysis we will consider the target SINR which has practical meaning instead of the minimum threshold SINR, i.e., we will use Γ^T instead of $\Gamma_{\min}$.

In matrix form Equation (5.15) can be rewritten as

$$[\mathbf{I} - \mathbf{\Gamma}^T \mathbf{H}]\mathbf{P} = \mathbf{u} \qquad (5.16)$$

where $\mathbf{u}$ is a vector with positive elements, and the ith element u_i is given

$$u_i = \frac{\mathbf{\Gamma}^T N_i}{G_{ii}}, \quad i = 1, \ldots, Q, . \qquad (5.17)$$

It can be shown that the linear system of equations in Equation (5.16) has a unique positive power vector solution if [5]

$$\mathbf{\Gamma}^T < \frac{1}{\rho(\mathbf{H})} \qquad (5.18)$$

In this case, the power vector $\mathbf{P}^*$

$$\mathbf{P}^* = [\mathbf{I} - \mathbf{\Gamma}^T \mathbf{H}]^{-1} \mathbf{u} \qquad (5.19)$$

is the solution of the optimization problem posed in Equations (5.1)–(5.3).

Exercise 5.4   Prove that in case of different target SINR values, condition (5.18) can be reformulated as

$$\rho(\mathbf{\Gamma}^T \mathbf{H}) < 1 \qquad (5.20)$$

where $\mathbf{\Gamma}^t = diag(\mathbf{\Gamma}_1^T, \mathbf{\Gamma}_2^T, \cdots, \mathbf{\Gamma}_Q^T)$.

There are no guarantees that $\mathbf{\Gamma}^T \leq \gamma^*$ or that the power vector $\mathbf{P}^*$ is within the constraints (Equation (5.3)). In this case *a removal algorithm* would be needed to reduce the number of users in the cell. Removal algorithms are discussed in Chapter 7.

Example 5.6

Three CDMA terminals are in connection with a single access point. The channel gain follows the relation $G_i = 1/d_i^4$, where d_i is the distance between terminal i and the access point, the noise figure of the access point is 5 dB. All terminals use same transmission power of 23 dBm. Calculate the achieved SINR for every terminal. If the processing gain $PG = 128$ and the minimum required E_b/N_0 is 15 dB discuss the system performance. The distance of every terminal is given in Table 5.3. The system bandwidth is 5 MHz.

Solution

The channel gains of terminals are $G_1 = 1.6 \times 10^{-7}$, $G_2 = 1.52 \times 10^{-8}$, and $G_3 = 4.82 \times 10^{-9}$. We should unify the numbers either in dB or linear scales. Let's work in the

Table 5.3   MS parameters of Example 5.6

	Terminal 1	Terminal 2	Terminal 3
Distance (m)	50	90	120

linear scale. The transmit power is $10^{23/10}mW = 0.2W$, $\Gamma_{min} = \frac{10^{1.5}}{128} = 0.247 = -6.07dB$, and the additive noise is $N = KFT_0B = 1.38 \times 10^{-23} \times 10^{0.5} \times 290 \times 5 \times 10^6 = 6.3 \times 10^{-14}$. Since all terminals use the same power value we can calculate the achieved SINR for every terminal as

$$\Gamma_1 = \frac{0.2 \times 1.6 \times 10^{-7}}{0.2 \times 1.52 \times 10^{-8} + 0.2 \times 4.82 \times 10^{-9} + 6.3 \times 10^{-14}} = 7.97 = 9dB$$

$$\Gamma_2 = \frac{0.2 \times 1.52 \times 10^{-8}}{0.2 \times 1.6 \times 10^{-7} + 0.2 \times 4.82 \times 10^{-9} + 6.3 \times 10^{-14}} = 0.0925 = -10.34dB$$

$$\Gamma_3 = \frac{0.2 \times 4.82 \times 10^{-9}}{0.2 \times 1.6 \times 10^{-7} + 0.2 \times 1.52 \times 10^{-8} + 6.3 \times 10^{-14}} = 0.0275 = -15.6dB$$

It is clear that only the first element can achieve the minimum required SINR. In other words, both the second and third terminals will not be able to access the network even though they are using their highest transmission power. In the next example we will compute the optimum transmission power for all terminals.

Example 5.7

Compute the optimum power values of Example 5.6 and discuss the results.

Solution

As we did in Examples 5.1 and 5.2, we start by calculating the **H** matrix.

$$\mathbf{H} = \begin{bmatrix} 0 & 0.095 & 0.030 \\ 10.498 & 0 & 0.316 \\ 33.178 & 3.161 & 0 \end{bmatrix}$$

We also calculate the **u** vector as

$$\mathbf{u} = \begin{bmatrix} 0.0097 \\ 0.1021 \\ 0.3227 \end{bmatrix} \times 10^{-5}$$

The optimum power vector can be now easily calculated using Equation (5.19) as:

$$\mathbf{P}^* = \left(\begin{bmatrix} 1 & 0 & 0 \\ 0 & 1 & 0 \\ 0 & 0 & 1 \end{bmatrix} - 0.247 \begin{bmatrix} 0 & 0.095 & 0.030 \\ 10.498 & 0 & 0.316 \\ 33.178 & 3.161 & 0 \end{bmatrix} \right)^{-1} \begin{bmatrix} 0.0097 \\ 0.1021 \\ 0.3227 \end{bmatrix} \times 10^{-5}$$

$$= \begin{bmatrix} 0.0192 \\ 0.2018 \\ 0.6377 \end{bmatrix} \times 10^{-5}W$$

The optimum power vector is now much smaller than the fixed maximum power values given in the previous example. However, all terminals now achieve the target SINR. This example shows the importance of power control to solve the near-far problem.

Exercise 5.5 Repeat the previous example if the terminals have different target SINRs such as $\Gamma_1^T = -5dB$, $\Gamma_2^T = -12dB$, and $\Gamma_3^T = -8dB$.

Example 5.8

Repeat Example 5.3 if the target $E_b/N_0 = 20\,\text{dB}$.

Solution

In this example we have $\Gamma^T = \frac{10^{2.0}}{128} = 0.7813 = -1.07dB$. When we solve for the optimum power vector as in the previous example, negative power values are obtained! Of course this solution is not accepted because the transmission power values must be nonnegative. This answer means that it is impossible to achieve the required SINR with the given network situation, i.e., there is no feasible solution. It is common to call this situation an infeasible power control problem.

5.3 Graphical Description of Power Control

The power control problem can be described graphically for a simple case. Consider two users within one cell. The first user has the link gain $G_1(t)$ and the second user the link gain $G_2(t)$. The link gains are functions in time due to the dynamical behavior of the mobile communication system. Assume that N is the average noise power. Recall the optimum power control problem Equations (5.1)–(5.3). The problem is to determine the minimum transmit power vector that satisfies the required QoS. We can write $\frac{P_1(t)G_1(t)}{P_2(t)G_2(t)+N} \geq \Gamma_{\min}$ for the first user and $\frac{P_2(t)G_2(t)}{P_1(t)G_1(t)+N} \geq \Gamma_{\min}$ for the second user. Solving the previous inequalities we obtain

$$P_1(t) \geq \Gamma_{\min} \frac{G_2(t)}{G_1(t)} P_2(t) + \frac{N\Gamma_{\min}}{G_1(t)}$$

$$P_2(t) \geq \Gamma_{\min} \frac{G_1(t)}{G_2(t)} P_1(t) + \frac{N\Gamma_{\min}}{G_2(t)}$$

$$(5.21)$$

In practice, the gains are random variables due to slow and fast fading behaviors.

To solve the system of linear equations in Equation (5.21), we need to assume that the gains are constant, i.e. they are frozen at time t. This is termed a *snapshot* assumption. Figure 5.3 illustrates the graphical interpretation of power control with the snapshot assumption at certain parameter values. It is interesting to observe that the power vector $\mathbf{P}_t$ in the figure is much larger than the optimum power vector; however, it cannot achieve even the minimum required SINR for both terminals. In multiuser environments, increasing the transmit power above the required is not helpful, because the others will increase their powers as well, so that the network will be congested and many links could be dropped. The shaded area shows the set of feasible power pair values to achieve the required QoS.

Exercise 5.6 From Figure 5.3 find the parameters: G_1, G_2, N, and $\Gamma_{\min}$.

As we indicated before, centralized power control is not practical because of its high computation and information requirements. Therefore it is common in practice to use a distributed power control technique.

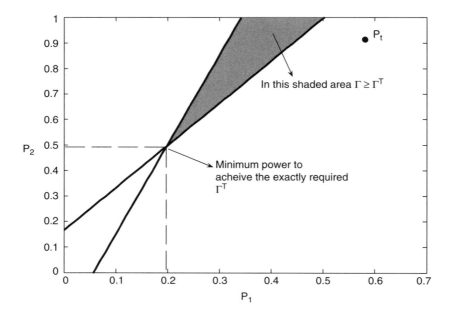

Figure 5.3   Graphical representation of two terminal transmission power

5.4 Distributed Power Control Algorithms

For distributed power control, only local information is needed for a specific transmitter to converge to the optimum power value. As we have seen from the previous section the centralized power control problem can be solved as a system of linear equations. The distributed power control is based on the concept of iterative methods to solve a system of linear equations. Although it is a straightforward to show the relations between the different distributed power control algorithms and iterative methods (e.g. Gauss–Saidel and over-relaxation methods), we will not do so here.

As we mentioned earlier local information is needed to update the transmit power in distributed power control. This information can be the received power, the total interference, the block error rate (BLER), or the SINR. The BLER is probably the most robust measure for the channel equality and hence the best indication to update the transmit power value. The data block consists of several time slots. However, it is not reasonable for fast power updates. The reason is that we will need a long time to achieve a robust measure for the real BLER. For example, if the length of the data block is 1 ms, and the target BLER is 10^{-2}, then we will need at least 100 ms to measure the channel quality efficiently. This means that the power update rate will be about 10 Hz. This is a very slow power control rate especially in highly dynamic channels. In the UMTS standard the power update rate is 1500 Hz.

The SINR is a very good indicator for the channel quality, and also it is relatively easy to estimate. Moreover, the SINR can be estimated based on the received signals' measurements much faster than the BLER, however, it has direct relations with the BLER. The SINR estimation is performed at the receiver and reported to the transmitter using the feedback channel. For example, if the power control is for the downlink, then the measurements of the

SINR is made at the mobile station. The results are reported back to the base station to update its transmit power. If the feedback channel is assumed perfect, then the actual measurement can be transmitted from the mobile to the base station. However, this is not practical: we would need large bandwidth (the scarcest resource) for the feedback channel. The available bandwidth is strictly limited, shared between all transmitters and used as a feedback channel to report the channel parameters, therefore the feedback part of the bandwidth should be kept at the minimum requirement. It is unwise to use a considerable part of the scarce resources to send measurements instead of user data. For this reason in practice we use quantized feedback channels, as will be discussed later.

If the power control is for uplink, then the measurement of the SINR has to be done at the base station. The results are reported back to the mobile, and the mobile will adjust its transmitted power in the next time slot. The optimum power values can also be calculated at the base station and then reported back to the mobile terminals. The power control in the UMTS standard will be discussed later.

One important property which must be checked for every distributed iterative power control algorithm is the convergence proofs. Of course, we are not interested in distributed algorithms which diverge to the maximum power or converge to the wrong power values, because this will lead to network congestion and instability or at least very bad network performance. From a linearity point of view there are two kinds of power control algorithms: linear and nonlinear distributed algorithms. It is relatively easy to prove the convergence of linear algorithms using linear algebra tools. Traditionally there were no standard tools to do the same for nonlinear distributed algorithms, however, Yates has introduced some relatively simple tools to prove the convergence of many kinds of linear or nonlinear power control algorithms [6]. We start by introducing this method along with some simple proofs.

The transmitted power of all users can be described mathematically as

$$\mathbf{P}(t+1) = \mathbf{\Psi}(\mathbf{P}(t)) \quad t = 0, 1, \ldots \tag{5.22}$$

where $\mathbf{\Psi}(\mathbf{P}(t)) = [\Psi_1(\mathbf{P}(t)), \ldots, \Psi_Q(\mathbf{P}(t))]'$ is the interference function. There are different types of interference functions in the literature. Other interference functions will be given later.

The interference function $\mathbf{\Psi}(\bullet)$ is called the standard when the following properties are satisfied for all components of the nonnegative power vector $\mathbf{P}$ [6]:

- positivity $\mathbf{\Psi}(\mathbf{P}) > 0$;
- monotonicity, if $\mathbf{P} \geq \bar{\mathbf{P}}$ then $\mathbf{\Psi}(\mathbf{P}) \geq \mathbf{\Psi}(\bar{\mathbf{P}}) > 0$;
- scalability, for all $\alpha > 1$, $\alpha\mathbf{\Psi}(\mathbf{P}) > \mathbf{\Psi}(\alpha\mathbf{P})$.

Theorem 5.1
If the standard power control algorithm (Equation (5.22)) has a fixed point, then that fixed point is unique [6].

Proof
Let's assume that a standard power control algorithm has two distinct fixed points (i.e., different power vector solutions) which are given by $\mathbf{P}$ and $\bar{\mathbf{P}}$. Now assume that for a jth power element $P_j \neq \bar{P}_j$, and let $P_j < \bar{P}_j$. Hence, there exists $\alpha > 1$ such that $\alpha\mathbf{P} \geq \bar{\mathbf{P}}$ and that for the

jth element $\alpha P_j = \bar{P}_j$. The monotonicity and scalability properties imply

$$\bar{P}_j = \Psi_j(\bar{\mathbf{P}}) \leq \Psi_j(\alpha\mathbf{P}) < \alpha\Psi_j(\mathbf{P}) = \alpha P_j \qquad (5.23)$$

Since $\alpha P_j = \bar{P}_j$, we have found a contradiction, implying that the fixed point must be unique [6].

Theorem 5.2

If $\Psi(\mathbf{P})$ is feasible, then for any initial power vector $\mathbf{P}(0)$, the standard power control algorithm converges to a unique fixed point $\mathbf{P}^*$ [6].

Proof
As we discussed before the feasibility means that all terminals achieve their target SINR within the possible power values $\mathbf{0} \leq \mathbf{P}^* \leq \mathbf{P}_{\max}$, where $\mathbf{P}^* = \Psi(\mathbf{P}^*)$.

First, if the initial power vector $\mathbf{P}(0) \geq \mathbf{P}^*$, the feasibility of $\mathbf{P}(0)$ implies that $\mathbf{P}(1) \leq \mathbf{P}(0)$. Suppose $\mathbf{P}(t+1) \leq \mathbf{P}(t)$. The montonicity property implies $\Psi(\mathbf{P}(t+1)) \leq \Psi(\mathbf{P}(t))$. That is $\mathbf{P}(t+1) \geq \Psi(\mathbf{P}(t+1)) = \mathbf{P}(t+2)$. Hence $\mathbf{P}(t)$ is a decreasing sequence of feasible power vectors. Since the sequence $\mathbf{P}(t)$ is bounded below by zero, Theorem 5.1 implies the sequence must converge to a unique fixed point $\mathbf{P}^*$ [6].

Second, if the initial power vector $\mathbf{P}(0) \leq \mathbf{P}^*$, for example let $\mathbf{P}(0) = \mathbf{z}$, where $\mathbf{z}$ is all zero vectors. Since $\Psi(\mathbf{P})$ is feasible, it is clear that $\mathbf{P}^* = \Psi(\mathbf{P}^*) \geq \Psi(\mathbf{P}(t)) \geq \Psi(\mathbf{P}(t-1)) \geq \cdots \geq \mathbf{P}(1) = \Psi(\mathbf{P}(0)) > \mathbf{z}$. Hence the sequence is non-decreasing and bounded above by $\mathbf{P}^*$. Theorem 5.1 implies that the sequence must converge to $\mathbf{P}^*$ [6].

Yates introduced a set of very useful theorems of power control in [6]. The definition of the standard interference function simplifies the convergence analysis mainly in nonlinear power control algorithms. The nonlinearity can be inherent in the distributed power control or according to the transmit power constraint. For example, even if the distributed power algorithm $\mathbf{P}(t+1) = \Psi(\mathbf{P}(t))$ is linear, the constraint version $\mathbf{P}(t+1) = \min\{P_{\max}, \Psi(\mathbf{P}(t))\}$ is nonlinear.

There are different ways of analyzing the convergence properties of distributed power control algorithms. We can apply the classical methods of fixed point iteration analysis in the distributed power control algorithms. Another method which can be used for the convergence test is to compare the convergence speed between different linear or nonlinear power control algorithms – this is presented next.

In this section, the power control algorithm will be defined as a contraction mapping.

Equation (5.22) is in the form of a contraction mapping. The convergence analysis of contraction mapping can then be applied to Equation (5.22). Suppose that, the power control algorithm is feasible, i.e. $\mathbf{P}^* = \Psi(\mathbf{P}^*)$ and $\Gamma_i = \Gamma^T \quad \forall \ i = 1, \ldots, Q$, and that the partial derivatives

$$d_{ij}(P) = \frac{\partial \Psi_i(P_i, \Gamma_i)}{\partial P_j} \qquad 1 \leq i, j \leq Q \qquad (5.24)$$

exist for $\mathbf{P} \in \mathfrak{R}$ where $\mathfrak{R} = \{\mathbf{P} : \| \mathbf{P} - \mathbf{P}^* \| < \varepsilon\}$ and ε is a positive constant. Let $\mathbf{D}(\mathbf{P})$ be $Q \times Q$ matrix with elements $d_{ij}(\mathbf{P})$. A necessary condition for Equation (5.22) to converge is that the spectral radius of $\mathbf{D}(\mathbf{P}^*)$, $\{\rho[\mathrm{D}(\mathbf{P}^*)]\}$, is less than or equal to one. Define a constant

$m > 0$ such that

$$\rho[\mathbf{D}(\mathbf{P})] \leq \mathrm{m} < 1 \quad \mathbf{P} \in \mathfrak{R} \tag{5.25}$$

The rate of convergence depends linearly on m, and we have [9]

$$\| \mathbf{P}(t+1) - \mathbf{P}^* \| \leq m \| \mathbf{P}(t) - \mathbf{P}^* \| \tag{5.26}$$

From Equations (5.25) and (5.26), we may use the spectral radius of the matrix $\mathbf{D}(\mathbf{P})$ as an indication of the asymptotic average rate of convergence. Similar results using a different methodology are presented in [4].

5.4.1 The Linear Iterative Method

We start with a general introduction to solve a set of linear equations with iterative methods. The power control problem including additive noise can be described as

$$[\mathbf{I} - \mathbf{\Gamma}^T \mathbf{H}]\mathbf{P} = \mathbf{u} \tag{5.27}$$

Now define

$$[\mathbf{I} - \mathbf{\Gamma}^T \mathbf{H}] = \mathbf{M} - \mathbf{N}, \tag{5.28}$$

where $\mathbf{M}$ and $\mathbf{N}$ are $Q \times Q$ matrices, and $\mathbf{M}$ is nonsingular. Then Equation (5.27) can be solved iteratively as

$$\mathbf{P}(t+1) = \mathbf{M}^{-1}\mathbf{N}\mathbf{P}(t) + \mathbf{M}^{-1}\mathbf{u}. \tag{5.29}$$

From Equation (5.29), $\mathbf{P}(1) = \mathbf{M}^{-1}\mathbf{N}\mathbf{P}(0) + \mathbf{M}^{-1}\mathbf{u}$, and $\mathbf{P}(2) = (\mathbf{M}^{-1}\mathbf{N})\mathbf{P}(1) + \mathbf{M}^{-1}\mathbf{u} \Rightarrow \mathbf{P}(2) = (\mathbf{M}^{-1}\mathbf{N})^2\mathbf{P}(0) + (\mathbf{M}^{-1}\mathbf{N})\mathbf{M}^{-1}\mathbf{u} + \mathbf{M}^{-1}\mathbf{u}$, in general at iteration t:

$$\mathbf{P}(t) = (\mathbf{M}^{-1}\mathbf{N})^t\mathbf{P}(0) + \sum_{k=0}^{t-1} (\mathbf{M}^{-1}\mathbf{N})^k\mathbf{M}^{-1}\mathbf{u}. \tag{5.30}$$

Let's define $\mathbf{A} = (\mathbf{M}^{-1}\mathbf{N})$, and using diagonal decomposition, matrix $\mathbf{A}$ can be decomposed as

$$\mathbf{A} = \mathbf{V}\mathbf{\Sigma}\mathbf{V}^H \tag{5.31}$$

where V is an orthonormal matrix consisting of the eigenvectors associated with their eigenvalues. The matrix $\mathbf{\Sigma}$ is a diagonal matrix containing the eigenvalues. Since $\mathbf{A}^t = \mathbf{V}\mathbf{\Sigma}^t\mathbf{V}^H$, then $\lim_{t \to \infty}\mathbf{A}^t \to 0$ if and only if the maximum absolute value of the eigenvalues (the spectral radius) is less than one. If $\rho(\mathbf{M}^{-1}\mathbf{N}) < 1$, then in Equation (5.30) $\lim_{t \to \infty}(\mathbf{M}^{-1}\mathbf{N})^t \to 0$ and also $\lim_{t \to \infty}\sum_{k=0}^{t-1}(\mathbf{M}^{-1}\mathbf{N})^k \to (\mathbf{I} - \mathbf{M}^{-1}\mathbf{N})^{-1}$, substitute this in Equation (5.30), we obtain

$$\lim_{t \to \infty} \mathbf{P}(t) \to (\mathbf{I} - \mathbf{M}^{-1}\mathbf{N})^{-1}\mathbf{M}^{-1}\mathbf{u} = (\mathbf{I} - \mathbf{\Gamma}^T\mathbf{H})^{-1}\mathbf{u}, \tag{5.32}$$

which is the solution of power control problem.

The foregoing analysis shows that for any initial power vector $\mathbf{P}(0)$, the linear iterative method converges to the fixed-point solution $\mathbf{P}^*$, providing that the spectral radius of $(\mathbf{M}^{-1}\mathbf{N})$ is less than one. There are many different algorithms in the literature, however, we will review just a few of the important ones.

5.4.2 The Distributed Balancing Algorithm (DBA)

Zander proposed a distributed balancing algorithm in [10]. The method is based on the power method for finding the dominant eigenvalue (spectral radius) and its corresponding eigenvector.

The DBA algorithm is as follows

$$\mathbf{P}(0) = \mathbf{P}_0 \qquad\qquad\qquad \mathbf{P}_0 > 0$$
$$\mathbf{P}_i(t+1) = \beta P_i(t)\left(1 + \frac{1}{\Gamma_i(t)}\right), \quad \beta > 0, \ t = 0, 1, \ldots, \ i = 1, \ldots, Q \tag{5.33}$$

The algorithm starts with an arbitrary positive vector $\mathbf{P}(0)$. The SINR level $\Gamma_i(t)$ is measured in link i.

Theorem 5.3

Using the DBA algorithm (Equation (5.33)) and for a noiseless case the system will converge to the SINR balance, i.e.,

$$\lim_{t \to \infty} P(t) = P^* \quad t = 0, 1, \ldots$$
$$\lim_{t \to \infty} \Gamma_i(t) = \gamma^* \quad i = 1, \ldots, Q \tag{5.34}$$

where γ^* is the maximum achievable SINR, which is equal to $1/\rho(\mathbf{H})$, and $\mathbf{P}^*$ is the corresponding eigenvector which represents the optimum transmitted power.

Proof
The DBA algorithm can be represented as

$$P_i(t+1) = \beta P_i(t) + \beta \frac{P_i(t)}{\Gamma_i(t)} = \beta P_i(t) + \beta \sum_{\substack{j=1 \\ j \neq i}}^{Q} \frac{G_{ij}}{G_{ii}} P_j(t)$$

In matrix format it can be reformulated as

$$\mathbf{P}(t+1) = \beta(\mathbf{I} + \mathbf{H})\mathbf{P}(t) \tag{5.35}$$

Referred to the non-zero initial power vector, Equation (5.35) can be represented as

$$\mathbf{P}(t) = \beta^t(\mathbf{I} + \mathbf{H})^t \mathbf{P}(0) \tag{5.36}$$

Let $\mathbf{A} = \mathbf{I} + \mathbf{H} \Rightarrow \mathbf{P}(t) = \beta^t \mathbf{A}^t \mathbf{P}(0)$, now representing the matrix $\mathbf{A}$ with diagonal decomposion, Equation (5.36) can be represented as $\mathbf{P}(t) = \beta^t \mathbf{V}\mathbf{\Sigma}^t \mathbf{V}^H \mathbf{P}(0)$. As we discussed earlier, matrix $\mathbf{\Sigma}$ is a diagonal matrix with the ith element as $1 + \lambda_i$, where λ_i is the ith eigenvalue of matrix $\mathbf{H}$. The matrix $\mathbf{V}$ consists of the Q eigenvectors such as $\mathbf{V} = [\mathbf{v}_1 \quad \mathbf{v}_2 \quad \cdots \quad \mathbf{v}_Q]$, and $\mathbf{V}^H$ is the Hermitian transpose of matrix $\mathbf{V}$. Assume that the largest element of the diagonal matrix $\mathbf{\Sigma}$ is the first element. Of course the largest element is $1 + \rho(\mathbf{H})$, where $\rho(\mathbf{H})$ is the spectral radius of $\mathbf{H}$. Now we can represent Equation (5.36) as

$$\mathbf{P(t)} = \beta^t(1 + \lambda_1)^t \mathbf{V}\hat{\mathbf{\Sigma}}^t \mathbf{V}^H \mathbf{P}(0) \tag{5.37}$$

where $\lambda_1 = \rho(\mathbf{H})$, and $\hat{\mathbf{\Sigma}}$ is normalized version of $\mathbf{\Sigma}$, and it is represented as

$$\hat{\mathbf{\Sigma}} = \begin{bmatrix} 1 & 0 & \cdots & 0 \\ 0 & (1+\lambda_2)/(1+\lambda_1) & & 0 \\ 0 & 0 & \ddots & \\ 0 & 0 & \cdots & (1+\lambda_Q)/(1+\lambda_1) \end{bmatrix} \tag{5.38}$$

The absolute value of all diagonal elements except the first one are less than 1, so that

$$\lim_{t \to \infty} \hat{\mathbf{\Sigma}}^t = \begin{bmatrix} 1 & 0 & \cdots & 0 \\ 0 & 0 & & 0 \\ 0 & 0 & \ddots & \vdots \\ 0 & 0 & \cdots & 0 \end{bmatrix} \tag{5.39}$$

The transmit power vector at iteration t becomes

$$\mathbf{P}(t) = \beta^t (1+\lambda_1)^t \mathbf{V} \hat{\mathbf{\Sigma}}^t \mathbf{a} \tag{5.40}$$

where $\mathbf{a} = \mathbf{V}^H \mathbf{P}(0)$, and since the $\hat{\mathbf{\Sigma}}^t$ matrix will approach zero diagonal except for the first element as shown in Equation (5.39), then for large t, Equation (5.40) becomes

$$\mathbf{P}(t) \cong \beta^t (1+\lambda_1)^t \mathbf{V}[c \quad 0 \quad \cdots \quad 0]' = \beta^t (1+\lambda_1)^t c \mathbf{v}_1 = \alpha \mathbf{v}_1 \tag{5.41}$$

where c is the first element in $\mathbf{a}$ vector, $\alpha = \beta^t (1+\lambda_1)^t$ is a parameter which depends on the time iteration so that β should be selected so that α does not become too large or too small, and $\mathbf{v}_1$ is the eigenvector associated with the largest eigenvalue (spectral radius) of matrix $\mathbf{H}$. Equation (5.41) indicates that the power will converge to the eigenvector which represents the optimum transmit power. The convergence speed depends on how fast the other values of the diagonal elements will vanish with time iterations. In other words the convergence speed is related to $|1 + \lambda_2|/|1 + \lambda_1|$ assuming that λ_2 is the second-largest eigenvalue after the spectral radius.

Exercise 5.7 If the noise is considered and the target SINR is Γ^T, prove that the adaptation parameter β should be selected as $\beta = \frac{\Gamma^T}{1+\Gamma^T}$.

Exercise 5.8 If additive noise is considered, prove that Equation (5.33) will converge to the solution of the optimum power vector in Equation (5.19).

It is clear that the DBA uses only local SINR information and utilizes an iterative scheme to control the transmitted power. The DBA requires a normalization procedure after each iteration (in the noiseless case) to constrain the transmitted power.

5.4.3 The Distributed Power Control (DPC) Scheme

It has been shown that the distributed power control scheme for satellite systems can be applied to cellular systems [11]. The results presented in [11] indicate that the DPC scheme has the potential to converge faster than the DBA scheme at high SINRs.

The power adjustment made by the ith mobile at the tth time slot is given by

$$P_i(t+1) = \beta(t)\frac{P_i(t)}{\Gamma_i(t)} \qquad i = 1, \ldots, Q, \quad t = 0, 1, \ldots \tag{5.42}$$

where $\beta(t)$ is some positive coefficient chosen to achieve the proper power control vector (not too large or too small). When the additive noise is considered then $\beta(t) = \Gamma^T$.

Theorem 5.4
For noiseless systems with $\beta(t) > 0$, the DPC algorithm will converge to the optimum solution where the SINR is the maximum achievable SINR and the power is the corresponding eigenvector.

Proof
The proof will be left for the reader. The procedure is very similar to the proof of Theorem 5.3.

Theorem 5.5
For feasible systems and when the additive noise is considered, prove that Equation (5.42) will converge to the optimum solution given in Equation (5.19).

Proof
The DPC algorithm, Equation (5.42), can be reformulated as

$$P_i(t+1) = \frac{\Gamma^T}{\Gamma_i(t)}P_i(t) = \frac{\Gamma^T}{P_i(t)G_{ii} \Big/ \left(\sum_{j\neq i}^{Q} P_j(t)G_{ji} + N_i\right)}P_i(t)$$

$$\Rightarrow P_i(t+1) = \Gamma^T\left(\sum_{j\neq i}^{Q} P_j(t)\frac{G_{ji}}{G_{ii}} + \frac{N_i}{G_{ii}}\right) \tag{5.43}$$

In matrix form Equation (5.43) becomes

$$\mathbf{P}(t+1) = \Gamma^T\mathbf{H}\mathbf{P}(t) + \mathbf{u} \tag{5.44}$$

Comparing Equation (5.44) with Equation (5.29) we can see that $\mathbf{M}^{-1} = \mathbf{I}$ and $\mathbf{N} = \Gamma^T\mathbf{H}$. The convergence proof for such system is given in Section 5.4.1.

5.4.4 The Distributed Constrained Power Control (DCPC) Algorithm

The transmitted power of a mobile station or a base station is limited by some maximum value P_{max}. The constrained power control generally takes the following form

$$P_i(t+1) = \min\{P_{\max}, \Psi_i(\mathbf{P}(t))\}, \quad t = 0, 1, 2, \ldots; i = 1, \ldots, Q \tag{5.45}$$

where $\Psi_i(\cdot), \quad i = 1, \ldots, Q$ is the standard interference function. The DCPC algorithm has the following form [7]

$$P_i(t+1) = \min\left\{P_{\max}, \Gamma^T\frac{P_i(t)}{\Gamma_i(t)}\right\}, \quad t = 0, 1, 2, \ldots; \quad i = 1, \ldots, Q \tag{5.46}$$

Theorem 5.6

Starting with any nonnegative power vector $\mathbf{P}(0)$, the DCPC scheme described in Equation (5.46) converges to the fixed point $\mathbf{P}^*$ of

$$\mathbf{P}(t+1) = \min\{\mathbf{P}_{max}, \mathbf{\Gamma}^T \mathbf{H} \mathbf{P}(t) + \mathbf{u}\}, \quad t = 0, 1, 2, \ldots \tag{5.47}$$

If the target SINR is greater than the maximum achievable SINR, i.e., $\mathbf{\Gamma}^T \geq \gamma^*$ then the fixed point $\mathbf{P}^*$ will converge to P_{max}.

Proof
The proof is left to the reader. You can use the concept of standard interference function to prove that Equation (5.47) will always converge to a fixed point.

Example 5.9

Assume a single-cell scenario with four terminals located at 60, 90, 110, and 250 m from the base station. The channel gain is given as $G_i = 1/d_1^4$, write a MATLAB® code to simulate the power updates for each terminal using DPC and DBA. The required target is -10 dB for all terminals. Assume that the noise figure N_F of the base station is 2 dB. The maximum transmit power for each terminal is 0.2 W. Show the transmitted power with time iterations using both techniques in log-log scale. Take the maximum power (0.2 W) as an initial power value, i.e., $P_i(0) = 0.2, \forall i = 1, \cdots, 4$.

Solution

The average additive noise at the input of the base station is $N_i = KT_0 N_F B = 1.38 \times 10^{-23} \times 290 \times 10^{0.2} \times 5 \times 10^6 \cong 3.2 \times 10^{-14}$ W. The channel gain vector is $G = \begin{bmatrix} 0.772 & 0.152 & 0.068 & 0.003 \end{bmatrix} \times 10^{-7}$.

Figures 5.4 and 5.5 show the transmit power values with time iterations for the DPC and DBA algorithms, respectively. Both algorithms have the same steady-state values for the transmit power. With these steady-state power values all terminals achieve the required SINR (check it!). However, the convergence behavior is not the same. Furthermore, it is clear that the transmit power of the worst user in terms of the distance with the base station (the fourth one) increases up to about 7 W before it converges to the optimum power value. This is not feasible because it states in the example that the maximum power of the terminals is 0.2 W. Repeating the same example with maximum power constraint we obtain the result shown in Figure 5.6, which shows the transmit power only for the fourth user. Figure 5.6 shows that, even if we use maximum constraint in the distributed power algorithms, it will converge to the optimum power value if it is feasible. We can see from same figure that the convergence speed of the DPC is faster than the DBA (try to prove that mathematically).

Exercise 5.9 Repeat the previous example if the target SINR is -3 dB. Discuss the feasibility and the achieved SINR for every mobile. Propose a method to relax the congestion situation.

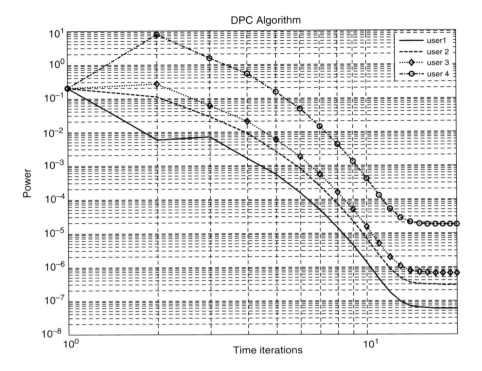

Figure 5.4   Transmit power with time iteration using DPC algorithm

5.4.5 The Foschini and Miljanic Algorithm (FMA)

Foschini and Miljanic proposed a simple and efficient distributed power control algorithm [12]. The algorithm is based on the following continuous time differential equation:

$$\dot{\Gamma}_i(\tau) = -\beta[\Gamma_i(\tau) - \Gamma^T], \quad \beta > 0, \tau \geq 0 \tag{5.48}$$

The steady-state solution of the above differential equation for user i is $\Gamma_i = \Gamma^T$. The speed of the convergence depends on the coefficient β.

Define the total interference of user i:

$$I_i(\tau) = \sum_{\substack{j \neq i}}^{Q} G_{ij}(\tau)P_j(\tau) + N_i \tag{5.49}$$

Then Γ_i from Equation (5.2) becomes

$$\Gamma_i(\tau) = \frac{G_{ji}(\tau)P_i(\tau)}{\sum\limits_{\substack{j \neq i}}^{Q} G_{ji}(\tau)P_j(\tau) + N_i} = \frac{G_{ji}(\tau)P_i(\tau)}{I_i(\tau)}, \quad i = 1, \dots, Q. \tag{5.50}$$

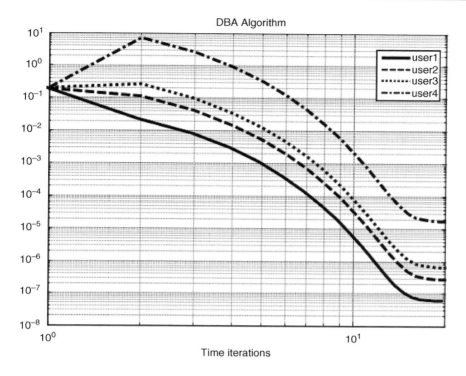

Figure 5.5   Transmit power with time iteration using DBA algorithm

Assuming that $I_i(\tau)$ and $G_{ki}(\tau)$ are constant, substituting Equation (5.50) into Equation (5.48) gives

$$\frac{G_{ji}\dot{P}_i(\tau)}{I_i} = -\beta\left[\frac{G_{ji}P_i(\tau)}{I_i} - \Gamma^T\right], \quad i = 1,\dots,Q. \tag{5.51}$$

Using Equation (5.49), this becomes

$$\dot{P}_i(\tau) = -\beta\left[P_i(\tau) - \frac{\Gamma^T}{G_{ii}}\left(\sum_{j\neq i}^{Q} G_{ji}(\tau)P_j(\tau) + N_i\right)\right], \quad i = 1,\dots,Q. \tag{5.52}$$

Using matrix notation we can rewrite Equation (5.52) as

$$\dot{\mathbf{P}}(\tau) = -\beta[\mathbf{I} - \Gamma^T\mathbf{H}]\mathbf{P}(\tau) + \beta\,\mathbf{u}. \tag{5.53}$$

At the steady state, we have

$$\mathbf{P}^* = [\mathbf{I} - \Gamma^T\mathbf{H}]^{-1}\mathbf{u}. \tag{5.54}$$

The discrete form of Equation (5.53) is

$$\mathbf{P}(t+1) = \beta\left[\left(\frac{1}{\beta} - 1\right)\mathbf{I} + \Gamma^T\mathbf{H}\right]\mathbf{P}(t) + \beta\,\mathbf{u}, \quad t = 0, 1, \dots \tag{5.55}$$

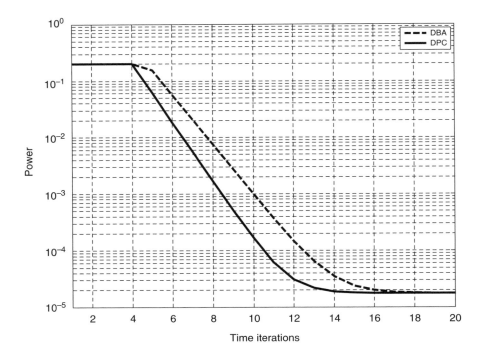

Figure 5.6   Convergence speed comparison between DPC and DBA

The iterative power control for each terminal i is

$$P_i(t+1) = (1-\beta)P_i(t)\left[1 + \frac{\beta}{(1-\beta)}\left(\frac{\Gamma^T}{\Gamma_i(t)}\right)\right], \quad t = 0, 1, 2, \ldots; \quad i = 1, \ldots, Q \quad (5.56)$$

The constraint algorithm Equation (5.45) can be directly applied with Equation (5.56). The resultant algorithm is called the constraint FMA (CFMA) algorithm.

Example 5.10

Prove that the FMA presented in Equation (5.55) will always converge to the optimum power solution in Equation (5.19). What are the necessary conditions for convergence?

Solution

Comparing Equation (5.55) with Equation (5.29) we can see that $\mathbf{M} = \frac{1}{\beta}\mathbf{I}$ and $\mathbf{N} = \left(\frac{1}{\beta} - 1\right)\mathbf{I} + \Gamma^T\mathbf{H}$. The convergence proof for such system is given in Section 5.4.1.

Example 5.11

Repeat Example 5.9 using the CFMA algorithm with $\beta = 0.2, 0.6, 1.0$ and 2. Show also the SINR convergence behavior with all β. Show the results for the fourth terminal only. The target SINR is $-10\,\text{dB}$.

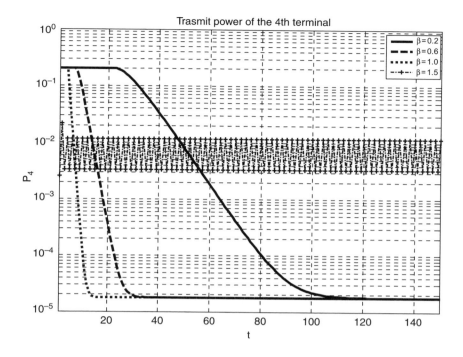

Figure 5.7   The convergence behavior of CFMA

Solution

The resultant transmit power of the fourth terminal is shown in Figure 5.7. It is clear that as β value increases the convergence speed will increase. However, when $\beta > 1$ the CFMA does not work, and no valid solution is obtained. This can be clarified by looking at the SINR value of the fourth terminal in Figure 5.8.

Exercise 5.10   Prove that at $\beta = 1$, the FMA algorithm becomes identical to the DPC algorithm in Equation (5.42).

Exercise 5.11   Since the convergence speed of the FMA reduces when β is selected to be less than one, explain in what situations we select $\beta < 1$?

5.4.6 The Constrained Second-Order Power Control (CSOPC) Algorithm

The second-order power control (SOPC) algorithm has been suggested to accelerate the convergence speed of distributed power control algorithms with a snapshot assumption [13]. The algorithm is based on utilizing the successive over-relaxation (SOR) method as follows

$$P_i(t+1) = a(t)\frac{\Gamma_i^T}{\Gamma_i(t)}P_i(t) + (1 - a(t))P_i(t-1) \qquad (5.57)$$

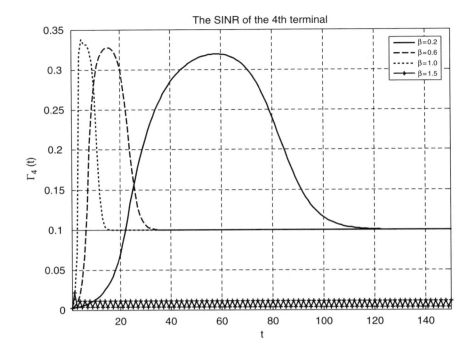

Figure 5.8   The achieved SINR for fourth terminal in Example 5.11

The relaxation factor $a(t)$ is a non-increasing sequence of control parameters, where $1 < a(1) < 2$ and $\lim_{t \to \infty} a(t) = 1$. The following relaxation factor is employed in [13]

$$a(t) = 1 + \frac{1}{1.5^t} \qquad (5.58)$$

To guarantee that the transmit power is always between $[0, P_{\max}]$, then Equation (5.57) can be reformulated as:

$$P_i(t+1) = \min\left\{ P_{\max}, \max\left\{ 0, a(t)\frac{\Gamma_i^T}{\Gamma_i(t)}P_i(t) + (1 - a(t))P_i(t-1) \right\} \right\}, \qquad t = 1, 2, \ldots$$

$$(5.59)$$

The convergence properties of the CSOPC algorithm are given in [13]. However, we recommend the reader to prove the convergence of the algorithm using the same techniques discussed in Section 5.4.

Exercise 5.12   Repeat Example 5.9 with the CSOPC algorithm.

5.4.7 *The Estimated Step Power Control (ESPC) Algorithm*

In all previous algorithms we assumed a perfect feedback channel. In existing cellular communication systems only a very limited feedback channel is available for the purpose of power control. To reduce the bandwidth of the feedback channel only one bit is used to update the transmit power (it can be coded with two or three bits using for example repetitive code). In the existing CDMA cellular system, the power control for the UL is performed as follows:

1. Measure and estimate the SINR of user i at its assigned BS.
2. Compare the estimated SINR with the target.
3. If the estimated SINR is less than the target SINR, then send $(+)$ command to ask the mobile to increase its transmitted power by one step.
4. If the estimated SINR is larger than the target SINR, then send $(-)$ command to ask the mobile to decrease its transmitted power by one step.

The target SINR is set by the outer-loop power control (OLPC). The OLPC update rate is very slow compared with the fast inner-loop power control. This is because the OLPC updates the target SINR based on BLER measurements. The optimum target SINR to achieve certain QoS depends on the channel and the network situation. For this reason the OLPC is important in UMTS systems. Figure 5.9 shows a simple block diagram for the power control in a UMTS network. From the previous discussion, it is clear that the mobile terminal does not know the actual SINR value at the BS. The mobile's transmit power follows the instructions of the base station blindly. This type of power control is called 'bang-bang' power control or fixed step

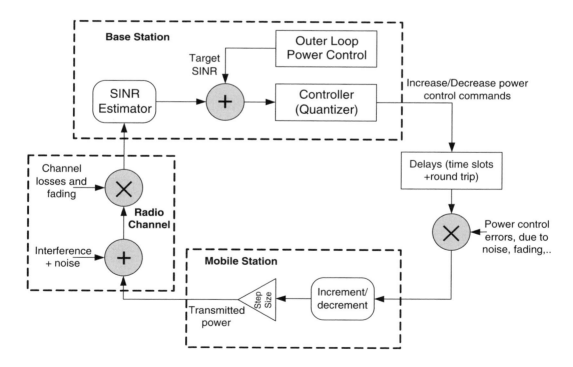

Figure 5.9   General block diagram for quantized power control system

power control (FSPC). Mathematically, this can be represented as (all values are in dB)

$$P_{i,FSPC}(t+1) = P_{i,FSPC}(t) + \delta sign(\Gamma^T - \Gamma_i(t)), \quad t = 0, 1, 2, \ldots; \quad i = 1, \ldots, Q \quad (5.60)$$

where $P_{min} \leq P_{FSPC}(t) \leq P_{max}$ is the transmitted power at time slot t, P_{min} and P_{max} are the minimum and maximum transmit powers respectively, δ is the step size of the power update, Γ^T is the target SINR which is determined by the outer-loop power control, and $\Gamma(t)$ is the measured SINR at time slot t. The sign function is given by

$$sign(x) = \begin{cases} +1, & x \geq 0 \\ -1, & x < 0 \end{cases} \quad (5.61)$$

It is clear from Equation (5.60) that the mobile terminal is commanded to increase or decrease its transmitted power without detailed information about the channel situation, i.e., the mobile terminal does not know how large the difference is between the target SINR and the measured SINR. If the measured SINR is much greater than the target SINR then it will take a relatively long time to adjust the transmitted power to the proper value to make the actual SINR close to the target SINR. This can reduce the performance and capacity of the system. Theoretically the performance of the FSPC algorithm is greatly affected by the step size δ. If δ is too small then it will take a long time to converge but with small steady-state error. On the other hand if δ is too large, the convergence can be fast but with large steady-state error. These concepts are shown in Figures 5.10 and 5.11. Figure 5.10 shows the transmit power for a two-mobile scenario when δ is large. Figure 5.11 shows the same scenario when δ is small.

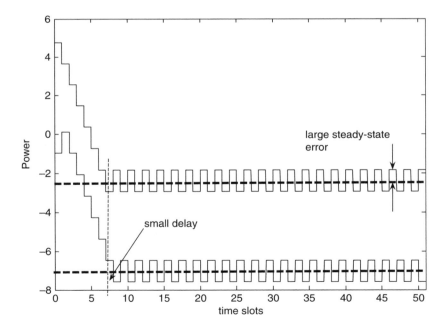

Figure 5.10   Quantized power control with large step size

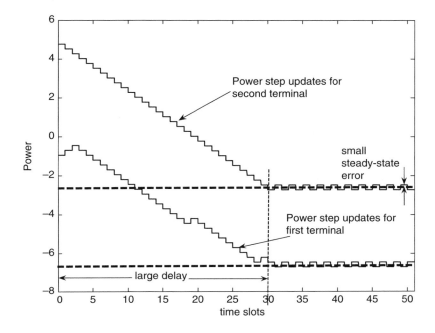

Figure 5.11   Quantized power control with small step size

The maximum transmit power value depends on the mobile manufacturer. However, according to the UMTS recommendations the maximum power of cell phones should be 23 dBm, i.e., about 0.2 W. The dynamic range of the mobile is about 70 dB. This means that the difference between the minimum transmit power and the maximum is 70 dB.

Example 5.12

What is the minimum transmitting power (theoretically) of a cell phone if it is switched on.

Solution

Since the maximum power is 23 dBm and the dynamic range is 70 dB. The minimum transmit power is $23\,\text{dBm} - 70\,\text{dB} = -47\,\text{dBm}$ or $2 \times 10^{-8}\,\text{W}$.

Example 5.13

In the UMTS standard, what is the required time for the cell phone to increase its transmit power from minimum to maximum if there are no power control command errors.

Solution

Since the power control rate is 1500 Hz, then the required time is $70/1500 = 46.7\,\text{ms}$.

Exercise 5.13   Repeat the previous example when there are power command errors, i.e., the BS sends $(+)$ command but it is received $(-)$, and the probability of error is 0.1.

The DCPC algorithm Equation (5.46) can be rewritten in dB scale as

$$P_{i,DPC}(t+1) = P_{i,DPC}(t) + (\Gamma^T - \Gamma_i(t)), \quad t = 0, 1, 2, \ldots; \quad i = 1, \ldots, Q \qquad (5.62)$$

where all values are in dB.

It is clear that the DCPC algorithm assumes no quantization distortion so more information about the channel is available. For this reason the DCPC performance is better than the FSPC algorithm in quasi-static channels. Now we introduce another power control algorithm based on the estimation of the difference $(\Gamma^T - \Gamma_i(t))$ using one-bit signaling. This is called the estimated step power control (ESPC) algorithm [14].

In what follows we consider only the uplink, but the proposed method is applicable also to the downlink. The ESPC algorithm is based on a simple tracking method, which uses one memory location for the previous BS power command.

Define for all users, $i = 1, \ldots, Q$, and time slots, $t = 0, 1, 2, \ldots$

$$e_i(t) = \Gamma^T - \Gamma_i(t), \qquad (5.63)$$

$$v_{tr,i}(t) = \text{sign}(e_i(t)), \qquad (5.64)$$

$$v_i(t) = -v_{tr,i}(t)E_{PC,i}(t), \qquad (5.65)$$

where $E_{PC,i}(t)$ is 1 with probability $P_{PCE,i}(t)$ and -1 with probability $1 - P_{PCE,i}(t)$. $P_{PCE,i}(t)$ is the probability of a bit error in the power control command transmission at time t.

Let the estimate of the error signal $e_i(t)$ be $\tilde{e}_i(t)$. We propose a simple form for the error estimation:

$$\tilde{e}_i(t) = \frac{1}{2}[1 + v_i(t)v_i(t-1)]\,\tilde{e}_i(t-1) + \delta_i v_i(t), \qquad (5.66)$$

where δ_i is the adaptation step size of user i. The ESPC algorithm is given by

$$P_{ESPC,i}(t+1) = P_{ESPC,i}(t) + \tilde{e}(t), \qquad (5.67)$$

where $Pc_{\min} \leq P_{ESPC,i}(t) \leq P_{\max}$, and all the values are in dB. Figure 5.12 shows the block diagram of the suggested algorithm.

If we define

$$a_i(t) = \frac{1}{2}[1 + v_i(t)v_i(t-1)] = \begin{cases} 1, & v_i(t) = v_i(t-1) \\ 0, & v_i(t) \neq v_i(t-1) \end{cases} \qquad (5.68)$$

then, solving Equation (5.66) recursively, we obtain

$$\tilde{e}_i(t) = \prod_{m=1}^{t} a_i(m)\tilde{e}_i(0) + \delta_i \sum_{k=1}^{t-1} v_i(t-k) \prod_{n=0}^{k-1} a_i(t-n) + \delta_i v_i(t). \qquad (5.69)$$

It is clear that the first term will be zero (after any zero crossing of $e_i(t)$ or if $\tilde{e}_i(0) = 0$). Then Equation (5.69) can be rewritten as

$$\tilde{e}_i(t) = \delta_i \left[\sum_{k=1}^{t-1} v_i(t-k) \prod_{n=0}^{k-1} a_i(t-n) + v_i(t) \right] \qquad (5.70)$$

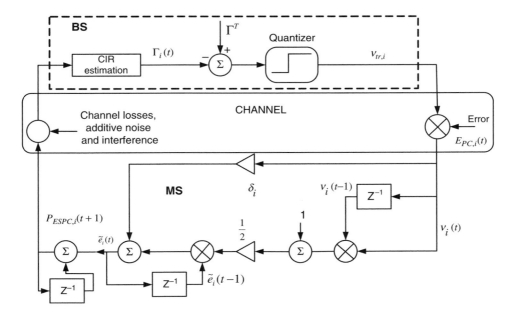

Figure 5.12 Block diagram of the ESPC algorithm

Define

$$c_i(t, k) = \prod_{n=0}^{k-1} a_i(t - n) \tag{5.71}$$

Substitute Equation (5.71) in Equation (5.70) to obtain

$$\tilde{e}_i(t) = \delta_i \left[\sum_{k=1}^{t-1} v_i(t - k) c_i(t, k) + v_i(t) \right] \tag{5.72}$$

The first part of Equation (5.72) can be seen as the convolution between the input $v_i(t)$ and a time-varying system $c_i(t, k)$. The statistical properties of $\tilde{e}_i(t)$ depend on the statistical properties of the channel, interference, and the additive noise. The performance of the ESPC algorithm can be further improved by using a variable step size. The idea is to increase the step size if the same command is received for example three consecutive times. The algorithm is called modified ESPC or ESPC-M.

5.4.8 The Multi-Objective Distributed Power Control (MODPC) Algorithm

All previously discussed power control algorithms were based on the idea of minimizing the transmit power and keeping the target SINR as a constraint. Another power control algorithm based on analytical multi-objective optimization is proposed in [15]. A brief overview of the multi-objective optimization and its application to fixed rate power control will be given in this section.

The MO optimization technique is a method to optimize between different (usually) conflicting objectives. In the MO optimization problem we have a vector of objective functions instead of one scalar objective. Each objective function is a function of the decision (variable) vector. The mathematical formulation of an MO optimization problem becomes [16]:

Find **x** which achieves

$$\min\{f_1(x), f_2(x), \ldots, f_m(x)\}, \qquad \text{subject to } x \in S, \tag{5.73}$$

where we have m (≥ 2) objective functions $f_i : \mathfrak{R}^n \to \mathfrak{R}$, **x** is the decision (variable) vector, $\mathbf{x} \in \mathbf{S}$, **S** is (nonempty) feasible region (set). The abbreviation $\min\{.\}$ means that we want to minimize all the objectives simultaneously. Since usually the objectives are at least partially conflicting and possibly incommensurable then there is no single vector **x** minimizing all the objectives. In the MO optimization we have different optimal solutions, which are called Pareto optimal solutions.

Note: A decision vector $\mathbf{x}^\bullet \in S$ is Pareto optimal if there is no other decision vector $\mathbf{x} \in S$ such that $f_i(\mathbf{x}) \leq f_i(\mathbf{x}^*)$ for all $i = 1,2, \ldots, m$ and $f_j(\mathbf{x}) < f_j(\mathbf{x}^*)$ for at least one index j.

The *Pareto optimal set* is the set of all possible Pareto optimal solutions. This set can be nonconvex and nonconnected [16].

After the generation of the Pareto set, we are usually interested in one solution of this set. This solution is selected by a decision maker. There are different techniques to solve the MO optimization problems. One way is to use soft-computing methods such as genetic algorithms. Here we will concentrate on the analytical solutions of the MO optimization problems. A technique to solve the MO optimization problem is by converting it to a single objective optimization problem as in the *weighting method*. The weighting method transforms the problem posed in Equation (5.73) into

$$\min \sum_{i=1}^{m} \lambda_i f_i(\mathbf{x}), \qquad \text{subject to } \mathbf{x} \in \mathbf{S}, \tag{5.74}$$

where the tradeoff factors $\lambda_i \geq 0$, $\forall i = 1, \ldots, m$ and $\sum_{i=1}^{m} \lambda_i = 1$.

The Pareto set can be obtained by solving the single objective (SO) optimization problem (Equation (5.74)) for different tradeoff factor values.

Another important method which is of special interest in the application of MO optimization in radio resource scheduling (RRS) is that of *weighted metrics*. If the optimum solution of each objective is known in advance then Equation (5.73) can be formulated as

$$\min \left(\sum_{i=1}^{m} \lambda_i |f_i(\mathbf{x}) - z_i^*|p \right)^{1/p}, \quad \text{subject to } \mathbf{x} \in \mathbf{S}, \tag{5.75}$$

where $1 \leq p \leq \infty$, z_i^* is the desired solution of the objective i, and the tradeoff factors $\lambda_i \geq 0$, $\forall i = 1, \ldots, m$ and $\sum_{i=1}^{m} \lambda_i = 1$.

In this section, we propose using MO optimization techniques to solve the single rate power control problem Equations (5.1)–(5.3).

The target QoS is not usually strict but it has some margin which is the difference between the target QoS and the minimum allowed QoS as described in Section 5.1. We call any QoS level inside the margin as the accepted QoS level. *The preferred power control is that an accepted QoS level can be achieved very quickly at low power consumption.* The proposed power control algorithm achieves an accepted QoS level at low average power consumption. If we have two power control algorithms, both of them will eventually converge to the same power value. We will select the one which can converge faster and at lower transmit power consumption. The distributed power control algorithms use the estimated SINR to update the power. The proposed algorithm achieves two objectives by applying the multi-objective optimization method. The first objective is minimizing the transmitted power and the second objective is achieving the target QoS, which is represented here by the target SINR. In the next formulation, the power control problem has been represented by two objectives as follows: (i) minimizing the transmitted power; (ii) keeping the SINR as close as possible to some target SINR value. In other words, the MO power control algorithm tracks the target SINR, while minimizing the transmitted power. The above statement can be interpreted mathematically for user i, $i = 1, \ldots, Q$, by the following error function

$$e_i(t) = \lambda_{1,i}|P_i(t) - P_{\min}| + \lambda_{2,i}|\Gamma_i(t) - \Gamma_i^T|, \quad t = 0, 1, \ldots \quad (5.76)$$

where $0 \leq \lambda_{1,i} \leq 1$; $\lambda_{2,i} = 1 - \lambda_{1,i}$ are tradeoff factors of user i, Γ_i^T is the target SINR of user i, $P_{\min}$ is the minimum transmitted power of the mobile station. The user's subscript i will be dropped from the tradeoff factors and the target SINR for simplicity. But generally each user can have different values of tradeoff factors as well as different target SINRs.

The above error function (Equation (5.76)) has been constructed from two objectives. The first objective is to keep the transmitted power $P_i(t)$ as close as possible to the minimum power $P_{\min}$. The second objective is to keep the SINR $\Gamma_i(t)$ as close as possible to the target SINR. It is clear that Equation (5.76) has the form of the weighted metrics method with $p = 1$ (Equation (5.75)).

To generalize the optimization over all users and for time window of N slots we define the optimization problem as the following.

Find the minimum of the cost function

$$J(\mathbf{P}) = \left[\sum_{i=1}^{Q} \sum_{t=1}^{N} \varsigma^{N-t} e_i^2(t) \right], \quad (5.77)$$

with respect to the power vector $\mathbf{P}$, where ς is an adaptation factor, and $\mathbf{P} = [P_1, P_2, \ldots, P_Q]'$.

Problem (5.76)–(5.77) is a non-smooth optimization problem because of the absolute function in (5.76). One of the advantages of using the cost function (5.76) is that it can be used for different tasks, for example, it can be applied to reduce transmitted power, achieve some target QoS, increase the throughput, reduce the packet delay, and increase the fairness levels. The new algorithm is called the multi-objective distributed power control (MODPC) algorithm. The derivation of the MODPC algorithm leads to the following simple form [15]:

$$P_i(t) = \frac{\lambda_1 P_{\min} + \lambda_2 \Gamma^T}{\lambda_1 P_i(t-1) + \lambda_2 \Gamma_i(t-1)} P_i(t-1), \quad i = 1, \ldots, Q; \quad t = 1, 2, \ldots \quad (5.78)$$

It is clear that setting $\lambda_1 = 0$ and $\lambda_2 = 1$ in Equation (5.78), the DPC algorithm of Equation (5.42) is obtained. This means that the DPC algorithm is a special case of the MODPC algorithm. At the other extreme, where $\lambda_1 = 1$ and $\lambda_2 = 0$, the handset transmits at the minimum power regardless of the SINR situation (no power control). The proper values of the tradeoff factors, which can be adaptive, can greatly enhance the performance of the algorithm depending on the scenario. The adaptation of the tradeoff factors makes the system more cooperative in a distributed manner. In terms of MO optimization the proper values of the tradeoff factors for certain network condition is selected by a decision maker, which determines the optimum point from a Pareto optimal set.

At steady state (i.e. $P_i(t+1) = P_i(t) = P_i^{ss}$), Equation (5.78) results in the steady-state SINR of user $i(\Gamma_i^{ss})$, which is given by

$$\Gamma_i^{ss} = \Gamma^T - \frac{\lambda_1}{\lambda_2}(P_i^{ss} - P_{\min}) \tag{5.79}$$

One of the key features of the MODPC algorithm can be observed in the steady-state solution given in Equation (5.79). It is clear that the steady-state SINR equals the target SINR when the steady-state power equals the minimum power. The penalty of using any excessive power is to reduce the steady-state SINR. The decision maker should select the values of the tradeoff factors in order to guarantee that all users achieve at least the minimum allowed SINR level. In the worst-case situation the steady-state power is the maximum allowed power ($P_{\max}$). The maximum allowed power is determined by the power amplifier of the handset. The MODPC algorithm with maximum power constraint is given by

$$P_i(t) = \min\left\{ P_{\max}, \frac{\lambda_1 P_{\min} + \lambda_2 \Gamma_i^T}{\lambda_1 P_i(t-1) + \lambda_2 \Gamma_i(t-1)} P_i(t-1) \right\}, \quad i = 1, \ldots, Q; \, t = 1, 2, \ldots \tag{5.80}$$

It is interesting to observe that the transmitted power of MODPC algorithm (5.78) is naturally upper bounded such as

$$P_i(t) \le P_{\min} + \frac{\lambda_2}{\lambda_1} \Gamma^T, \quad i = 1, \ldots, Q \tag{5.81}$$

We assume that the maximum allowed power ($P_{\max}$) is less than the natural upper bound of the MODPC algorithm (5.81).

Solving for (λ_1 and λ_2) using Equation (5.79) and the fact that $\lambda_1 + \lambda_2 = 1$ (two equations with two unknowns), the values of tradeoff factors are derived as (assuming $P_{\min} = 0$)

$$\lambda_2 = \frac{P_{\max}}{P_{\max} + \Gamma^T - \Gamma_{\min}}, \quad \lambda_1 = \frac{\Gamma^T - \Gamma_{\min}}{P_{\max} + \Gamma^T - \Gamma_{\min}} \tag{5.82}$$

The convergence properties of the MODPC algorithm are discussed in [15].

Theorem 5.7
For any $\mathbf{P}(0) > 0$, the MODPC algorithm (Equation (5.78)) with $\lambda_1 > 0$ will always converge to a unique fixed point $\hat{\mathbf{P}}$. At $\lambda_1 = 0$ the feasibility condition is necessary for convergence.

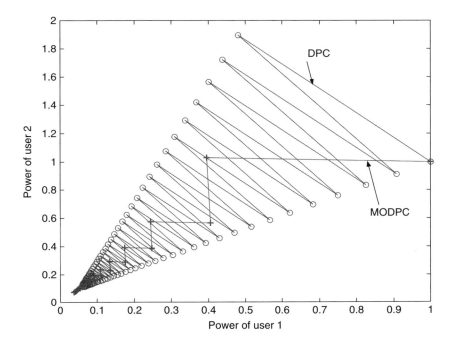

Figure 5.13   Comparison of convergence rates of DPC and MODPC

Proof
We leave the proof for the reader. The idea is to check the positivity, monotonicity, and scalability conditions of the MODPC algorithm. The detailed proof is given in [15].

As indicated in Equations (5.76) and (5.79) there is a penalty if using an extra transmit power. For this reason the MODPC algorithm goes through a fewer number of iterations than other DPC algorithms in order to converge to the accepted solution. Figure 5.13 shows this property of the MODPC algorithm. The power path trajectory of the MODPC algorithm and DPC algorithm for two users are shown in the figure. It is clear that the MODPC algorithm converges faster than the DPC algorithm to reach the feasible region.

Exercise 5.14   Repeat Example 5.9 using the MODPC algorithm.

5.4.9 The Kalman Filter Distributed Power Control Algorithm

In this section we use the Kalman filter in a distributed power control algorithm. It is a well-known fact that the Kalman filter is the optimum linear tracking device on the basis of second-order statistics. This feature motivates us to apply the Kalman filter in this type of application. The Kalman filter has been proposed in the literature recently for use in different applications related to power control such as interference estimation and prediction [18,19], power estimation [17], and channel gain prediction for power control [19]. In this section we use the Kalman filter directly to estimate the best transmitted power in a distributed way.

Although the MODPC algorithm outperforms Kalman filter based power control in terms of convergence speed, Kalman filter power control has a well-known linear behaviour which may make it preferred in some applications.

The power control is considered as a linear time-variant first-order Markov model [20]. The transmitted power of user i at time slot t is given by

$$P_i(t) = w_i(t-1)P_i(t-1) \tag{5.83}$$

where the weight vector (states) $\mathbf{w}(t) = [w_1(t), \ldots, w_Q(t)]'$ can be estimated by solving the following state-space equations

$$\mathbf{w}(t) = \mathbf{F}(t-1)\mathbf{w}(t-1) + \mathbf{q}(t-1), \tag{5.84}$$

$$\mathbf{y}(t) = \mathbf{G}(t)\mathbf{w}(t) + \mathbf{v}(t), \tag{5.85}$$

where $\mathbf{F}(t)$ is the transition matrix, the state vector $\mathbf{w}(t)$ represents the tap-weight vector at time slot t, $\mathbf{q}(t)$ is the process noise, $\mathbf{y}(t)$ is the desired QoS response, $\mathbf{G}(t)$ is the measurement matrix, and $\mathbf{v}(t)$ is the measurement noise. $\mathbf{q}(t)$ and $\mathbf{v}(t)$ are assumed to be zero-mean white noise with covariance matrices $\mathbf{Q}(t) = q_o\mathbf{I}$, and $\mathbf{R}(t) = v_o\mathbf{I}$ respectively.

To solve the problem in a distributed manner (i.e. each user updates its power based on local information), we have designed the matrices $\mathbf{F}$ and $\mathbf{G}$ in diagonal forms:

$$\mathbf{G}(t) = \begin{bmatrix} g_1(t) & 0 & \cdots & 0 \\ 0 & g_2(t) & 0 & \vdots \\ \vdots & 0 & \ddots & 0 \\ 0 & 0 & \cdots & g_Q(t) \end{bmatrix} \tag{5.86}$$

where $g_i(t)$ is the SINR of user i at time slot t, i.e. $g_i(t) = \frac{P_i(t)}{I_i(t)}$.

The transition matrix $\mathbf{F}(t)$ is given by

$$\mathbf{F}(t) = \begin{bmatrix} f_1(t) & 0 & \cdots & 0 \\ 0 & f_2(t) & 0 & \vdots \\ \vdots & 0 & \ddots & 0 \\ 0 & 0 & \cdots & f_Q(t) \end{bmatrix} \tag{5.87}$$

where $f_i(t-1) = \frac{g_i(t-1)}{g_i(t)}$.

Assuming that all users have the same QoS, the desired system response is given by

$$\mathbf{y}(t) = [\Gamma^T, \Gamma^T, \cdots, \Gamma^T]' \tag{5.88}$$

To explain the proposed state-space representation of the power control, assume the measurement noise and the process noise equal zero. For one user i (scalar form), Equations (5.84) and (5.85) are given by

$$w_i(t) = \frac{g_i(t-1)}{g_i(t)} w_i(t-1) \tag{5.89}$$

$$\Gamma^T = g_i(t)w_i(t) \tag{5.90}$$

In power control, the optimum transmitted power is determined to achieve the target SINR in the next time slot. From Equation (5.83), the next time slot SINR can be predicted as

$$\tilde{g}_i(t+1) = g_i(t)w_i(t) \tag{5.91}$$

From Equations (5.89)–(5.91), the adaptation weight is computed to achieve the target SINR in the next time slot. The modeling error (process noise) and the measurement noise should be taken into consideration to complete the state-space modeling. The Kalman filter is used to estimate the optimum adaptation weight $w_i(t)$ in order to make the next time slot SINR very close to the target SINR.

5.4.9.1 Kalman Filter Algorithm

The Kalman filter algorithm is specified by the following equations [21].

Let $\mathbf{C}(0)$ be the initial error covariance and $\mathbf{w}(0)$ the initial weight vector. Then for $t = 0,1,2,\ldots$.

$$\hat{\mathbf{w}}^-(t) = \mathbf{F}(t-1)\hat{\mathbf{w}}^+(t-1) \quad \text{(weight extrapolation)}, \tag{5.92}$$

$$\mathbf{C}^-(t) = \mathbf{F}(t-1)\mathbf{C}^+(t-1)\mathbf{F}'(t-1) + \mathbf{Q}(t-1) \text{ (error covariance extrapolation)}, \tag{5.93}$$

$$\mathbf{K}(t) = \mathbf{C}^-(t)\mathbf{G}'(t)[\mathbf{G}(t)\mathbf{C}^-(t)\mathbf{G}'(t) + \mathbf{R}(t)]^{-1} \text{ (Kalman gain)}, \tag{5.94}$$

$$\hat{\mathbf{w}}^+(t) = \hat{\mathbf{w}}^-(t) + \mathbf{K}(t)[\mathbf{y}(t) - \mathbf{G}(t)\hat{\mathbf{w}}^-(t)] \text{ (weight update)}, \tag{5.95}$$

$$\mathbf{C}^+(t) = [\mathbf{I} - \mathbf{K}(t)\mathbf{G}(t)]\mathbf{C}^-(t) \text{ (error covariance update)}, \tag{5.96}$$

The algorithms in (5.92)–(5.96) can be solved in scalar form, since all the matrices are in diagonal form.

Theorem 5.8

In feasible systems with snapshot assumption, using the proposed Kalman filter power control, all users will approach their SINR targets.

Proof

From Equations (5.92)–(5.95), the weight update equation for user i can be represented as

$$\begin{aligned}\hat{w}_i^+(t) &= f_i(t-1)\hat{w}_i^+(t-1) + K_i(t)[\Gamma^T - g_i(t)f_i(t-1)\hat{w}_i^+(t-1)] \\ &= f_i(t-1)\hat{w}_i^+(t-1) + K_i(t)[\Gamma^T - g_i(t-1)\hat{w}_i^+(t-1)].\end{aligned} \tag{5.97}$$

At steady state, let

$$\hat{w}_i^+(t) = \hat{w}_i^s, \quad K_i(t) = K_i^s, \quad g_i(t) = g_i^s, \quad \text{and } f_i(t) = f_i^s. \tag{5.98}$$

Then

$$\hat{w}_i^s = f_i^s\hat{w}_i^s + K_i^s[\Gamma^T - g_i^s\hat{w}_i^s] = \frac{K_i^s\Gamma^T}{1 - f_i^s + K_i^s g_i^s}. \tag{5.99}$$

Since at steady state, $f_i^s = 1$, the steady-state weight becomes

$$\hat{w}_i^s = \frac{\Gamma^T}{g_i^s}. \tag{5.100}$$

For user i, the SINR becomes

$$\Gamma_i^s = \frac{\hat{w}_i^s P_i^s}{I_i^s} = \Gamma^T. \tag{5.101}$$

The convergence properties of the Kalman filter depend on the values of the covariance matrices. From Equation (5.96) the error covariance equation can be represented as

$$C_i^+(t) = (1 - K_i(t)g_i(t))(f_i^2(t)C_i^+(t-1) + q_0), \tag{5.102}$$

where the Kalman gain Equation (5.94) in scalar form is represented as

$$K_i(t) = \frac{(f_i^2(t-1)C_i^+(t-1) + q_0)g_i(t)}{g_i^2(t-1)C_i^+(t-1) + g_i^2(t)q_0 + v_0} \tag{5.103}$$

By substituting Equation (5.103) into Equation (5.102), we obtain

$$C_i^+(t) = \frac{f_i^2(t-1)C_i^+(t-1)v_0 + q_0 v_0}{g_i^2(t-1)C_i^+(t-1) + g_i^2(t)q_0 + v_0} \tag{5.104}$$

At steady state:

$$C_i^s = -\frac{q_0}{2} + \frac{q_0}{2}\sqrt{1 + 4\frac{v_0}{q_0(g_i^s)^2}} \tag{5.105}$$

We are interested in the nonnegative solution because the variance $C_i^+(t)$ of uncertainty is, by definition, nonnegative [21]. From Equation (5.105), if the factor v_0/q_0 is small then faster convergence can be achieved [21].

5.4.9.2 Convergence

The convergence properties of the Kalman filter depend on the values of the covariance matrices. The convergence speed is an important issue in power control due to the dynamical behavior of the mobile communication system. Unfortunately, it is difficult to find an analytical expression for the convergence speed of the Kalman filter because of the time-varying nature of the system. The performance analysis of the Kalman filter depends on solving the Riccati equations. It is possible to simulate the Riccati equations without computing the state estimates themselves. This gives us a good indication of the convergence speed of the algorithm. The discrete-time Riccati equation is the solution of the following equation [21]

$$\mathbf{C}^-(t+1) = \mathbf{A}(t+1)\mathbf{B}^{-1}(t+1), \tag{5.106}$$

where

$$
\begin{bmatrix} \mathbf{A}(t+1) \\ \mathbf{B}(t+1) \end{bmatrix} = \begin{bmatrix} (\mathbf{F}(t) + \mathbf{Q}(t)(\mathbf{F}^{-1}(t))'\mathbf{G}'(t)\mathbf{R}^{-1}(t)\mathbf{G}(t)) & \mathbf{Q}(t)(\mathbf{F}^{-1}(t))' \\ (\mathbf{F}^{-1}(t))'\mathbf{G}'(t)\mathbf{R}^{-1}(t)\mathbf{G}(t) & (\mathbf{F}^{-1}(t))' \end{bmatrix} \begin{bmatrix} \mathbf{A}((t)) \\ \mathbf{B}(t) \end{bmatrix}
$$

$$(5.107)$$

We recommend the interested reader to go through reference [21] for more details.

References

[1] J. Proakis, *Digital Communications*, 3rd edition, McGraw-Hill, 1995.

[2] S. Haykin, *Communication Systems*, John Wiley & Sons Ltd, 2001.

[3] J. Cheng and N. Beaulieu, 'Accurate DS-CDMA bit-error probability calculation in Rayleigh fading', *IEEE Transactions on Wireless Communications*, **1**, January, 2004.

[4] R. Jäntti, Power control and transmission rate management in cellular radio systems, Ph.D. thesis, Control Engineering Laboratory, Helsinki University of Technology, Finland, 2001.

[5] F. Gantmacher, *The Theory of Matrices*, Vol. 2, Chelsea Publishing Company, 1964.

[6] R. Yates, 'A framework for uplink power control in cellular radio systems', *IEEE Journal on Selected Areas in Communications*, **13**, 1341–7, 1995.

[7] S. Grandhi and J. Zander, 'Constrained power control in cellular radio systems', *Proceedings of the IEEE Vehicular Technology Conference*, vol. 2, pp. 824–8, Piscataway, USA, June, 1994.

[8] H. Holma and A. Toskala, *WCDMA for UMTS*, 3rd edition, John Wiley & Sons Ltd, 2004.

[9] G. Dahlquist, *Numerical Methods*, Prentice-Hall, 1974.

[10] J. Zander, 'Distributed co-channel interference control in cellular radio systems', *IEEE Transactions on Vehicular Technology*, **41**, 305–11, 1994.

[11] S. Grandhi, R. Vijayan and D. Goodman 'Distributed power control in cellular radio systems', *IEEE Transactions on Communications*, **42**, 226–8, 1994.

[12] G. Foschini and Z. Miljanic, 'A simple distributed autonomous power control algorithm and its convergence' *IEEE Transactions on Vehicular Technology*, **42**, 641–6, 1993.

[13] R. Jäntti and S. Kim, 'Second-order power control with asymptotically fast convergence', *IEEE Journal on Selected Areas in Communications*, **18**, 447–57, 2000.

[14] M. Elmusrati, M. Rintamäki, I. Hartimo and H. Koivo, 'Fully distributed power control algorithm with one bit signaling and nonlinear error estimation', *Proceedings of the IEEE Vehicular Technology Conference*, Orlando, USA, October, 2003.

[15] M. Elmusrati, R. Jantti, and H. Koivo, 'Multi-objective distributed power control algorithm for CDMA wireless communication systems', *IEEE Transactions on Vehicular Technology*, March, 2007.

[16] K. Miettinen, *Nonlinear Multiobjective Optimization*, Kluwer, 1998.

[17] T. Jiang, N. Sidiropoulos and G. Giannakis, 'Kalman filtering for power estimation in mobile communication', *IEEE Transactions on Wireless Communications*, **2**, 151–61, 2003.

[18] K. Leung, 'Power control by interference prediction for broadband wireless packet networks', *IEEE Transactions on Wireless Communications*, **1**, 256–65, 2004.

[19] K. Shoarinejad, J. Speyer, and G. Pottie, 'Integrated predictive power control and dynamic channel assignment in mobile radio systems', *IEEE Transactions on Wireless Communications*, **2**, 976–88, 2003.

[20] J. Candy, *Signal Processing – The Modern Approach*, McGraw-Hill, 1987.

[21] M. Grewal and A. Andrews, *Kalman Filtering*, Prentice-Hall, 1993.

6

Power Control II: Control Engineering Perspective

6.1 Introduction

Fast and accurate power control plays the main role in radio resource management of wireless communications. The first issue with a base station and the mobile units in the same cell is to overcome the *near-far effect*, that is, the nearest mobile unit being closest to the base station overshadows those mobile units which are far away. This is intuitively clear but is simple to deduce from distance power law as discussed in Chapter 5. The task of *uplink power control*, from MS to BS, is to offset the channel variations, so that the received power per bit of all MS is equalized at BS.

In *downlink power control*, when the BS transmits a signal, it is received by many MS and the signal experiences the same attenuation, before reaching the MS. This implies that power control is not as useful in downlink as it is in uplink.

In this chapter, we will concentrate on discussing fast uplink power control. In Chapters 3 and 5 a static, 'snapshot' approach was used to gain a basic understanding of power control issues. In practice, the channel variations are quick, so the control policy and decisions must also be high speed. Uplink power control can be classified into three classes: *open loop*; *closed loop*; and *outer loop*. We will also concentrate on treating values in logarithmic scale. This leads to a log-linear model of power control algorithms, which is a more natural setting for treating dynamic power control problems.

Open-loop power control: the principle is as follows. Estimate the path loss, seen as a time-varying power gain $G(t)$, from the downlink beacon signal. To compensate the variations, the transmitted power should be the inverse of the estimated power gain. This idea was used in early wireless point-to-point links.

In practice open-loop power control does not work well, because the estimate is inaccurate. The reason is that channel fading in downlink is quite different from fading in uplink as explained above. Open-loop power control is used, however, to give a ballpark estimate for the initial MS power setting.

Systems Engineering in Wireless Communications   Heikki Koivo and Mohammed Elmusrati
© 2009 John Wiley & Sons, Ltd

Closed-loop power control (inner-loop power control): an important issue in the transmission is how many bit errors occur in a transmitted frame. This is described by the *bit error rate* (BER) after decoding. Another way to look at quality of service is the *frame error rate* (FER) or block error rate (BLER). Estimates of BER and FER require some time to be formed. Therefore they cannot be used as variables to be controlled in the fast loop, since channel variations are rapid. This is why SINR is chosen as the variable to be controlled.

In closed-loop power control the BS estimates the received SINR at each time slot and compares it with the set target SINR. Based on the error difference the controller makes the control decision in order to decrease the error at the next time slot. Currently, the most common controller is the one-bit, relay controller: if the error difference is positive, then the measured SINR is lower than the target. This implies that the BS will ask the MS to increase the transmission power by a fixed step. If the error is negative, then the measured SINR is higher than the target, and the command from BS to MS is to decrease the transmission power. A block diagram of closed-loop power control is represented in Figure 6.1.

Outer loop power control: the outer loop power control can be considered in two ways. In both cases it provides the inner loop power control the SINR target value $\Gamma^{\text{target}}(t)$. In a *hierarchical sense*, the outer loop power control can be considered to be at a higher and slower level than the inner loop power control as seen in Figure 6.2. In practice, the radio network controller (RNC), at a higher level, collects information about the individual radio links in the cell aiming to guarantee the quality of transmission. The quality is often defined by setting a target for either BER or FER, which are easy to observe using the cyclic redundancy check (CRC). The outer loop then adjusts the SINR target $\Gamma^{\text{target}}(t)$, so that a desired FER is guaranteed (Figure 6.2). The other way of viewing outer loop power control from a control point of view is to say that this is a typical *cascade control system*. The inner loop is a fast adjusting loop, and the outer loop slower, making decisions based on the quality of service. The outer loop continuously advises the inner loop what SINR target value $\Gamma^{\text{target}}(t)$ to use at each time slot (Figure 6.3).

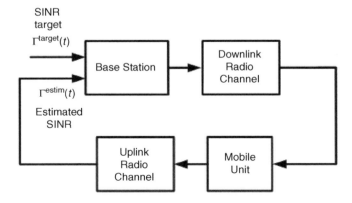

Figure 6.1   Block diagram of fast closed-loop power control

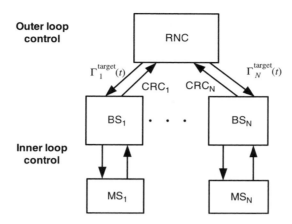

Figure 6.2 Hierarchical view of the outer power control loop. The RNC receives information about the CRC. If the FER value shows that the transmission quality is decreasing, then the RNC advises the base station to increase the SINR target value appropriately

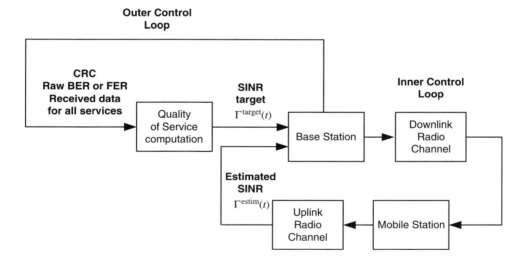

Figure 6.3 Cascade control view of the outer control loop. In the outer loop, the RNC computes the SINR target value for the fast inner loop

6.2 Issues in Uplink Power Control

In cellular systems the coverage of one base station extends to the other cells causing interference. In addition to this *inter-cell interference,* users interfere with each other within a cell leading to *intra-cell interference.* Ideally, each cell would have its own frequency channel to reduce the inter-cell interference, but this would severely limit the system capacity. The cells using the same frequency channel are called *co-channel cells*. The co-channels belong to different clusters. The interference between co-channel cells is designated as *co-channel interference I*. The inter-cell interference is due to both non-co-channel cells and co-channel cells. Since non-co-channel cells use a different frequency, their interference is negligible compared with the co-channel interference.

Recall the SINR definition given by Equation (5.2) in Chapter 5.

$$\bar{\Gamma}_i(t) = \frac{\bar{x}_i(t)}{\bar{I}(t)} = \frac{\bar{G}_{ii}(t)\bar{P}_i(t)}{\bar{I}(t)} \, i = 1, \ldots, Q \tag{6.1}$$

where $\bar{x}_i(t)$ is the received power, $\bar{G}_{ii}(t)$ the channel gain, $\bar{P}_i(t)$ the transmitted power, and $\bar{I}(t)$ the *total interference power* including the noise. Equation (6.1) is written in linear scale (bars are used above variables). In logarithmic scale this becomes

$$\Gamma_i(t) = G_{ii}(t) + P_i(t) - I(t), \, i = 1, \ldots, Q \tag{6.2}$$

A simplified closed-loop upper link power control model is shown in Figure 6.4. The subscript i has been dropped off, since only one control loop is considered.

Bit errors occur only in the standard one bit, relay power control loop, which is treated in Section 6.3. For the other power control algorithms this does not happen and the phenomenon should be ignored. The bit error probability in the transmission channel at time t is given as $\mathbb{P}(t)$. The bit errors are modeled with signal $E_{PC}(t)$, which multiplies the transmitted downlink

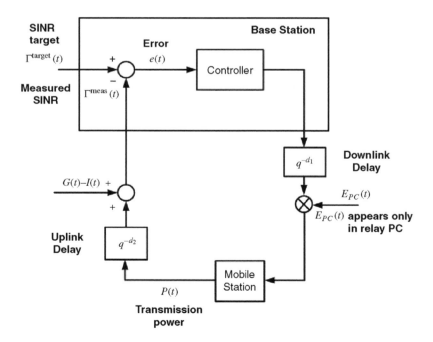

Figure 6.4 Upper link power control model

commands:

$$E_{PC}(t) = \begin{cases} 1 \text{ with probability} & \mathbb{P}(t) \\ -1, \text{ with probability} & 1 - \mathbb{P}(t) \end{cases} \tag{6.3}$$

The *loop delay* $d = d_1 + d_2$ is the overall delay of the control system in Figure 6.4. It consists of measuring, transmission and processing the information to form the power control command. When the MS sends the slot to the base station, it is received after a propagation delay. Some time is consumed by the BS in forming the SINR estimate Γ^{est} and then the power control command. The command is received by the MS after a transmission delay. This is then used in the next time slot to correct the transmitted power. The procedure can be carried out in one power control period. If the estimation together with the control algorithm takes more time, it is possible that the total delay is longer. This implies that the loop delay can be two power control periods. The delays are known. In the following, delays d_1 and d_2 are not treated separately. They are combined into loop delay d.

6.2.1 Information feedback

The BS receives the exact SINR measurements and thus the error

$$e(t) = \Gamma^{\text{target}}(t) - \Gamma^{\text{meas}}(t) \tag{6.4}$$

The error signal is processed in various ways depending on the control algorithm chosen.

6.2.2 Decision feedback

The error signal is processed via the sign function resulting in either $+1$ or -1 (multiplied by the correction step Δ).

$$\chi(t) = \Delta sign(\Gamma^{\text{target}}(t) - \Gamma^{\text{meas}}(t)) = \begin{cases} \Delta \\ -\Delta \end{cases} \tag{6.5}$$

In the following a variety of different control algorithms are applied to the dynamic power control problem. The same example is used throughout, so that the comparison is easier. The most used algorithm is relay (one-bit) power control, but here we demonstrate that in the BS we could apply more powerful algorithms advantageously. Figure 6.5 shows the possible benefits of this. It must be borne in mind that the examples and simulations are simplified, but similar results have been achieved with more complete wireless network simulators.

6.3 Upper Link Power Control with a Relay Controller

The relay controller was introduced in Equation (6.5). In UMTS and IS-95 systems only one bit is used for power control. Assume for simplicity that $\Gamma(t) = \Gamma^{\text{meas}}(t)$. The one-bit relay controller in Figure 6.6 becomes

$$P(t+1) = P(t) + \Delta sign(e(t)) = P(t) + \Delta sign(\Gamma^{\text{target}}(t) - \Gamma(t)) \tag{6.6}$$

where Δ is the fixed power control step size, varying between 0.5 dB and 2 dB uplink or in multiples of 0.5 dB downlink. The signum function in this context is defined as

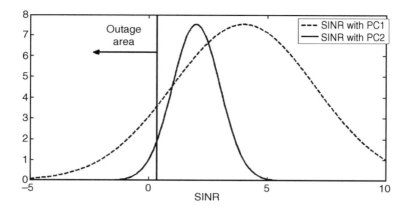

Figure 6.5 Variance of SINR for two different power control algorithms (PCs). The variance with PC1 has a wider spread and poorer performance than with PC2. Significant savings can be achieved if PC2 is used. This could be possible with more advanced PCs

$$sign(\alpha) = \begin{cases} +1, \text{ if } \alpha \geq 0 \\ -1, \text{ if } \alpha < 0 \end{cases} \qquad (6.7)$$

This is also called a *fixed-step power control* (FSPC) algorithm. Equation (6.6) represents the relay controller in discrete, recursive form. Equation (6.6) is a discrete form of an integral controller. The control system is depicted in Figure 6.6.

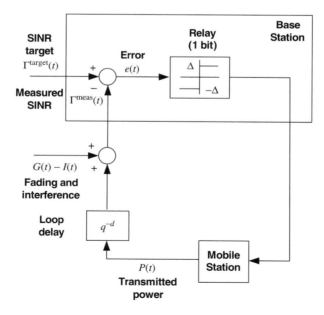

Figure 6.6 The upper link power control with one-bit relay controller with step size Δ

Consider control algorithm (6.6). This can be rewritten as (recall that $qP(t) = P(t+1)$)

$$P(t+1) - P(t) = (q-1)P(t) = \Delta sign(e(t)) \tag{6.8}$$

or

$$P(t) = \frac{1}{(q-1)}\Delta sign(e(t)) \tag{6.9}$$

Operator q corresponds to z in the z-transform. The difference is that in the z-transform the initial value is taken into account, in q-operator formulation it is not. In Simulink$^{®}$ the term $1/(q-1)$ is thus replaced by the discrete transfer function term $1/(z-1)$.

Example 6.1   Relay control

Consider the uplink power control model in Figure 6.6 with control algorithm (6.6), where step size $\Delta = 0.75$ dB. Assume two cases: (a) a constant $\Gamma^{target}(t) = 1.5$ and (b) step function $\Gamma^{target}(t) = 1.5, 0 \leq t \leq 200$ and $\Gamma^{target}(t) = 3, 201 \leq t \leq 500$, carrier frequency $f_c = 1.8$ GHz sampling rate $f_s = 1$ kHz. Here $t = 1, 2, \ldots$ (number of time steps), where the step size is $t_{step} = 0.002$ s. This same convention is used in all the examples of this chapter. Two mobile speeds, 5 km/h (pedestrian walk) and 100 km/h are studied. For channel gain and interference, $G(t) - I(t)$ use the Rayleigh fading model (Rayleigh fading generation was discussed in Section 3.3.2 of Chapter 3) and for shadowing a Gaussian-distributed random signal with zero mean and variance one. Simulate the system behavior using MATLAB$^{®}$/Simulink$^{®}$.

Solution

A Simulink$^{®}$ configuration of the control system is displayed in Figure 6.7.

The loop delay is one, representing the sum of uplink and downlink delay. Channel gain and interference, $G(t) - I(t)$, are shown at the upper side of the configuration. This will be repeated in the upcoming examples. Data for *tot* block comes from a function generating Rayleigh fading. Shadowing is shown here explicitly. Usually it would be in the same code as Rayleigh fading. If the seed number in the *Shadowing* block is changed, the simulation will not be exactly the same as with a different seed number. Bit error probability is also taken into account. The controller block implements Equation (6.9). One resulting set of simulations is shown in Figure 6.8. At the slow speed of 5 km/h the relay controller does a satisfactory job, but at high MS speeds, such as 100 km/h, the performance is not as good.

One question that arises is how to compare the results of the possible control responses. There are a number of ways of doing this, such as computing outage probability. Here we use the loss function $V = \sum e^2(i)$. The value of the loss function for the step function response at 5 km/h is $V = 1943.2$ and at 100 km/h it is 4013.0. If a constant value of SINR target $\Gamma^{target}(t) = 1.5$ is used, the loss functions are slightly smaller, at 5 km/h $V = 1540.4$ and at 100 km/h $V = 3782.7$. □

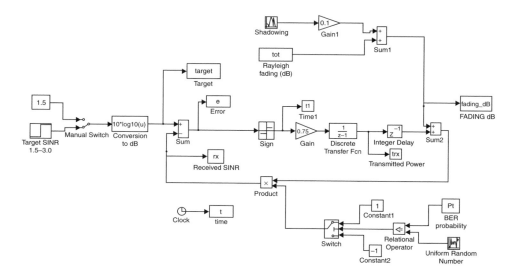

Figure 6.7 Uplink control system, when a relay controller is used. For channel gain and interference the Jakes' model is applied

Since shadowing involves random numbers, the simulations results will not be exactly the same when they are repeated! Many simulation runs should be carried out and the results should be statistically analyzed. Here only one sample run is shown. Rayleigh fading is used as an example. Such a model does not cover all situations appropriately because the environment

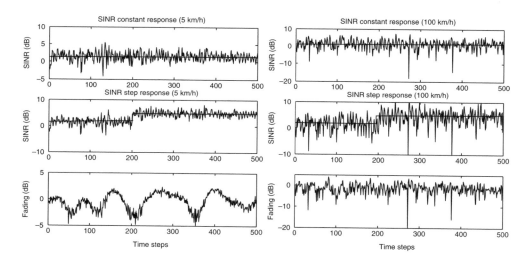

Figure 6.8 Simulation results of uplink power control: on the left responses for speed of 5 km/h and on the right the corresponding responses for speed of 100 km/h. Below the corresponding fading. Observe that the scales are not the same in all drawings

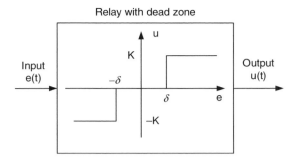

Figure 6.9   Characteristics of a relay with a dead zone

changes. The examples treated in this chapter demonstrate the different power control laws and their strengths. More sophisticated network simulators must be used when designing power control for larger networks.

Two bits may be included in the frame for power control. This makes it possible to implement a relay with a dead zone controller as shown in Figure 6.9.

When the input amplitude is small, the output is zero meaning that the controller eliminates noise with small amplitude. If the absolute value of the signal is larger than the dead zone width δ, the output is either $+K$ or $-K$. Written mathematically the dead zone characteristics are given as

$$u(t) = \begin{cases} K, & \text{if } e(t) > \delta \\ 0, & \text{if } 0 \le e(t) \le \delta \\ -K, & \text{if } e(t) < \delta \end{cases} \qquad (6.10)$$

There are numerous ways of implementing this in Simulink®, which does not have a dead zone block of Figure 6.9. One way is to use the *Sign* and *Dead Zone* blocks as shown in Figure 6.10.

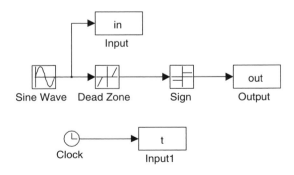

Figure 6.10   Implementation of a relay with a dead zone in Simulink®

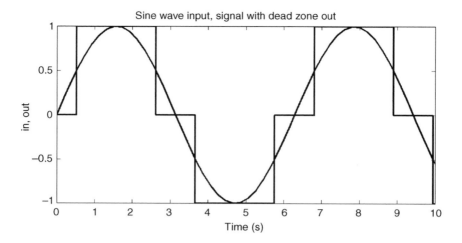

Figure 6.11 The output of the system with dead zone parameter $\delta = 0.5$. As expected the output is zero when the input amplitude is smaller than 0.5. If the input is bigger than 0.5 or smaller than -0.5, the output is either $+1$ or -1, correspondingly

The combination shown in Figure 6.10 produces the desired result as seen in Figure 6.11. Replace the relay control in Figure 6.6 by that in Figure 6.12.

Repeating Example 6.1 with the dead zone as a controller, the result of Figure 6.13 is achieved. Visual comparison is not easy, but computing the loss function for the two cases reveals the situation.

The loss function in the case of constant response at 100 km/h takes value, $V = 1537.8$ for dead zone parameter $\delta = 1.7$. The corresponding value for relay control is $V = 1812.7$. The latter figure is now smaller than in the previous example, because the number of samples is only 500. It was previously 1000. Clear advantages can be gained at high speeds. Once more, it should be emphasized that these values were for one run only.

The comparative runs in Figure 6.13 have been carried out with the Simulink® configuration of Figure 6.14, where parallel runs of the dead zone controller and relay controller were performed simultaneously with the same fading and shadowing data. Similar results were also obtained for step responses.

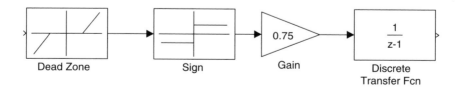

Figure 6.12 Simulink® implementation of a relay with a dead zone

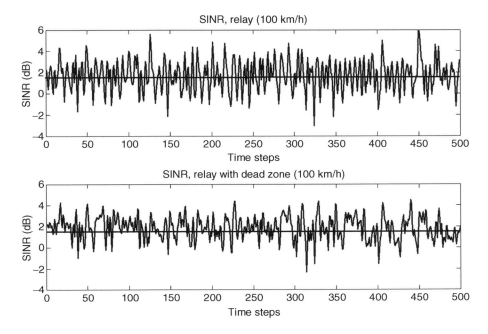

Figure 6.13 The uplink control with a dead zone and relay controller. On the left the speed is 5 km/h and on the right 100 km/h. The legend is the same in both

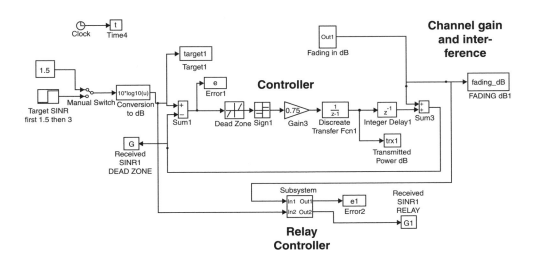

Figure 6.14 The parallel Simulink® configuration of a dead zone and relay controller with the same inputs and same fading data. This makes the comparison easier and will also be used with the other controllers

6.4 PID Control

PID control was discussed in detail in Chapter 2. Since it is used extensively in all sorts of applications, it is the first choice after the relay controller. Replace the controller shown in Figure 6.6 with a controller in Figure 6.15. It is otherwise the same except that a PID controller has been added in front of the sign block. This is a decision feedback form as in Equation (6.5) except that the error signal is processed through a PID controller.

The PID algorithm is applied in its velocity form as given in Equation (2.32) repeated here with slightly different notation ($k \mapsto t$ and difference operator Δ is not used):

$$u(t+1) = u(t) + K_P(e(t+1) - e(t)) + K_I e(t) + K_D \frac{e(t+1) - 2e(t) + e(t-1)}{h} \quad (6.11)$$

Because the signum function is followed with an integrator, it is expected that the integral term is not going to have very much effect.

An issue to resolve is how to determine good parameter values K_P, K_I and K_D for the PID controller. If a model (simulation model suffices) is available, then optimization can be applied. Typical cost functions are (given in continuous form for simplicity):

$$J_{ISE}(\mathbf{p}) = \int_0^\infty e^2(t)dt$$

$$\quad (6.12)$$

$$J_{ITAE}(\mathbf{p}) = \int_0^\infty t|e(t)|dt$$

Here $e(t)$ is the error shown, e.g. in Figure 6.4. For notational purposes define the control parameter vector as $\mathbf{p} = [K_P, K_I, K_D]$.

The abbreviations in Equation (6.12) are *ISE = Integral Square Error* and *ITAE = Integral Time Absolute Error*. The idea in *ITAE* is that often the error is large in the beginning, so multiplying it with time t balances the cost over time better.

The idea in the optimization process is as follows. First set up the Simulink® diagram of the system. Give the initial, numerical value for $\mathbf{p}$, $\mathbf{p}^0 = [K_P^0, K_I^0, K_D^0]$ on the MATLAB® command side. With this value of the control parameter vector, the error $e(t)$ can be computed with the Simulink® model and therefore also the corresponding cost function value. Next apply an optimization algorithm which uses only cost function values. In basic MATLAB® *fminsearch* uses a Nelder–Mead direct search technique. Another command that could be used is *fminunc*.

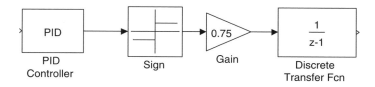

Figure 6.15 Decision feedback type of control, where the error signal is first fed to a PID controller before entering the sign function. The Discrete Transfer Fcn block acts as an integrator

Both of these update the control parameter vector **p** and the process can be repeated until the error tolerance in the cost function is satisfied. More formally:

1. Configure the Simulink® diagram of the system, *sys*, leaving $\mathbf{p} = [K_P, K_I, K_D]$ as a parameter.
2. Set initial value for control parameter vector $\mathbf{p}^0 = [K_P^0, K_I^0, K_D^0]$ in the MATLAB® Command window. Set $i = 0$.
3. Set $i \mapsto i + 1$. Simulate the system with parameter vector **p**, $\mathbf{p}^i = [K_P^i, K_I^i, K_D^i]$ and compute error $e^i(t)$.
4. Compute the value of cost function $J^i = J(\mathbf{p}^i)$ with the Simulink® model.
5. Apply the Nelder–Mead minimization algorithm (*fminsearch*) to obtain new, improved parameter vector **p**
6. Compute $|J^i - J^{i-1}|$. If this is bigger than the given error tolerance, go to step 2, otherwise to step 7.
7. Stop

An observation about the cost function calculation helps to combine step 4 with step 3. Consider the ITAE cost function in Equation (6.12). Replace the upper limit of integration ∞ with *t* and rewrite ITAE mathematically in a more exact way (the dummy integration variable is α)

$$J_{ITAE}(t; \mathbf{p}) = \int_0^t \alpha |e(\alpha)| d\alpha \qquad (6.13)$$

Compute the derivative of $J_{ITAE}(t; \mathbf{p})$ with respect to *t* using the Leibnitz rule:

$$\frac{dJ_{ITAE}(t; \mathbf{p})}{dt} = t|e(t)| \qquad (6.14)$$

Thus the cost function can be set up in Simulink® using a clock to generate *t* and then after multiplication integrating the product on the right-hand side. This is computed simultaneously with the rest of the system simulation. The upper limit need not go to infinity. It must be adjusted so that the system reaches its steady state. It is easy to adjust the upper limit to a sufficiently large value.

Observe that the formulated problem is a finite dimensional optimization problem. There are only three or fewer parameters in $\mathbf{p} = [K_P, K_I, K_D]$. The difference here, compared to curve fitting in Chapter 3, is that the parameters are embedded in a dynamical system. The response in practice is almost impossible to compute analytically by hand, but meaningful practical problems can be solved using simulators. Sophisticated, complex simulators often involve analytical expressions including dynamical equations, but they might also include software procedures, which are analytically impossible to handle, e.g. gradients cannot be computed. These present no problems in simulation, which provides a value for the cost function, or for the following optimization.

Example 6.2   PID control tuning with optimization

Let the open-loop system transfer function be

$$G(s) = \frac{0.4}{5s + 1} e^{-s} \qquad (6.15)$$

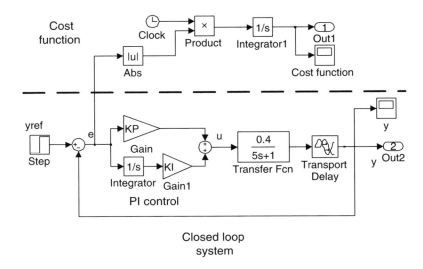

Figure 6.16 Below the dashed line is the PI controlled feedback system. Above the dashed line is the realization of the ITAE cost computation simultaneously with system simulation

The controller is a PI controller with $[K_P, K_I] = \mathbf{p}$. The overall Simulink® configuration of the feedback system is shown in Figure 6.16. Compute the minimal cost and the optimal $\mathbf{p} = [K_P, K_I]$, which minimizes the ITAE criterion and plot the output response and the cost function.

Solution

The simplest way to solve the problem is as follows. Start by setting up the Simulink® configuration for the system together with the cost ITAE. The result. *PItuning*, is presented in Figure 6.16. Next define a MATLAB® M-file *cost*.

```
function [c]=cost(p)
global p;
T_final=50; %Simulation time
Kp=p(1);
Ki=p(2);
[t,x,y]=sim('PItuning',T_final); %PItuning is simulated with
  values [Kp, KI]
c=max(y(:,1)); % value of cost function
```

Define K_P, K_I to be global variables. Then assign them starting values and apply optimization command *fminsearch*. In this command *cost* is defined as the function to be minimized. The value of the cost function is computed by the function defined above. It simulates (*sim* command) the specified system *PItuning* in Simulink® with the given control parameter values. Optimization proceeds with *fminsearch*, which receives the value of the computed cost function. Based on this information K_P, K_I values are improved according to the procedure described above.

In the MATLAB® Command window this proceeds as follows

```
global Kp Ki; % Minimize 'cost'
fminsearch('cost',[1 1]) % Initial values Kp = 1, Ki = 1
% Simulate for plotting purposes until T_final=50 s - This can be
   longer or shorter
[t,x,y]=sim('PItuning',50); % Output y includes both the output
   step response and cost
% Plot output step response and cost figure; plot(t,y)
% Compute the value of the cost
cost([Kp Ki])
% ans = 5.3644; minimum of cost function
% result of optimization Kp = 6.7910, KI = 1.3560
```

The optimized step response is shown in Figure 6.17, which is quite good. Note that the unit step begins at time 1 s. It was assumed that $T\_final = 50$ s would suffice for the cost function to settle. This is true as seen in the figure.

It is also possible to formulate constrained problems. For such problems command *fmincon_* is available, but will not be discussed here.

Optimization is a very powerful design tool, when used properly. The designer may easily become too confident in using this approach. A number of issues have to be kept in mind. Does the cost function include all the necessary specifications which need to be satisfied? When minimization is performed, several initial values should be tried, because usually the problem is likely to have many local minima. Optimization here is performed with respect to a step response. What can be done if the input is not a step function but takes different

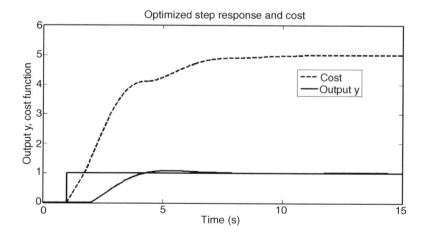

Figure 6.17   Optimized step response and cost function are shown. There is very little overshoot in the output step response. The cost function reaches the steady-state value well before 50 s, which is used in the optimization

forms? Multiobjective optimization provides possibilities in resolving some of these issues.

Example 6.3   PID power control

Consider the uplink control model in Figure 6.6, where the relay controller is replaced with the controller depicted in Figure 6.15. Repeat Example 6.1 with the same data parameter values and Jakes' fading model. Study the system behavior using Simulink®. Consider a slow MS speed of 5 km/h (pedestrian walk) and fast speed of 100 km/h (vehicle moving on an expressway).

Solution

The Simulink® configuration of the power control system is shown in Figure 6.18. It provides many alternatives for simulation. Either a constant value or a step function can be chosen as the target SINR. The PID control block includes a PID controller or a PID controller with relay as displayed in Figure 6.19. The basic relay controller is shown for comparison. There is a separate block for the cost function calculation, which follows the same procedure as that in Figure 6.16 and is not repeated here.

If only slow speeds are considered, then relay control performs slightly better than PID control, but if the speed of MS increases, then the PID controller becomes significantly better. For a speed of 100 km/h a computation over 500 time steps results in loss function value $V_{PID} = \sum e^2(i) = 5723$ for the PID controller and in $V_{relay} = 6576$ for the relay controller. Simulation results for the PID controller are displayed in Figure 6.20 where the MS speed is 100 km/h.

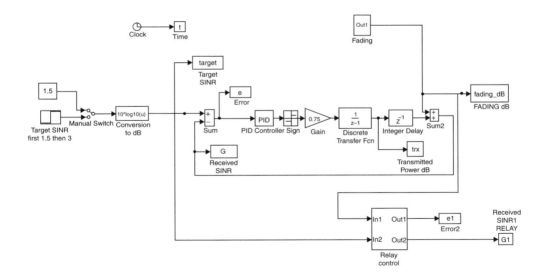

Figure 6.18   Simulink® configuration of decision feedback type of PID control

PID control block

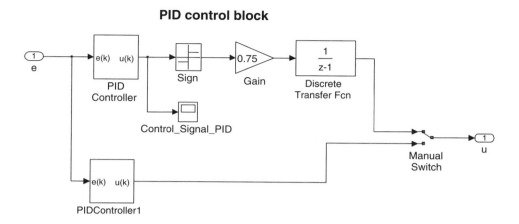

Figure 6.19 Expanded view of PID control block in Figure 6.16. Above: PID controller with relay control; below: without relay control, decided by a switch

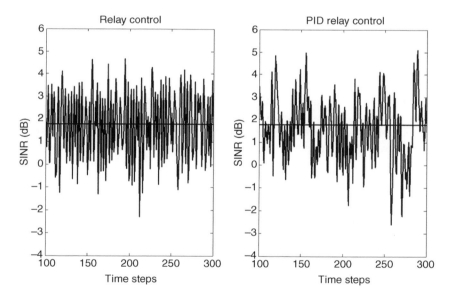

Figure 6.20 Simulated SINR responses. On the left-hand side relay, on the right-hand side relay PID. The legend is the same in all SINR figures. The loss function calculation shows that for a slower speed (5 km/h) the relay controller gives smaller loss function values and for higher speeds PID results in smaller loss functions for this particular run. Only part of the simulation interval [100 300] is shown

6.5 The Self-Tuning Predictive Power Control Algorithm

Self-tuning predictive power control is based on the self-tuning predictor presented in Section 4.9. The principle of the feedback control loop is displayed in Figure 6.21. The major change

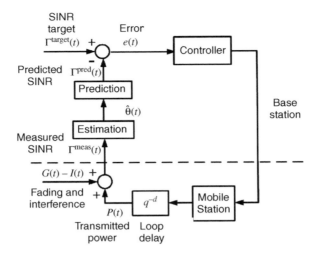

Figure 6.21 Basic structure of the predictive self-tuning predictive control algorithm

compared with Figure 6.4 is that estimation and prediction of SINR are explicitly carried out in the base station.

The task of the predictor is to minimize prediction error variance. The main goal of the fast power control is to keep the SINR as close to the target as possible (or above it). Because the control system contains a nonlinear element, the *separation principle* does not apply as such, that is, the predictor and the controller cannot be designed separately as is done in linear stochastic control. In addition to the unknown predictor parameters, the control system includes many other constants and parameters. These need special attention, so that the overall system will perform well. Simulation studies provide a good way of harnessing the most usable parameter values.

6.5.1 Predictor Structure

The predictor structure (Equation (4.107)), rewritten here, is defined by (n_c, n_g, k) where k is the prediction horizon.

$$
\begin{aligned}
\hat{\Gamma}(t+k|t) &= -\widehat{c}_1\hat{\Gamma}(t+k-1|t-1) - \cdots - \widehat{c}_{n_c}\hat{\Gamma}(t+k-n_c|t-n_c) \\
&\quad + \widehat{g}_0\Gamma(t) + \widehat{g}_1\Gamma(t-1) + \cdots + \widehat{g}_{n_g}\Gamma(t-n_g) \\
&= \boldsymbol{\phi}^T(t)\hat{\boldsymbol{\theta}}(t-1)
\end{aligned}
\tag{6.16}
$$

where the data vector $\boldsymbol{\phi}(t) = [-\hat{\Gamma}(t+k-1|t-1), \ldots, -\hat{\Gamma}(t+k-n_c|t-n_c), \Gamma(t),$ $\ldots, \Gamma(t-n_g)]^T$ and the vector of parameter estimates $\hat{\boldsymbol{\theta}}(t-1) = [\widehat{c}_1, \ldots, \widehat{c}_{n_c}, \widehat{g}_0, \ldots, \widehat{g}_{n_g}]^T$.

If the integers in (n_c, n_g) are large, a longer data history is needed and therefore more computations. Since the environmental conditions change quite fast, simpler structures, such as (2,1) at low speed and (3,4) at high speed, are found useful with one or two step prediction steps ahead.

6.5.1.1 Self-Tuning Predictive Power Control Algorithm

Define first the structure and the prediction horizon of the estimator: (n_c, n_g, k). Then

1. Read the new SINR measurement $\Gamma(t)$.
2. Update the data vector $\boldsymbol{\phi}(t) = [-\hat{\Gamma}(t+k-1|t-1), \ldots, -\hat{\Gamma}(t+k-n_c|t-n_c), \Gamma(t),$
   $\ldots, \Gamma(t-n_g)]^T$.
3. Calculate the new parameter estimate vector from RELS as explained in Chapter 4:

$$\hat{\boldsymbol{\theta}}(t) = \hat{\boldsymbol{\theta}}(t-1) + \mathbf{K}(t)[\Gamma(t) - \boldsymbol{\phi}^T(t)\hat{\boldsymbol{\theta}}(t-1)] \tag{6.17}$$

$$\mathbf{K}(t) = \mathbf{P}(t)\boldsymbol{\phi}(t) = \frac{\mathbf{P}(t-1)\boldsymbol{\phi}(t)}{\lambda(t) + \boldsymbol{\phi}^T(t)\mathbf{P}(t-1)\boldsymbol{\phi}(t)} \tag{6.18}$$

$$\mathbf{P}(t) = \frac{1}{\lambda(t)}\left[\mathbf{P}(t-1) - \frac{\mathbf{P}(t-1)\boldsymbol{\phi}(t)\boldsymbol{\phi}^T(t)\mathbf{P}(t-1)}{\lambda(t) + \boldsymbol{\phi}^T(t)\mathbf{P}(t-1)\boldsymbol{\phi}(t)}\right] \tag{6.19}$$

4. Use the estimated parameters to compute the new k step-ahead predictor from

$$\begin{aligned}
\hat{\Gamma}(t+k|t) &= \boldsymbol{\phi}^T(t)\hat{\boldsymbol{\theta}}(t-1) \\
&= -\widehat{c}_1\hat{\Gamma}(t+k-1|t-1) - \cdots - \widehat{c}_{n_c}\hat{\Gamma}(t+k-n_c|t-n_c) \\
&\quad + \widehat{g}_0\Gamma(t) + \widehat{g}_1\Gamma(t-1) + \cdots + \widehat{g}_{n_g}\Gamma(t-n_g)
\end{aligned} \tag{6.20}$$

5. Set $t \to t+1$.
6. Power control decision is made based on the sign of the error (relay control).
7. Apply new power control command.
8. Go back to step 1.

The initial value of error covariance $\mathbf{P}(0)$ is chosen to be large, if little or no knowledge of it is available. As indicated in Chapter 4, the initial value of the error covariance is taken as a diagonal matrix $\mathbf{P}(0) = \delta\mathbf{I}$, where the scalar $\delta = 100 - 1000$. The forgetting factor $\lambda(t)$ is usually given a constant value between 0.9 and 0.995.

Example 6.4   Self-tuning predictive power control

Consider the uplink control model in Figure 6.21 with self-tuning predictive control. Repeat Example 6.1 with the same data values and Jakes' fading model. Study the system behavior using Simulink® with different predictor structures (n_c, n_g, k). For prediction horizon k try values $k = 1, 2$. Consider a slow MS speed of 5 km/h (pedestrian walk) and a fast speed of 100 km/h (vehicle moving on an expressway).

Solution

The Simulink® configuration of the power control system is shown in Figure 6.22. When the self-tuning predictive power control algorithm is implemented, we need to remember the old data $\boldsymbol{\phi}(t) = [-\hat{\Gamma}(t+k-1|t-1), \ldots, -\hat{\Gamma}(t+k-n_c|t-n_c), \Gamma(t), \ldots, \Gamma(t-n_g)]^T$ and previous parameter estimates $\hat{\boldsymbol{\theta}}(t-1) = [\widehat{c}_1, \ldots, \widehat{c}_{n_c}, \widehat{g}_0, \ldots, \widehat{g}_{n_g}]^T$. Implementation could be done in Simulink®, e.g. using delay blocks, but it would be fairly clumsy. Therefore a MATLAB®

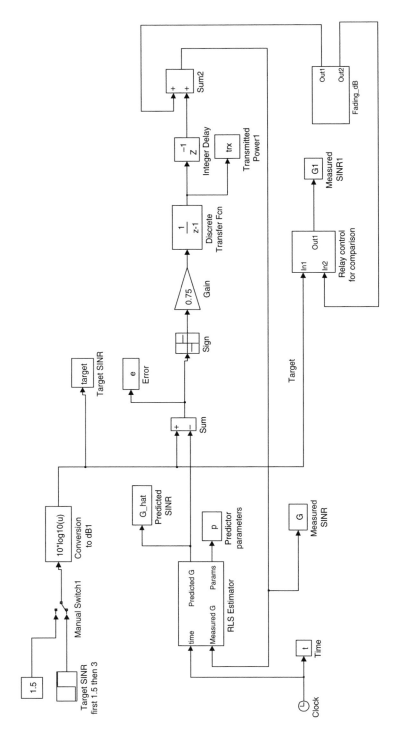

Figure 6.22 Self-tuning predictive power control implementation. Note the joint *Estimator and Predictor* MATLAB function in the feedback loop. The relay control loop, which is used for comparison, has its own block as does fading. Compare with Figure 6.14, where the relay controller is displayed in detail

function block RLS estimator is set up with an M-file, which computes the recursive least-squares estimates:

```
function y = RLS_estimator(u)
global Gh G TH_prev P;
% STATIC PARAMETERS
ng = 2; % Number of g-parameters
nc = 1; % Number of c-parameters
k = 1; % Prediction horizon
L = 0.99; % Forgetting factor
T_delay = (1 + ng) + (1 + nc) + k;

% CURRENT VALUES
curr_t = u(1); %Current time
curr_G = u(2); %Current measurement
if curr_t > T_delay

% Update measurement data
G = [G(2:end) curr_G];
% Update estimation
phi = [-Gh(end:-1:end-nc+1) G(end-1:-1:end-ng)];
K = P*phi'/(L + phi*P*phi');
TH = TH_prev + K*(curr_G - phi*TH_prev);
P = (1/L)*(P - K*phi*P);
curr_Gh = sum([-Gh(end:-1:end-nc+1) G(end:-1:end-ng+1)].*TH');

% Update variables for the next round
TH_prev = TH;
Gh = [Gh(2:end) curr_Gh];
else
% Collect data (and output zero)
Gh = [Gh 0];
G = [G curr_G];
TH_prev = [zeros(1,nc) zeros(1,ng)]';
P = 10000*eye(length(TH_prev));
end

% Set latest prediction as the first element of the output vector
% The following elements contain nc+ng parameters
y = [Gh(end) TH_prev'];
```

Here the structure of predictor is given as $n_g = 2, n_c = 1, k = 1$, which was found to be quite efficient. These have to be changed when other predictor structures are considered.

When the loss function value for is computed for step input and STPPC, $V_{\text{Predictive}} = \sum e^2(i) = 2822$ and for relay controller $V_{\text{Relay}} = \sum e^2(i) = 3451$. Here the speed is 100 km/h and prediction horizon $k = 1$. An overall picture of different situations is shown in Figure 6.23.

More detailed step responses of STPPC and relay power control are displayed in Figure 6.24 between time steps 120 and 280. It can be seen that although STPPC performs well and better

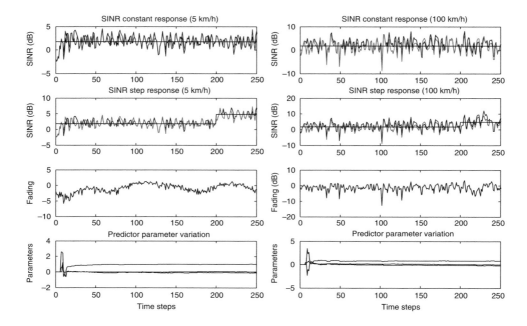

Figure 6.23 Overall simulation result. On the left-hand side MS speed is 5 km/h, on the right-hand side 100 km/h. From top to bottom: SINR constant response; SINR step response (here both relay and STPPC are drawn in the same figure); fading; and variation in predictor parameters. Observe that for clarity the time from 0 steps 90 to 250 steps 90

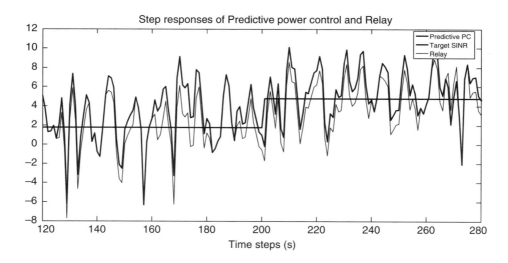

Figure 6.24 Step responses of STPPC and relay control. Computing the loss function shows that for this run, STPPC has a smaller loss function than relay. This is not taken from the simulation shown in Figure 6.23

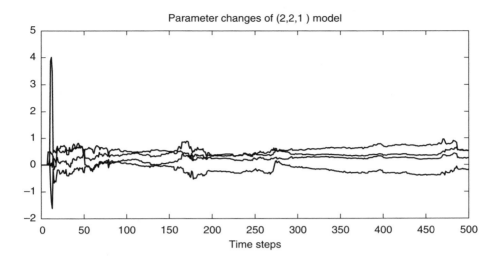

Figure 6.25 Parameter changes of the predictor in the run of Figure 6.24 as function of time. Strong changes occur at the early learning, but after that the variations are fairly small

than the relay controller, it has a slight bias. The corresponding parameters behave rather nicely in Figure 6.25, except in the beginning when learning is going on. At that time a backup power control, such as a relay controller should be used.

Figure 6.26 compares a typical measured SINR and the predicted value (one step ahead). The proposed prediction works quite well, and when it is part of a feedback loop minor variations disappear. This is depicted in Figure 6.27.

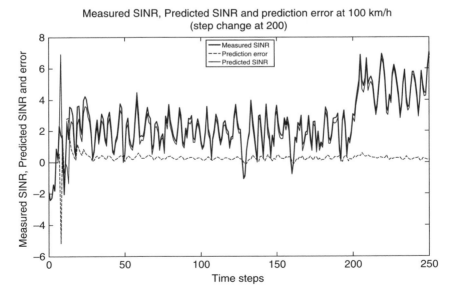

Figure 6.26 Comparison of a typical measured SINR and predicted SINR. The prediction error is also shown. Agreement is quite good

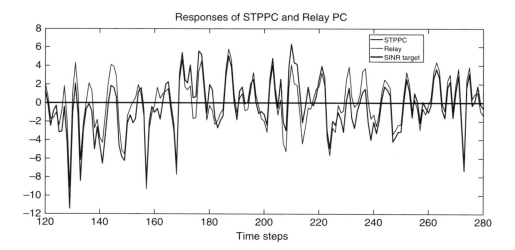

Figure 6.27 Responses of STPPC (thick line) and relay PC (thin line). We can see visually the same as the loss functions indicated. The STPPC is clearly better

STPPC is quite powerful when a proper structure for the predictor is found. Parameters settle down fairly quickly after the initial transient. In the beginning, during the learning period, the relay controller should be applied as a back-up controller until STPPC parameter variations stabilize sufficiently.

6.6 Self-Tuning Power Control

The purpose of using a self-tuning power control is similar to a self-tuning predictor – constant parameter controllers cannot keep up with the time-varying nature of the channel or with nonlinearities. The control parameters have to be updated according to the changing environment. The main idea is illustrated in Figure 6.28.

The starting point for deriving the controller is the constant parameter ARMAX model for the system

$$A(q^{-1})e(t) = B(q^{-1})u(t-k) + C(q^{-1})\xi(t) \tag{6.21}$$

where the polynomial operators are given as

$$A(q^{-1}) = 1 + a_1 q^{-1} + \cdots a_n q^{-n_a} \tag{6.22}$$

$$B(q^{-1}) = b_1 q^{-1} + \cdots b_{n_b} q^{-n_b} \tag{6.23}$$

$$C(q^{-1}) = 1 + c_1 q^{-1} + \cdots c_n q^{-n_c} \tag{6.24}$$

Here $e(t) = \Gamma^{\text{target}}(t) - \Gamma(t)$ is the SINR error at time step t, $u(t)$ the input (control variable) of the system, $\xi(t)$ is the white noise process, and k delay of the process. From this point on to simplify notation q^{-1} is left out as the argument in the polynomial operators.

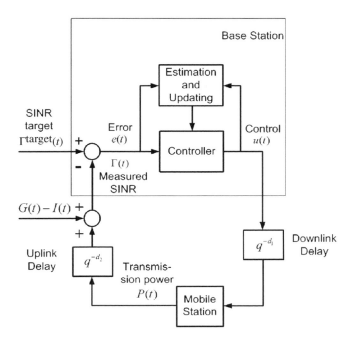

Figure 6.28   Principle of self-tuning controller in the BS

Determine the controller, which minimizes

$$J(u) = E\{e^2(t+k)|Y_t\} = E\{(\Gamma^{\text{target}}(t) - \Gamma(t))^2|Y_t\} \qquad (6.25)$$

where $Y_t = \{\Gamma(t), \Gamma(t-1), \ldots, u(t-1), u(t-2), \ldots\}$ is the measurement history. If $\Gamma^{\text{target}}(t) = 0$, the resulting optimal controller is the *minimum variance* controller. Here a slightly more general cost function with time-varying $\Gamma^{\text{target}}(t)$ (or constant) SINR target is used, resulting in the tracking controller, which is more typical in uplink power control.

The solution is found in a similar way as in the case of minimum variance predictor in Chapter 4, Section 4.8. The first form is

$$e(t+k) = \frac{B}{A}u(t) + \frac{C}{A}\xi(t+k) \qquad (6.26)$$

Recall the Diophantine equation

$$C = AF + q^{-k}G \qquad (6.27)$$

where

$$F = 1 + f_1 q^{-1} + \cdots f_{k-1} q^{-k+1} \qquad (6.28)$$

$$G = g_0 + g_1 q^{-1} + \cdots g_{n-1} q^{-n_g+k} \qquad (6.29)$$

Substituting Equation (6.27) in Equation (6.26) gives

$$e(t+k) = \frac{B}{A}u(t) + \frac{G}{A}\xi(t) + F\xi(t+k) \tag{6.30}$$

Solving for $\xi(t)$ in Equation (6.21) results after some manipulation in

$$e(t+k) = \left[\frac{C - q^{-k}G}{AC}\right]Bu(t) + \frac{G}{C}e(t) + F\xi(t+k) \tag{6.31}$$

When the Diophantine equation is applied to the term in brackets, we obtain

$$e(t+k) = \frac{BF}{C}u(t) + \frac{G}{C}e(t) + F\xi(t+k) \tag{6.32}$$

The last term on the right-hand side consists of future noise terms and is therefore independent of the first two terms. The best prediction of $e(t+k)$ at time t becomes

$$\hat{e}(t+k|t) = \frac{BF}{C}u(t) + \frac{G}{C}e(t) \tag{6.33}$$

Using this Equation (6.32) gives the output prediction error as

$$F\xi(t+k) = e(t+k) - \hat{e}(t+k|t) \tag{6.34}$$

Solving for $e(t+k)$ in Equation (6.34) and substituting in the cost function (Equation (6.25)) we obtain

$$\begin{aligned}J(u) &= E\{e^2(t+k)|Y_t\} = E\{[\hat{e}(t+k|t) + F\xi(t+k)]^2|Y_t\} \\ &= E\{\hat{e}^2(t+k|t)\} + (1 + f_1^2 + \cdots f_{k-1}^2)\sigma_\xi^2\end{aligned} \tag{6.35}$$

The last term on the right-hand side of the equation is constant, so the minimum is achieved when

$$\hat{e}(t+k|t) = 0 \tag{6.36}$$

The resulting control law is therefore

$$BFu(t) = -Ge(t) \tag{6.37}$$

or

$$u(t) = -\frac{G}{BF}e(t) \tag{6.38}$$

The cost function does not penalize against using large values of control, so the control signal may grow without limit. In power control the control signal is bounded by quantization, so this makes the developed controller more applicable in such an application. Adding b_1 to both sides of Equation (6.37) and recalling from Equation (6.23) the definition of operator B and from Equation (6.28) operator F, control $u(t)$ can be written as

$$u(t) = \frac{1}{b_1}[Ge(t) + (b_1 - BF)u(t)] \tag{6.39}$$

Note that the first term on the right-hand side contains terms of the error history (including current value) and the second term contains the control history. When implementing algorithm (6.39), care has to be taken that b_1 will not become zero. Often it is taken as a constant, e.g. $b_1 = 1$.

The adaptive form of Equation (6.39) can be implemented *indirectly* by first estimating the parameter values of A and B, then computing F and G from Equations (6.28)–(6.29) and then substituting them in the algorithm (6.39). This is also called the *explicit* form of a self-tuning controller. In a *direct* or *implicit* self-tuning controller the parameters of the controller are estimated directly as in a PID controller, therefore saving time in computing. It is the latter self-tuning control algorithm that is used here.

Substituting the operator polynomials from Equation (6.23) and Equations (6.27)–(6.29) we can write after some manipulation

$$u(t) = \frac{1}{b_1}\left[-\alpha_1 u(t-1) - \ldots - \alpha_{n_u} u(t-n_u) + g_0 e(t) + \cdots + g_{n_g - k} e(t+k-n_g)\right] \quad (6.40)$$

Here α_i are functions of b_i and f_i. Also n_u comes from the product of BF. In constructing the direct self-tuning controller, the structure is chosen first, that is, the number of history terms of errors ($e(t)$ is current) and controls are taken into account. Once that is done, the controller is hooked into the system and parameters in Equation (6.40) are estimated in real time, so there is no need to write down exact expressions for α_i. Equation (6.40) can be written in the form

$$u(t) = \frac{1}{b_1}\boldsymbol{\phi}^T(t)\hat{\boldsymbol{\theta}}(t-1) \quad (6.41)$$

where the data vector is

$$\boldsymbol{\phi}(t) = [-u(t-1), \ldots, -u(t+k-n_u), e(t), \ldots, e(t+k-n_g)]^T \quad (6.42)$$

and the parameter vector is

$$\hat{\boldsymbol{\theta}}(t-1) = [\widehat{\alpha}_1, \ldots, \widehat{\alpha}_{n_u}, \widehat{g}_0, \ldots, \widehat{g}_{n_g}]^T \quad (6.43)$$

The following procedure is analogous to the self-tuning predictive controller.

6.6.1.1 Self-Tuning Power Control (STPC) Algorithm

Define first the structure of the controller: (n_u, n_g, k).

1. Read the new SINR measurement $\Gamma(t)$ and compute the error $e(t) = \Gamma^{\text{target}}(t) - \Gamma(t)$.
2. Update the data vector

$$\boldsymbol{\phi}(t) = [-u(t-1), \ldots, -u(t+k-n_u), e(t), \ldots, e(t+k-n_g)]^T \quad (6.44)$$

3. Calculate the new parameter estimate vector

$$\hat{\boldsymbol{\theta}}(t-1) = [\widehat{\alpha}_1, \ldots, \widehat{\alpha}_{n_u}, \widehat{g}_0, \ldots, \widehat{g}_{n_g}]^{TT} \quad (6.45)$$

from RELS Equations (6.17)–(6.19).

4. Use the estimated new parameters to compute the control signal from

$$u(t) = \frac{1}{b_1} \boldsymbol{\phi}^T(t)\hat{\boldsymbol{\theta}}(t-1) \tag{6.46}$$

5. Set $t \rightarrow t + 1$
6. Power control decision is made based on the sign of the error (relay control).
7. Apply new power control command.
8. Go back to step 1.

The data vector $\boldsymbol{\phi}(t)$ is updated at every time step, but in practice parameter updating could be done more infrequently, especially if the parameters do not vary very much. The update frequency is a design parameter and has a direct effect on the parameter convergence and thus on the prediction accuracy. Too short an update period leads to non-convergence of the parameters and too long an update period leads to a more inaccurate estimation.

Example 6.5   Self-tuning power control

Consider the uplink control model in Figure 6.28 with self-tuning control. Repeat Example 6.1 with the same data values and Jakes' fading model. Study the system behavior using Simulink® with different self-tuning control structures (n_u, n_g). Consider a slow MS speed of 5 km/h (pedestrian walk) and a fast speed of 100 km/h (vehicle moving on an expressway). Study the effect of the updating period.

Solution

The Simulink® configuration of the power control system is shown in Figure 6.29. The relay controller is configured in parallel with the STPC for comparison.

The same recursive least-squares MATLAB® function block *RLS_estimate* as before is used. A small modification is done in the code by setting up an *update pulse*. The lines that have been changed are given in bold.

```
function y = RLS_estimator(u)

% CURRENT VALUES
curr_t = u(1); %Current time
curr_G = u(2); %Current measurement
pulse = u(3); %update pulse


if curr_t > T_delay && pulse
% Update measurement data
G = [G(2:end) curr_G];
```

In Simulink® the MATLAB® function block is modified by adding a *Pulse Generator* input shown in Figure 6.30.

The pulse counter is used to give the updating period. In STPPC the updating of the parameters is performed each time a new measurement becomes available. The pulse counter

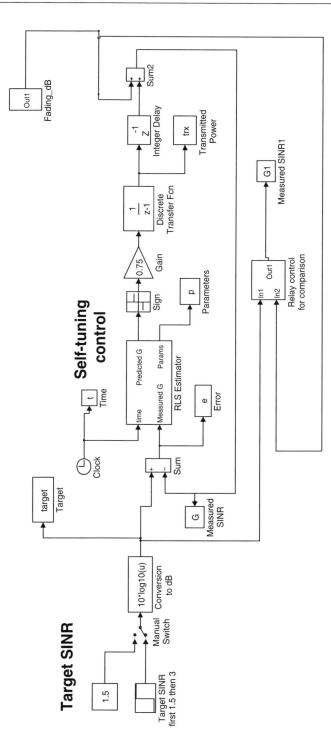

Figure 6.29 Above Simulink® configuration of STPC and relay control below, so that comparison can be performed easier

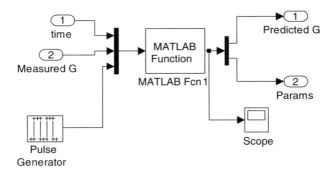

Figure 6.30   Pulse generator has been added to the MATLAB Function input, so as to choose the update period

allows choice of update: each time, waiting for two measurements, three measurements, and so on, before updating. The pulse counter is given integer values: *pulse* = 1,2,3, ...

After simulating the results with a number of different structures, STPC (2,1) and (3,2) gave good results. The updating period was also included as a design parameter. The best results were obtained with updating period = 3. Loss function calculations were used in comparisons of STPC and relay control. The initial learning period was dropped in these and the calculation of the loss function started e.g. after 50 time steps. In the following figures the scales are not always the same, so direct comparison is not straightforward. Another reason for this is that fading will vary each time a new run is generated due to in-built randomness. The main idea is to see that the concept works.

Figure 6.31 displays results for structure (2,1) for a constant target and a slow speed of 5 km/h. The controller follows the target quite well. The corresponding STPC parameters are also shown together with fading. The parameter variation is moderate.

In Figure 6.32 STPC performs better than the relay power control after the initial learning period. The loss function $V = \sum_{100}^{500} e^2(i)$ is 500.0 for STPC and 549.6 for relay control. These figures apply for this run, but STPC is not consistently better in other runs. One feature of STPC is that it has a tendency to learn and improve its performance as time progresses.

Two choices can be tried to improve the consistency of STPC: updating the STPC parameters more infrequently or considering a more complicated structure. If updating is done after three measurements, the situation improves significantly. A typical response is seen in Figure 6.33. A visual comparison of Figures 6.32 and 6.33 shows a better performance of STPC with slower updating.

When the speed increases to 100 km/h, STPC structure (2,1) is still applicable as shown in Figure 6.34, but in different runs it is not on average better than the relay controller (when computing the corresponding loss functions). Observe that there is much more parameter variation, which is expected because the MS speed is faster.

Using less infrequent parameter updating does not change the situation dramatically. Faster MS speed requires a more complex STPC structure, although STPC (2,1) gives satisfactory results.

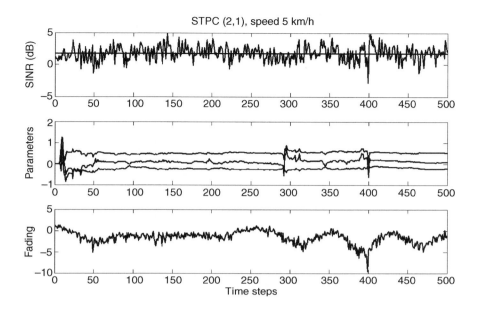

Figure 6.31   Response of STPC, structure (2,1), when the speed is 5 km/h and the SINR target is constant together with the corresponding control parameters and fading

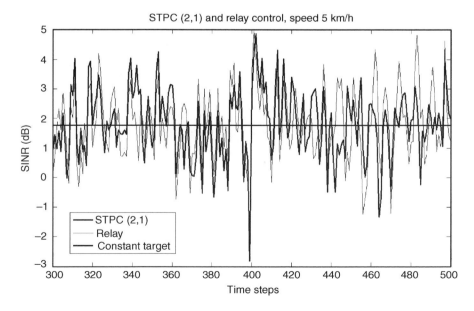

Figure 6.32   Comparison of STPC (2,1) and relay control, when speed 5 km/h and SINR target constant. A typical interval [300, 500] is chosen for closer examination

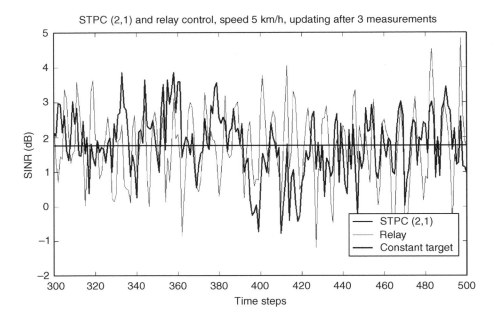

Figure 6.33 Responses of STPC (2,1) and relay power control, when the speed is 5 km/h. STPC parameters are updated after three measurements

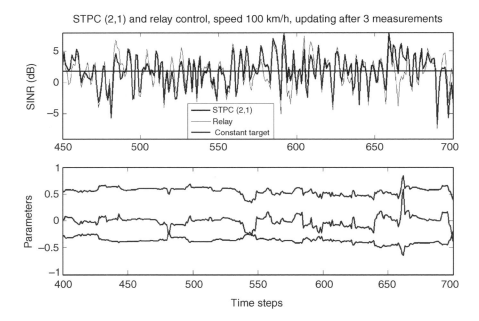

Figure 6.34 Above responses of STPC (2,1) and relay power control, when the speed is 100 km/h and the SINR target is constant. Below: parameters of STPC, which vary more than with slower speed

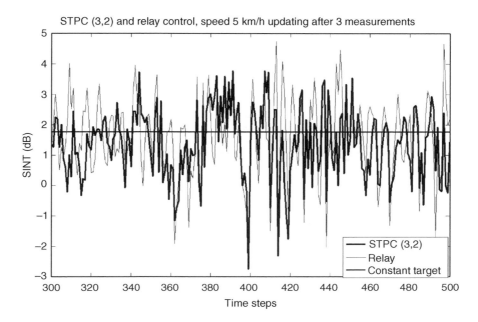

Figure 6.35 Above: responses of STPC (2,1) and relay power control, when the speed is 5 km/h, the SINR target is constant. Updating of STPC parameters is performed after three measurements

Another way to improve the result is to use a more complicated STPC structure (3,2). This proves to be useful especially at higher speeds. Again updating can be delayed, e.g. by three measurements with improved results. STPC (3,2) also works well with slower speeds as seen in Figure 6.35 with updating, although it also works well without.

For the speed of 100 km/h the performance of STPC (3,2) is good. This is depicted in Figure 6.36. Figure 6.37 shows a comparison between STPC (3,2) and the relay controller. For this run STPC gives better results (loss function is smaller than for the relay control). If updating is carried out after three measurements, the results are similar.

Step responses in different cases can also be studied. These do not provide any surprises. They follow the same line of results as before. A couple of examples are given, one for STPC (2,1) and the other for STPC (3,2). Figure 6.38 shows how STPC (2,1) recovers quickly from the step. The parameters change around the time that the step occurs.

Figure 6.39 displays the results for MS speed of 100 km/h.

Finally, it should emphasized that the results are sample runs only and further evidence must be obtained by having a greater number of runs and then computing averages.   □

The simulated results show that significant gains can be made by using more sophisticated controllers. Already the simplest STPC structure (2,1) performs quite well. STPC can also handle servo types of problems, such as step target. In general, STPC requires a relay controller for back-up, in order to guarantee performance if the STPC is unable to function properly.

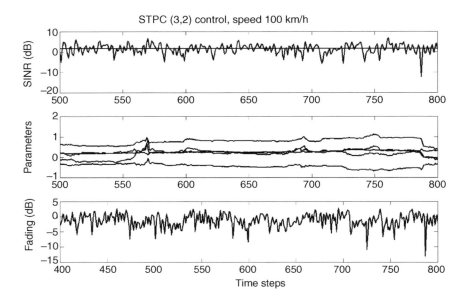

Figure 6.36   Response of STPC (3,2), when the MS speed is 100 km/h, and the SINR target constant. In the middle the corresponding STPC parameters are shown and fading is shown below

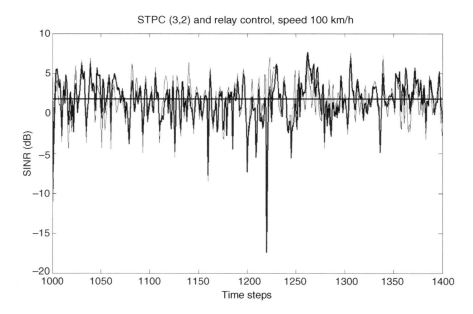

Figure 6.37   Responses of STPC of structure (3,2) and relay power control, when the speed is 100 km/h and the SINR target is constant. Updating is performed after three measurements

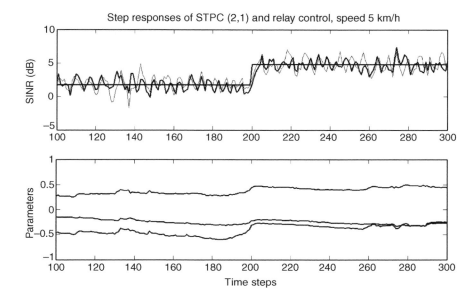

Figure 6.38 Step responses of STPC (2,1) and relay control above, when the SINR target is a step and MS speed is 5 km/h. STPC recovers more quickly from the step and performs better. STPC parameters are shown below

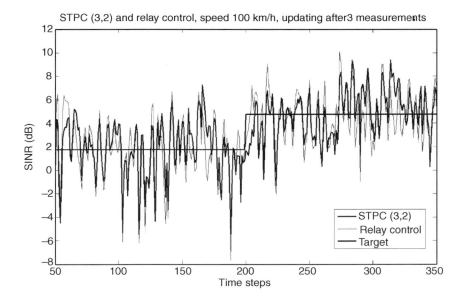

Figure 6.39 Step responses of STPC (3,2) and relay control, when MS speed is 100 km/h and updating is performed after three measurements. STPC works better based via visual observation. This is confirmed performing a loss function comparison

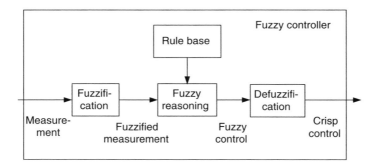

Figure 6.40   Structure of a fuzzy controller

6.7 Fuzzy Power Control

Fuzzy control has proven to be a very powerful method in many applications. It has also been applied to uplink power control. For a beginner, the design of a fuzzy controller may seem complicated at first. Recall from Chapter 2 the structure of the fuzzy controller, which is repeated here for convenience in Figure 6.40.

The controller consists of fuzzification, rule base, fuzzy reasoning, and defuzzification Each of these contains numerous parameters and a number of choices. Therefore the most logical avenue in the design process is to mimic a controller that is simpler and more familiar, that is, a PID controller.

The most crucial step in the design is generating the rule base. To grasp the basic ideas of this, let us generate a typical step response of a closed-loop system. Consider the system in Figure 6.41, where the open-loop plant transfer function in discrete form is

$$\frac{Y(z)}{U(z)} = z^{-2}\frac{0.1142}{z - 0.9048} \tag{6.47}$$

The sampling time $h = 1$ s and $K_p = 2$.

The step response of the closed-loop system is shown in Figure 6.42. Studying this will help us in constructing a rule base for our fuzzy controller.

The error $e(t) = y_{ref} - y(t)$ at time $t = 4$ s is positive (as seen in Figure 6.42). Suppose a human operator at that time needs to decide about the next control action in order to reach the reference value. The information available is the value of the error, but the previous error value is also measured. Therefore error $e(t)$ and difference of the error $\Delta e = e(t) - e(t - 1)$ are used

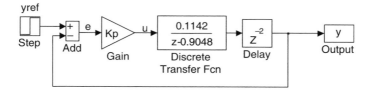

Figure 6.41   A first-order discrete plant with an information delay

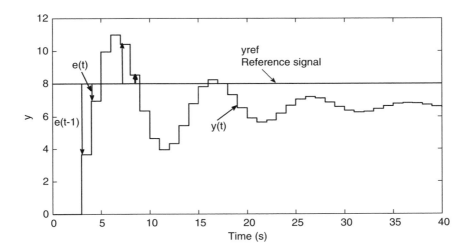

Figure 6.42 Step response of a closed-loop first-order system with a delay

in constructing the rule base. This is the same information used in the discrete PID controller in Chapter 2, Equation (2.29) in a linear manner. Here our fuzzy controller is a PD type controller acting nonlinearly.

Heuristically then, if error $e(t)$ is positive at time $t = 4$ s, as in Figure 6.42, increase the control effort, especially if at the same time the difference in the error $\Delta e(t) = e(4) - e(3)$ is negative. *Negative* and *positive* are now interpreted as fuzzy sets. Since the response is going in the right direction, only a moderate change in $u(t)$ should be made. If $e(t)$ is *positive* and $\Delta e(t)$ is *positive*, as is the case when $t = 8$ s then the response is going in the wrong direction and therefore $u(t)$ should be 'large'.

The following simple rule base, which follows velocity form of PI controller in Equation (2.35), is constructed

$$
\begin{aligned}
&- \; \textit{If } e(t) \textit{ is positive and } \Delta e(t) \textit{ is positive, then } \Delta u(t) \textit{ is positive} \\
&- \; \textit{If } e(t) \textit{ is positive and } \Delta e(t) \textit{ is negative, then } \Delta u(t) \textit{ is zero} \\
&- \; \textit{If } e(t) \textit{ is negative and } \Delta e(t) \textit{ is positive, then } \Delta u(t) \textit{ is zero} \\
&- \; \textit{If } e(t) \textit{ is negative and } \Delta e(t) \textit{ is negative, then } \Delta u(t) \textit{ is negative}
\end{aligned}
\tag{6.48}
$$

Positive, negative and *zero* are all fuzzy sets. Further fuzzy sets can be used, e.g. *no change* for $\Delta e(t)$. If further sets are added, then the rule base consists of more rules.

Recall that fuzzy control design begins by typing *fuzzy* on the MATLAB® Command window. From the Edit menu choose *Add Variable* to the input side. Then set up the fuzzy sets for the input and output variables. A rough design produces an initial design, which uses three sets for each of the variables: error e; difference of the error Δe; and output u. The range of the variables is determined by studying how each of the variables changes when simple feedback control is applied. Normalization of variables is advisable, although not applied here. Triangular symmetric membership functions are chosen.

Figure 6.43 shows what the system looks like – two inputs and one output. The design then proceeds by opening one of the variable blocks. In Figure 6.44 the variable is error e. The range

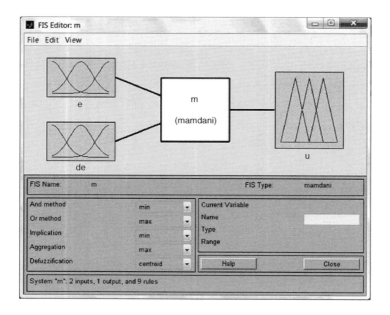

Figure 6.43 The FIS Editor. Two input variables e and de (Δe) are seen at the input and u at the output. The final result is exported to file m

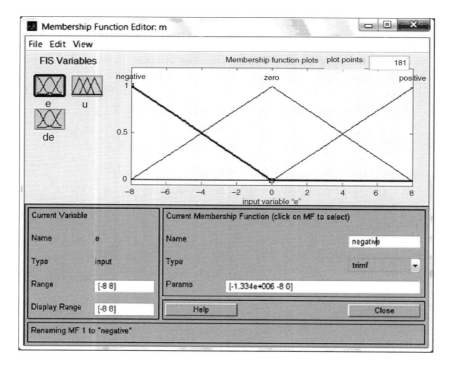

Figure 6.44 The Membership Function Editor is shown for input e. Three triangular membership functions are chosen: negative, zero, and positive. The range of the input variable is $[-6, 6]$

of the variable, number of membership functions and their type are determined. The fuzzy sets (=membership functions) of *e* are given linguistic names. The details are depicted in Figure 6.44.

Once the variables and their membership functions have been defined, the determination of the rule base begins. Choose *Edit* from the menu and then *Rules*. Adding *zero* membership function to both *e* and Δe expands the earlier rule base of Equation (6.48) to the following

1. *If $e(t)$ is positive and $\Delta e(t)$ is positive, then $\Delta u(t)$ is positive*
2. *If $e(t)$ is positive and $\Delta e(t)$ is zero, then $\Delta u(t)$ is positive*
3. *If $e(t)$ is positive and $\Delta e(t)$ is negative, then $\Delta u(t)$ is zero*
4. *If $e(t)$ is zero and $\Delta e(t)$ is positive, then $\Delta u(t)$ is zero*
5. *If $e(t)$ is zero and $\Delta e(t)$ is zero, then $\Delta u(t)$ is zero*   (6.49)
6. *If $e(t)$ is zero and $\Delta e(t)$ is negative, then $\Delta u(t)$ is zero*
7. *If $e(t)$ is negative and $\Delta e(t)$ is positive, then $\Delta u(t)$ is zero*
8. *If $e(t)$ is negative and $\Delta e(t)$ is zero, then $\Delta u(t)$ is negative*
9. *If $e(t)$ is negative and $\Delta e(t)$ is negative, then $\Delta u(t)$ is negative*

This is displayed in Figure 6.45.

The rule base can be viewed with a Rule Viewer, but it is more instructive in this case to use the Surface Viewer of Figure 6.46. This shows that the control surface is slightly nonlinear and its present form has not been optimally designed (look at the surface boundaries). This kind of view provides a basis for improving the design, e.g. so that the surface could be smoother. It should, however, perform adequately.

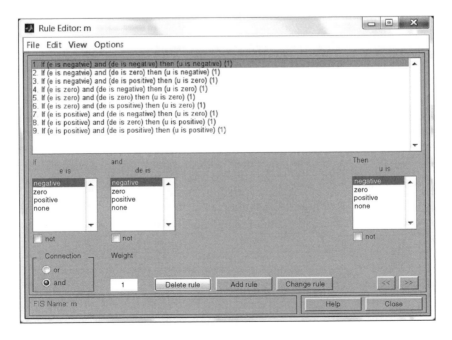

Figure 6.45 The Rule editor shows the final rule base in linguistic form

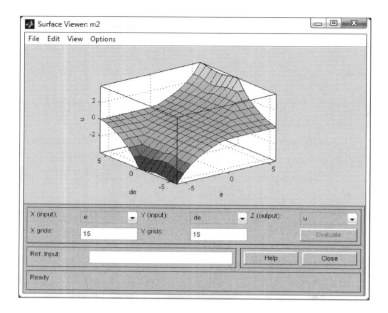

Figure 6.46 The Surface Viewer of the controller: the x-axis corresponds to e, the y-axis to de and the z-axis to u

Note that the fuzzy controller we have designed needs to have $u(t)$ at the output. Our rule base produces $\Delta u(t)$ at the output. This is fixed by the following observation (this is similar to the treatment in Equation (6.8)):

$$\Delta u(k) = u(k) - u(k-1) = u(k) - z^{-1}u(k) = (1 - z^{-1})u(k) \qquad (6.50)$$

or

$$u(k) = \frac{1}{1 - z^{-1}} \Delta u(k) \qquad (6.51)$$

Call the developed file m and *Export* it to a *File* (or *Workspace*).

We are now ready to study fuzzy power control using the same example as before.

Example 6.6   Fuzzy power control

Consider the uplink control model in Figure 6.21 with fuzzy control. Repeat Example 6.1 with the same data values and Jakes' fading model. Study the system behavior using Simulink®.

Solution

The fuzzy power control system is displayed in Figure 6.47. The Fuzzy logic controller block is found in the Simulink® block library under Fuzzy Logic Toolbox. The FIS file m developed

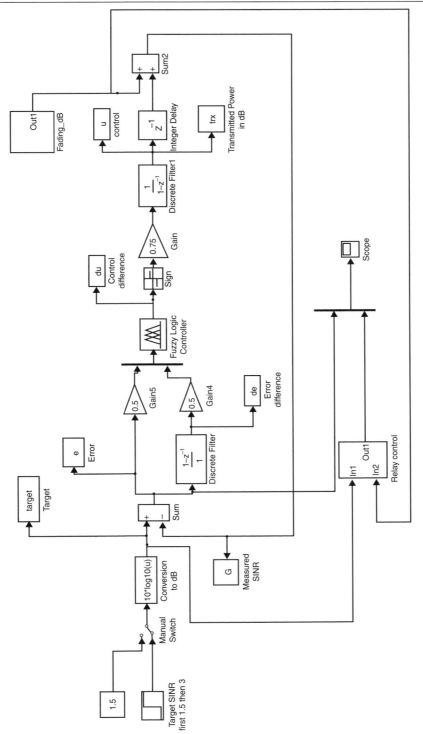

Figure 6.47 Fuzzy logic power control system

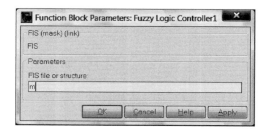

Figure 6.48 The Fuzzy Logic Controller block in SIMULINK. The name of the FIS file developed in the Fuzzy Logic Toolbox Editor is *m*. When the simulation is carried out, this FIS file has to be *Imported* from *File* with the FIS Editor and then *Exported* to *Workspace*

above is used in the block. The connection from FIS Editor to Simulink® Fuzzy logic controller block is achieved by opening this block. The result is shown in Figure 6.48.

Observe that the signum function is added after the fuzzy logic controller, and the transfer function (Equation (6.51)) replaces the previously used transfer function of Equation (6.9). If you *clear* the workspace between simulations, you need to *Export* FIS file *m* for each simulation.

Simulation results are depicted in Figure 6.49 for speed of 5 km/h and in Figure 6.50 for speed of 100 km/h. In both cases fuzzy control performs well. In the latter case fading is also shown.

All other cases considered also give very good results when FLPC is applied.

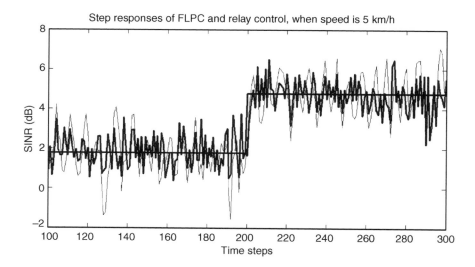

Figure 6.49 Step responses of FLPC and relay control (speed is 5 km/h and SINR target is a step). FLPC is superior

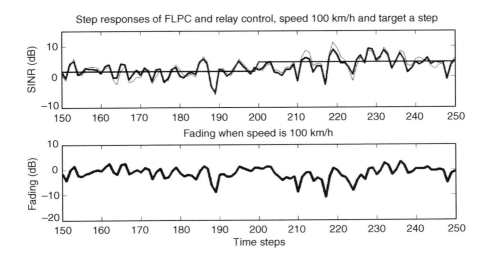

Figure 6.50   Above: step responses of FLPC and relay control (speed is 100 km/h and SINR target is a step). Below: fading

6.8 Handover

Handover and power control are closely tied together as essential parts in radio resource management of cellular systems. In radio resource management the first thing for the user *MS* is to be admitted to the network. After a request the network management refers the user to the *admission controller*. This checks the load in the network from the *load controller* and admits the user if the load is nominal. Admission and load control are discussed in more detail in Chapter 7.

Once the user is admitted, he is then referred to *power control*, where the *MS* is assigned a *BS*, a pair of channels (uplink and downlink), and transmitter powers for both the *BS* and the *MS*. The quality of transmission has to be satisfactory and this is constantly monitored by the power control as we have seen in this chapter. Since the *MS* typically moves around, it may cross to a different cell, where another *BS* will then take over. The transfer of the *BS* control is called *handover control*. If the *MS* leaves the home network, it enters a foreign network. The *MS* registers with the home network via the foreign agent and in this manner the home network has knowledge of the *MS* location.

Many reasons for handover have been identified. All of them cannot be listed here. One case arises when an MS moves out of range of a particular BS. The received signal weakens and the QoS set by the RNC drops below the minimum requirement. Another reason is due to heavy traffic in the cell. Then the MS is moved to another cell with less traffic load.

There are a number of different handover scenarios in a cellular system. Here we concentrate on *inter-cell, intra-BS*, which was described above. The basic issue in handover control is as an MS enters the coverage area of a new BS, when to establish a new connection and release the old one. The handover should not cause a cut-off, which is also called *call drop*. Typical handoff mechanisms are *hard handoff*, *soft handoff* and *softer handover*.

A handover can be divided into three phases: *measurement phase*, *decision phase*, and *execution phase*. In the measurement phase, the link quality is observed using the pilot signal strength, but also the signal strength of the neighboring cells is observed. These are reported to the RNC in the WCDMA system. Concentrating on the pilot signal only, if it becomes too weak, a threshold is reached at which point the change to the decision phase is made. In the decision phase the decision of the handoff is made by the RNC. This is natural, since it has the knowledge of the current state of the network. It receives the information from different MS and BS. The RNC uses different handover algorithms. Once the decision has been made execution follows by informing the new BS of the new resource allocation.

Hard handoff: in hard handoff the decision of releasing the old link before establishing a new one is abrupt. When the signal quality requirement falls below the threshold, a handover procedure is immediately initiated. Mathematically this implies that a relay is applied in decision making. When the MS is close to the border of two cells, hard handoff can easily lead to the so-called ping-pong effect causing exceedingly many switches from one BS to another. This creates unnecessary signaling and a severe load in the network. The following example illustrates hard handoff.

Example 6.7   Hard handoff

Consider a simple example of hard handoff shown in Figure 6.51. The MS is moving along a straight line from BS_1 to BS_2. It is first controlled by BS_1. The distance between the base stations is $d = d_1 + d_2 = 500$ m. Initially the distance of the MS to BS_1 is $d_0 = 10$ m (and to BS_2, $500 - d_0 = 500 - 10$ m). The base stations are assumed identical.

(a) Assume a large-scale path loss (Equation (3.9)), where $n = 3$ and $\alpha = 1$. Let the speed of the MS be $v = 23$ m/s. The handoff decision is based only on the signal strength. This means that the BS, whose signal is the strongest, is the winner. Set up a Simulink® model for the hard handoff of the system. Simulate and plot both received signal strengths (in dB) and the result of the handoff decision.
(b) Repeat part (a) when shadow fading is added.

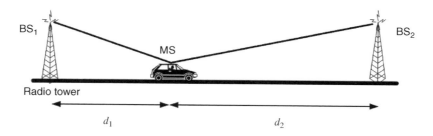

Figure 6.51   The base stations BS_1 and BS_2 use the same channel to transmit their signals. No shadowing is assumed

Solution

(a) Figure 6.52 displays the Simulink® configuration. From left to write, first distance d is formed, $d = vt + d_0$ (m). For BS_1 this becomes $= 23*t + 10$ and for $BS_2 = 23*t + (500 - 10)$. The information is fed to the parallel path-loss blocks. Path loss is assumed to be proportional to d^3. In the simplest handoff strategy the received signal strengths are compared by forming their difference and feeding it to the *Sign*-function block. If the difference is positive, then BS_1 wins. The handoff decision is then $+1$. If the difference is negative, then BS_2 is the winner and the decision is -1.

   The simulated signals are displayed in Figure 6.53 and the corresponding handoff decisions in Figure 6.54.

(b) In logarithmic scale, shadowing becomes a normally distributed variable. Figure 6.55 shows the required changes in the Simulink® configuration. Change the seed number in one of shadowing blocks so that a different random variable is generated. Either a Random Number or Band-Limited White Noise block suffices for this sort of qualitative study, where the main interest is in the handoff.

The simulation results corresponding to part (a) are displayed in Figures 6.56 and 6.57.

Figure 6.55 gives the same result for the shorter (BS_1) and larger distances (BS_2) as does the result in Figure 6.53. In the middle there is an area where the straightforward handover rule does not produce very good results. The ping-pong effect results. The handover decision function is displayed in Figure 6.57.

Softer handover: in soft handover the control is not immediately switched from one BS to another as in hard handover. Once a connection to a new BS is established, the old connection of the MS still remains intact for a while. With clever handoff algorithms, the ping-pong effect seen in Figure 6.57 can be significantly reduced.

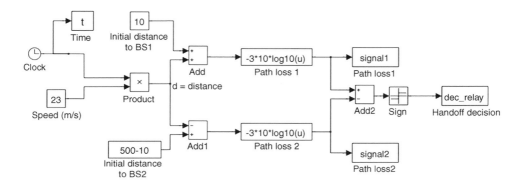

Figure 6.52 Simulink® configuration for hard handoff. The decision is formed by feeding the signal strength difference to a *sign*-function

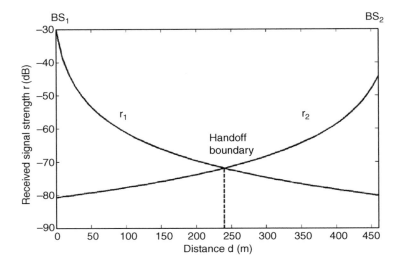

Figure 6.53   The received signals $\mathbf{r}_i$ are shown as a function of the distance d. When distance d is small, $\mathbf{r}_1 > \mathbf{r}_2$ and MS remains connected to BS_1. At $d = 240$ m, $\mathbf{r}_1 = \mathbf{r}_2$ and immediately when $\mathbf{r}_1 < \mathbf{r}_2$ decision is made to switch the control to BS_2. Finally, the decision function is given in Figure 6.54

Example 6.8   Soft handoff

Repeat Example 6.7(b), where the shadowing created a problem of excessive switching in the handoff decision. Apply hysteresis as the handoff decision algorithm. Simulate the result and compare especially the decision variable with that in Figure 6.57.

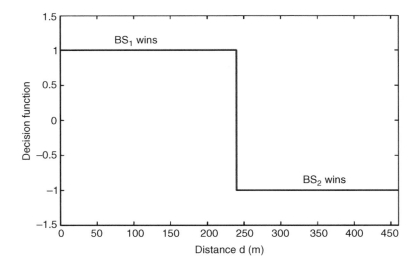

Figure 6.54   The decision in this trivial case is easy. When the MS is close to BS_1, the connection remains with it. When the distance becomes shorter to BS_2, it will receive the control of MS

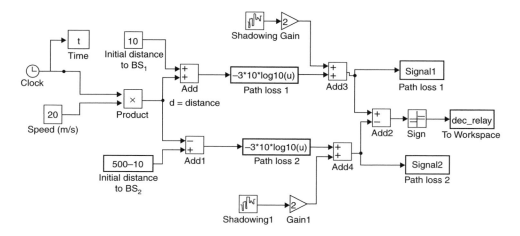

Figure 6.55 Simulink® configuration, where shadowing has been added to both links. Remember to change the seed number, so that they are not the same in both blocks

Solution

The Simulink® configuration is exactly the same as in Figure 6.55. Replace *sign*-function with a hysteresis. This in Simulink® is implemented with the *relay* block. The block parameters are changed, so that *eps* in *Switch on point* is replaced with 5 and *eps* in *Switch off point* with −5. In

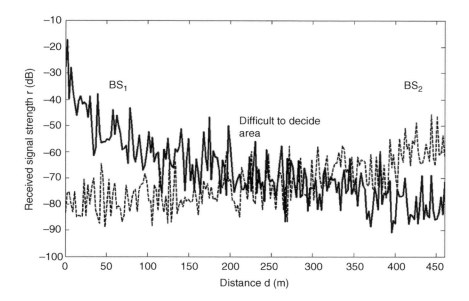

Figure 6.56 Shadowing makes the handoff decision more difficult. If the same handoff rule is applied, it will lead to too many, unnecessary switches

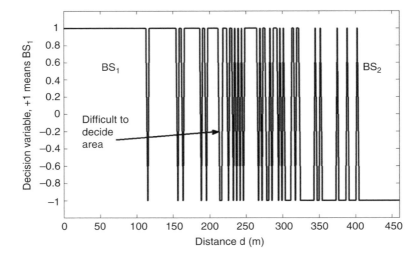

Figure 6.57 The handover decision becomes more complicated, when shadowing is introduced. Again at both ends there are not many switches, but in the middle, the decision making is difficult resulting in many switches and ping-pong effects. Here $+1$ implies control of BS_1 and -1 control of BS_2

addition, use -1 in output when off. In order to make comparisons, keep also the *Sign*-function. The change is shown in Figure 6.58.

The result of comparing the decision variables is depicted in Figure 6.59. It clearly shows how the hysteresis will decrease the switching and therefore the handoff probability/rate is significantly improved.

Softer handoff: this is a handoff in which the existing radio link is replaced by a new radio link and the old one deleted within the different sectors of the same cell. Softer handover is not discussed any further here.

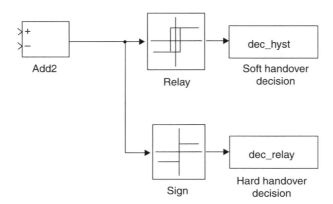

Figure 6.58 Hysteresis (represented by *Relay* block) is used for soft handoff and Sign function block for hard handoff

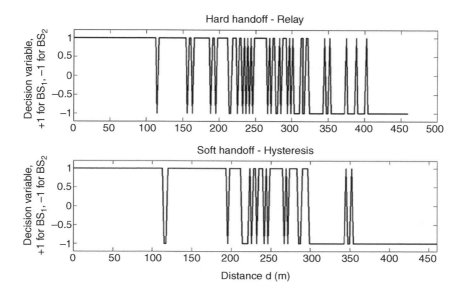

Figure 6.59   Comparison of the decision variables in the hard handoff case, when Sign function is used, and in the soft handoff case, when Relay with hysteresis is applied. Note the reduction in the number of switches in the soft handoff case

There are numerous other more sophisticated handoff algorithms in the literature. Often they involve some form of prediction of how the situation will develop. Fuzzy logic has been applied in this context.

In this section the objective has been to illustrate only the basic concepts of handoff, hard handoff and soft handoff. Simulink® provides a transparent way of doing this. Handoff algorithms also benefit from recent developments of GPS and digital maps.

When radio resource management is considered, better situation awareness can lead to more efficient power control. When load or transmission changes, *gain scheduling* offers a way to improve radio resource management. This means a change in control parameters, or even changing to different controller completely. The basic relay control offers a good backup controller, if performance with more sophisticated controllers becomes unacceptable.

6.9 Summary

In this chapter dynamic power control using various algorithms has been studied. The basic relay, one-bit controller works reasonably well, but use of two bits allows a dead-zone controller, which shows better performance. Each of the presented control algorithms has its advantages, but none works uniformly well in all situations.

When radio resource management is considered, better situation awareness can lead to more efficient power control. When load or transmission changes, *gain scheduling* offers a way to improve radio resource management. This means a change in control parameters, or even changing to different controller completely. The basic relay control offers a good backup controller, if performance with more sophisticated controllers becomes unacceptable.

The chapter concludes with a look at handoff in a cellular system. The basic mechanisms, hard handover and soft handover are discussed and illustrated with simulations.

Exercises

6.1 Set up uplink control system of Example 6.1 shown in Figure 6.7.
   (a) Study the effect of step size on the result. Try 0.5 dB, 1 dB and 1.5 dB.
   (b) Study the effect of Rayleigh fading and shadowing on received signal. In Figure 6.7 the gain value of shadowing is 0.1 and that of fading 0.5. Set both gain values to 0.1, 0.5 and 1.
   (c) BER probability P_t is assumed to have value 0.01. Study how its increase up to 0.1 affects the performance.

6.2 When a relay with dead zone (Equation (6.10)) is applied as A PC algorithm in the system shown in Figure 6.14, one of the key tuning parameters is the dead-zone width, K. Study its effect on the control performance by considering first small values of K, which are below 2 and then increasing its value. Take into account both slow, pedestrian type of speed 5 km/h and fast automobile speed of 100 km/h.

6.3 A PID controller can serve as an alternative controller in wireless communication. Consider the system in Figure 6.18. Ignore cost function calculations. Try to set up good parameter values for the PID controller. Gain of the integral controller is set $= 0$. The derivative gain is set to a small value. Start with proportional gain that is below one. Increase the derivative gain only slightly. Experiment with control parameters to understand how difficult it is to determine two control parameters in a satisfactory manner.

6.4 Example 6.3 introduces a powerful predictive power control. Set up the system in Figure 6.22. The structure of the predictor contains parameters determining its structure: n_g and n_c.
   (a) Test different structures of the predictor starting with $n_g = 2$ and $n_c = 1$. Is it beneficial to have a high order structure?
   (b) Test both low speed case of 5 km/h and high speed case of 100 km/h.
   (c) Study how the predictor parameters vary. If they converge and do not vary much after a while, test the resulting constant values as control parameters.

6.5 Repeat the structure examination of the self-tuning controller of Example 6.4. Investigate also the use of the updating period. Test both low speed case of 5 km/h and high speed case of 100 km/h.

6.6 Set up the fuzzy PC control system in Figure 6.48.
   (a) Experiment with scaling gains (in the figure they are both 0.5).
   (b) Investigate the effect of the step size, try 0.5, 1.0 and 1.5.
   (c) Study (a) and (b) when the speed is 5 km/h and 100 km/h.

6.7 In Chapter 4 the basics of neural network predictors was presented. Formulate a neural network predictor and use it in the control loop analogously to Example 6.3. Implement and simulate it.

7

Admission and Load Control

7.1 Introduction to Admission Control (AC)

Admission control (AC) is one of the important functions of radio resource management. Its main task is to accept or reject (deny) any new call or service request or even any higher requirement request (e.g., higher throughput) of current active terminals. When a new service or call is admitted to the network, it will need to use some of the network resources in the uplink as well as in the downlink. The AC should accept the new request if there are enough resources to handle at least its minimum requested QoS. Otherwise it may cause network congestion. When the network is congested, it may become unstable and several connections could be dropped. Load control is used to handle congestion as will be described later. Another important objective of the AC is to guarantee the continuity of current active connections. In other words the new call request should not be accepted if it degrades the performance of one or more of the active connections to be less than their minimum QoS requirements. The AC function is within the responsibility of the network layer (third layer of the OSI seven-layer model) and it should be done by interaction with power and rate control which is part of the link layer (second layer of OSI). This indicates that AC is one application of cross-layer optimization. The AC algorithm should be designed very carefully, otherwise the overall network performance could be considerably degraded. There are two types of decision errors that can be made by the AC algorithm:

1. *Accepting error:* the AC algorithm decides to accept a new service or call when it should be rejected. In this case the QoS of at least one current active connection will be degraded or even dropped. If the accepting error rate (AER) is high the network will be frequently congested. This error appears when the AC underestimates the required resources by the new request.
2. *Rejecting error:* the AC algorithm decides to reject a new service or call when it should be accepted. If the rejecting error rate (RER) is high the blocking rate of the new calls will be high and this reduces the performance and utilization of the network. Therefore, the

Systems Engineering in Wireless Communications   Heikki Koivo and Mohammed Elmusrati
© 2009 John Wiley & Sons, Ltd

operator will receive lower revenue and may lose its reputation. When the blocking rate of a certain operator is very high the customers may start to migrate to other operators (churn behaviour). The network must be optimized to work with full efficiency and this can be done by utilizing all network resources without congestion or losing the QoS. One reason for the RER is the overestimation of the resources required by the new request.

Perfect AC means that AER = 0 and RER = 0. There are many difficulties in obtaining the perfect AC algorithm. Some of these are:

1. Generally it is not possible to know precisely the impact of the new service or call on the network before it is accepted. For this reason the AC may predict its impact on the network and then make a decision based on that prediction. The AC may also temporarily accept the new request (for a test period) to evaluate its impact on the network. During the test period only probing packets are transmitted to and from the new terminal. After this testing period the AC may estimate more precisely the required resources of the new service request and this will reduce the probability of decision error. One problem of this technique is that it may reduce the efficiency of the network because some of the network resources will be reserved during the test period.
2. The wireless network is usually dynamic which makes future estimation more difficult. For example, if the channel quality for a certain active connection becomes bad (e.g. because of shadowing or fading), then this terminal will need more network resources (such as downlink power) to maintain its target QoS requirement. This means that the available resource is a random process depending on wireless channel fluctuations, network operations, and end-user requirements.
3. In order to make reliable decisions, the AC needs to have accurate measurements about the interference, noise, available resources (such as DL power), and so on. It also needs to know the statistical properties of these parameters to be able to accurately predict their possible future values. These measurements are not always easy to obtain especially if the channel has low time correlation.

When scalable QoS is not required, i.e., QoS is provided based on the best effort, then there is no need to use AC algorithms. In this case no service is denied because of the shortage of network resources. However, increasing the network load degrades the QoS of active terminals rapidly. Such systems are usually not suitable for real-time applications such as online voice and video chatting, because a certain QoS cannot be guaranteed. When a customer signs a contract with the mobile operator, he expects some QoS will be guaranteed for his calls. Conventional customers usually do not care about the detailed QoS parameters, however, they may use fuzzy measures to assist the network performance. Some of these fuzzy measures are: the call is usually dropped (or lost) when I am in my car; the download speed is very slow; the sound quality is not good, and so on. The operator should optimize its network parameters to satisfy its subscribers in terms of service qualities as well as costs. The AC is one of the important network entities that should be carefully optimized. The AC checks different available resources as well as the requested QoS before it makes the decision to accept or reject the new service request. These resources include DL power, bandwidth, maximum possible throughput, buffer size, user rank (or priority) and current active connections.

7.2 Theoretical Analysis of Centralized Admission Control

In this section we will discuss the main theoretical background of the AC. We will introduce the problem from a purely theoretical point of view. However, this type of analysis is not necessarily applicable in practice. The reason for its limited applications is that we need to consider some assumptions to be able to have perfect AC decisions. These assumptions may not be possible to justify in all real systems and situations.

At a certain time instant assume Q active users with different QoS requirements simultaneously use the network resources and the network is feasible (i.e., every terminal achieves at least its minimum QoS requirements).

In the uplink for all $i = 1, 2, \ldots, Q$ we have

$$\frac{P_i G_{ii}}{\sum_{\substack{j=1 \\ j \neq i}}^{Q} P_j \theta_{ij} v_j G_{ij} + N_i} \geq \Gamma_i^{\min} \quad \text{and} \quad 0 < P_i \leq P_{\max} \tag{7.1}$$

The parameters are defined in Chapter 5. As we showed in Chapter 5, the QoS of the active connection is determined by its $\Gamma_i^{\min}$. Without loss of generality we will assume that the voice activity factor $v_j = 1$ unless something else is specified. When $P_i = 0$, then the ith terminal is inactive. Since the network is feasible, then at least the following two conditions have been satisfied:

1. The spectral radius $\rho(\mathbf{\Gamma H}) < 1$, where $\mathbf{H}$ is the normalized channel matrix (see Chapter 5) and

$$\mathbf{\Gamma} = \begin{bmatrix} \Gamma_1^{\min} & 0 & \cdots & 0 \\ 0 & \Gamma_2^{\min} & & 0 \\ \vdots & & \ddots & \vdots \\ 0 & 0 & \cdots & \Gamma_Q^{\min} \end{bmatrix}.$$

2. The transmit power of every terminal is less than the maximum allowed, i.e., $\max(\mathbf{P}) \leq P_{\max}$, where $\mathbf{P} = [P_1, P_2, \cdots, P_Q]$.

For the downlink part we have

$$\frac{\hat{P}_i \hat{G}_{ii}}{\sum_{\substack{j=1 \\ j \neq i}}^{Q} \hat{P}_j \hat{\theta}_{ij} \hat{G}_{ij} + \hat{N}_i} \geq \hat{\Gamma}_i^{\min} \quad \text{and} \quad \sum_{i=1}^{Q} \hat{P}_i \leq \hat{P}_{\max} \tag{7.2}$$

where 'hats' refer to the downlink. In both formulations (Equations (7.1) and (7.2)), we have assumed that other networks' interferences are included in the additive noise part. At least the following two conditions must be satisfied for a feasible network (from the downlink point of view):

1. The spectral radius $\rho(\hat{\mathbf{\Gamma}}\hat{\mathbf{H}}) < 1$, where $\mathbf{H}$ is the normalized channel matrix and

$$
\hat{\mathbf{\Gamma}} = \begin{bmatrix} \hat{\Gamma}_1^{\min} & 0 & \cdots & 0 \\ 0 & \hat{\Gamma}_2^{\min} & & 0 \\ \vdots & & \ddots & \vdots \\ 0 & 0 & \cdots & \hat{\Gamma}_Q^{\min} \end{bmatrix}.
$$

2. The total required downlink power should be less than the total available power at the base station, i.e., $\sum_{i=1}^{Q} \hat{P}_i \leq \hat{P}_{\max}$.

In the text we sometimes use DL and UL instead of downlink and uplink, respectively.

Example 7.1

Represent Equation (7.2) for a single-cell case.

Solution

In a single-cell case the SINR at the terminal input is

$$
\frac{\hat{P}_i}{\sum\limits_{\substack{j=1 \\ j \neq i}}^{Q} \hat{P}_j \hat{\theta}_{ij} + \hat{N}_i/\hat{G}_{ii}} \geq \hat{\Gamma}_i^{\min} \quad \text{and} \quad \sum_{i}^{Q} \hat{P}_i \leq \hat{P}_{\max}.
$$

Let's define $\hat{P}_i = \alpha_i \hat{P}_{\max}$, $0 \leq \alpha_i \leq 1$, and $\sum_{i=1}^{Q} \alpha_i = 1$, then the above equation becomes (assuming $\hat{\theta}_i = \theta \, \forall i$)

$$
\frac{\alpha_i}{\theta(1 - \alpha_i) + \bar{N}_i} \geq \hat{\Gamma}_i^{\min} \tag{7.3}
$$

where $\bar{N}_i = \hat{N}_i/(P_{\max}\hat{G}_{ii})$.

Now assume that a new terminal requests to be accepted in the network. The AC needs to check the impact of this new terminal on the network (both uplink and downlink). Generally the new request can be accepted if the following four conditions are achieved:

1. $\rho(\mathbf{\Gamma}_{Q+1}\mathbf{H}_{Q+1}) < \varepsilon$, where $\mathbf{H}_{Q+1}$ is the new normalized channel matrix when the new

   terminal is added, and $\mathbf{\Gamma}_{Q+1} = \begin{bmatrix} \mathbf{\Gamma}_1^{\min} & 0 & \cdots & 0 \\ 0 & \mathbf{\Gamma}_2^{\min} & & 0 \\ \vdots & & \ddots & \vdots \\ 0 & 0 & \cdots & \mathbf{\Gamma}_{Q+1}^{\min} \end{bmatrix}$. $0 < \varepsilon < 1$ is the uplink

   protection value.
3. $\max(\mathbf{P}) \leq P_{\max}$, where $\mathbf{P} = [P_1, P_2, \cdots, P_Q, P_{Q+1}]$.
4. $\rho(\hat{\mathbf{\Gamma}}_{Q+1}\hat{\mathbf{H}}_{Q+1}) < \hat{\varepsilon}$, where $0 < \hat{\varepsilon} < 1$ is the downlink protection value.
5. $\sum_{i=1}^{Q+1} \hat{P}_i \leq \hat{P}_{\max}$

The above conditions assume a static environment, which is difficult to justify in cellular systems. To include random fluctuations of the channel matrix $\mathbf{H}$, we have used a protection value in both uplink and downlink. The main purpose of this protection value is to ensure the feasibility and stability of the cellular system even when some changes (regarding channel gains, noise levels, interferences, etc.) have occurred for the current active connections. Selecting ε as very small will ensure the feasibility and stability of the cellular system but at the expense of low system capacity and high service rejection rate and vice versa.

For random dynamic channels, it is possible to make admission decisions based on probabilistic models. For example, condition 3 is passed if $P_r(\rho(\hat{\mathbf{\Gamma}}_{Q+1}\hat{\mathbf{H}}_{Q+1}) > \hat{\varepsilon}) \leq 0.01$. This concept can be applied to all four conditions. However, we will not discuss the probabilistic models in this chapter.

Example 7.2

Assume a single cell with three mobile phones as shown in Figure 7.1. The channel gains (assumed to be the same for UL and DL) are $G_1 = -80dB$, $G_2 = -100dB$, and

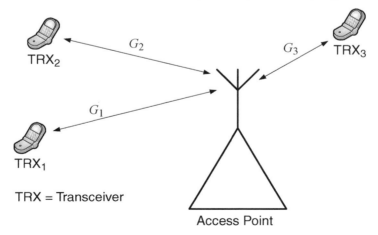

Figure 7.1 Single-cell environment for Example 7.2

$G_3 = -70dB$. The target SINR for every connection (the same in uplink and downlink) are $\Gamma_1^{tar} = -8dB$, $\Gamma_2^{tar} = -5dB$, and $\Gamma_3^{tar} = -4dB$. The minimum required SINR for all connections (the same in uplink and downlink) is $\Gamma_i^{min} = -10dB \ \forall i = 1, 2, 3$. The additive noise average power is -80 dBm for both uplink and downlink. Assume a unity orthogonality factor for the uplink and downlink. The maximum power of the base station is 1 W. The maximum transmit power of the terminals is 0.2 W.

(a) Check the feasibility of current system.
(b) A new terminal (with $G_4 = -90dB$) requests to get admitted to the network with target QoS represented by $\Gamma_4^{tar} = -3.5dB$, and minimum QoS $\Gamma_4^{min} = -7dB$. Should the AC accept it?

Solution

(a) We can check the feasibility by calculating the required uplink and downlink transmit power as well as the spectral radius.

The required transmit power to achieve the target QoS is (see Chapter 5)

$$\mathbf{P}_{UL} = \begin{bmatrix} P_1 \\ P_2 \\ P_3 \end{bmatrix} = \left[\begin{bmatrix} 1 & 0 & 0 \\ 0 & 1 & 0 \\ 0 & 0 & 1 \end{bmatrix} - \begin{bmatrix} 10^{-0.8} & 0 & 0 \\ 0 & 10^{-0.5} & 0 \\ 0 & 0 & 10^{-0.4} \end{bmatrix} \begin{bmatrix} 0 & 10^{-2} & 10 \\ 10^2 & 0 & 10^3 \\ 10^{-1} & 10^{-3} & 0 \end{bmatrix} \right]^{-1}$$

$$\times \begin{bmatrix} 10^{-0.8} \times 10^{-11}/10^{-8} \\ 10^{-0.5} \times 10^{-11}/10^{-10} \\ 10^{-0.4} \times 10^{-11}/10^{-7} \end{bmatrix} = \begin{bmatrix} 0.4 \\ 71 \\ 0.084 \end{bmatrix} mW$$

We can see that all three terminals' power values are less than the maximum possible power. From the uplink point of view the system is feasible and stable.

From the downlink point of view

$$\mathbf{P}_{DL} = \begin{bmatrix} P_1 \\ P_2 \\ P_3 \end{bmatrix} = \left[\begin{bmatrix} 1 & 0 & 0 \\ 0 & 1 & 0 \\ 0 & 0 & 1 \end{bmatrix} - \begin{bmatrix} 10^{-0.8} & 0 & 0 \\ 0 & 10^{-0.5} & 0 \\ 0 & 0 & 10^{-0.4} \end{bmatrix} \begin{bmatrix} 0 & 1 & 1 \\ 1 & 0 & 1 \\ 1 & 1 & 0 \end{bmatrix} \right]^{-1}$$

$$\times \begin{bmatrix} 10^{-0.8} \times 10^{-11}/10^{-8} \\ 10^{-0.5} \times 10^{-11}/10^{-10} \\ 10^{-0.4} \times 10^{-11}/10^{-7} \end{bmatrix} = \begin{bmatrix} 9.9 \\ 41.2 \\ 20.4 \end{bmatrix} mW$$

It is clear that the DL is also feasible and stable in this system.
(b) When the AC receives the new request, it should evaluate the impact of the acceptance of the new service. This can be done by following the previous four checking conditions.

Let's follow the previous four steps to assist the new service:

1. The feasibility test for UL:

$$\rho(\mathbf{\Gamma H}) = \rho\left(\begin{bmatrix} 10^{-0.8} & 0 & 0 & 0 \\ 0 & 10^{-0.5} & 0 & 0 \\ 0 & 0 & 10^{-0.4} & 0 \\ 0 & 0 & 0 & 10^{-0.35} \end{bmatrix} \begin{bmatrix} 0 & 10^{-2} & 10 & 10^{-1} \\ 10^2 & 0 & 10^3 & 10 \\ 10^{-1} & 10^{-3} & 0 & 10^{-2} \\ 10 & 10^{-1} & 10^2 & 0 \end{bmatrix}\right) = 0.96 < 1$$

This means that the system is feasible but since the spectral radius is close to one, the system will not be robust for possible changes in channel gains or other system parameters. However, since there is no specified protection value in this example, the AC should not reject this service because of that.

2. Transmit power test: Now we calculate the required transmit power of every terminal in the UL: the required transmit power vector is $\mathbf{P}_{UL} = \begin{bmatrix} 4.6 & 816.4 & 1 & 104.9 \end{bmatrix}' mW$. To achieve the target QoS, the second terminal will need to transmit at higher power than it is allowed (the maximum power is 200 mW). For this reason the fourth terminal cannot be accepted with its target QoS. The AC may decide to test the system again with reduced QoS from the target to the minimum required QoS for the new service. If it works, the UL test is passed otherwise it starts to reduce the target QoS of other active terminals. By reducing the target QoS of the new service to its minimum required QoS we obtain $\mathbf{P}_{UL} = \begin{bmatrix} 0.8 & 139.8 & 0.2 & 9.7 \end{bmatrix}' mW$. Now the power limits are satisfied. The UL admission is passed with minimum required QoS for the new terminal (SINR $= -7\,$dB).

3. We repeat the calculations for the downlink. For a single cell with the same parameters the spectral radius of uplink and downlink are the same (prove it). This means that $\rho_{DL}(\mathbf{\Gamma H}) = 0.96$.

4. The required power from the base station for all terminals is $\mathbf{P}_{UL} = \begin{bmatrix} 126.9 & 246.7 & 264 & 289.3 \end{bmatrix}' mW$, the total required power is about 927 mW which is less than the maximum available power at the base station, but it is very close to the maximum available power. The network could be congested by possible changes in channel gains or system parameters. When there is no specified protection factor the AC will accept the new service with UL QoS represented by SINR $= -7\,$dB and DL QoS represented by SINR $= -3.5\,$dB.

Exercise 7.1   After the AC accepted the new fourth terminal, the channel gain of the first terminal dropped to $G_1 = -110dB$ because of shadowing. Test the system feasibility now. Try to relax the system by reducing the target QoS for some terminals. If it is not possible to relax the system even with setting all terminals at their minimum requirements what should you do?

The previous AC scheme was centralized, i.e., the AC should collect all the information of the channel gains at the uplinks and downlinks as well as additive noises' power in order to make a correct decision to accept or reject the new request. This could be possible for a single-cell scenario. However, for a multi-cell scenario as in mobile networks, the centralized solution is avoided because of its high computation and measurement requirements. Thus the centralized AC is not a preferred option for practical systems. Next we introduce some distributed and more practical AC algorithms.

7.3 Non-Interactive Distributed Admission Control (NIDAC) Algorithm

This algorithm is based on a simple and fast method to make the admission decision [1]. The new terminal uses the pilot signal of the cell to measure the SINR at the mobile. If the SINR at the mobile (DL) is greater than the target SINR then the DL portion of the AC is passed and the mobile will request the service from the base station. The base station also estimates the SINR of the new terminal, and if it is greater than the requested uplink target, and there is enough power budget at the base station the AC accepts the new request. We can summarize the required condition to accept the new request as:

$$\text{For UL}: \quad \Gamma_{Q+1}^{tar} \leq \frac{G_{Q+1,Q+1}P_{\max}}{I_{Q+1}}$$

$$\text{For DL}: \quad \hat{\Gamma}_{Q+1}^{tar} \leq \frac{\hat{G}_{Q+1,Q+1}\hat{P}_{DL,Q+1}}{\hat{I}_{Q+1}} \text{ and } \sum_{i=1}^{Q+1} \hat{P}_i \leq \hat{P}_{\max}$$

where I_{Q+1} is the total interference experienced by the $(Q + 1)^{\text{th}}$ terminal and is given by:

$$I_{Q+1} = \sum_{j=1}^{Q} G_{Q+1,j}P_j + N_{Q+1} \tag{7.4}$$

The NIDAC is a simple, distributed, and fast AC. However, it is not preferred in practice because of its high RER as well as AER.

Exercise 7.2   Apply the NIDAC for Example 7.2.

7.4 Interactive Distributed Admission Control (IDAC) Algorithm

The IDAC algorithm has been proposed in [1]. The principal idea is to use a constrained power control scheme with a tighter power constraint on the new arrival. Its power constraint is gradually relaxed until the moment where it can be either rejected or accepted [1]. This AC scheme is similar to another algorithm called distributed power control with active link protection (DPC-ALP) [2]. The IDAC algorithm is applied with the distributed constraint power control (DCPC) algorithm (Chapter 5).

The DCPC algorithm (for UL) is given by

$$P_i(k+1) = \min\left(P_{\max}, \frac{\Gamma_i^{tar}}{\Gamma_i(k)}P_i(k)\right) = \min\left\{P_{\max}, \Gamma_i^{tar}\left(\sum_{\substack{j=1 \\ j\neq i}}^{Q} \frac{G_{ij}}{G_{ii}}P_j(k) + \frac{N_i}{G_{ii}}\right)\right\} \tag{7.5}$$

The $(Q + 1)^{\text{th}}$ new mobile can be accepted if there is a positive fixed-point solution (where all target QoS requirements are achieved) to

$$P_i = \min \left\{ P_{\max}, \Gamma_i^{tar} \left(\sum_{\substack{j=1 \\ j \neq i}}^{Q+1} \omega_{ij} P_j + \eta_i \right) \right\} \tag{7.6}$$

where $\omega_{ij} = \frac{G_{ij}}{G_{ii}}$, $\eta_i = \frac{N_i}{G_{ii}}$, and the iteration symbol (k) has been dropped to simplify the notation and also to indicate that the algorithm should converge even with asynchronous updates. In the IDAC algorithm it has been assumed that the new mobile terminal can estimate its channel gain as well as the additive noise term at the base station. The concept of the IDAC algorithm is to accept the new request but with small maximum power constraint. We then iterate the conventional DCPC algorithm and test the convergence behaviour of active links as well as the new mobile arrival. If all mobiles converge to an acceptable power vector as well as the required QoS we accept the new arrival. If not we increase the maximum allowed transmission power of the new arrival and repeat the previous step. We reject the new arrival if at least one active terminal falls bellows its minimum QoS or its power exceeds the maximum allowed. Also we reject the new arrival mobile if no solution is obtained for all iteration phases until we reach the actual maximum allowed transmission power of the new terminal. The IDAC algorithm can be summarized as follows [1].

Define the maximum power constraint of the new $(Q + 1)^{\text{th}}$ arrival terminal at phase m as $\bar{P}_{\max}(m)$ and also P_i^* is the stationary power value of terminal i. Moreover, assume that the maximum allowed power for all terminals is the same as $P_{\max}$.

1. Initialization: set DCPC phase $m = 0$, and $\mathbf{P}^*(0) = [P_1^* \quad \cdots \quad P_Q^* \quad 0]$, $\vartheta_0 = \min_{1 \leq i \leq N} \left(\frac{N_i}{G_{i,Q+1}} \right)$, and $\bar{P}_{\max}(0) = 0$.
2. Increment: $m + 1 \to m$.
3. Update: in this stage we update the maximum power constraint of the new arrival terminal:
   (a) If $\bar{P}_{\max}(k-1) \geq P_{\max}$, then reject mobile $(Q + 1)$, stop IDAC, and continue with DCPC only with the old active set of mobiles $\{1, 2, \ldots, Q\}$. Otherwise continue with IDAC.
   (b) Set $\varepsilon_m = \min_{1 \leq i \leq Q} \left(\frac{P_{\max}}{P_i^*(m-1)} - 1 \right)$. If $\varepsilon_m = 0$, then reject mobile $Q + 1$, stop IDAC, and continue with DCPC only with the old active set of mobiles $\{1, 2, \ldots, Q\}$. Otherwise, set the variables below and continue with IDAC: $\bar{P}_{\max}(m) = \bar{P}_{\max}(m-1)(1 + \varepsilon_m) + \vartheta_0 \varepsilon_m$ and the initial power vector as $\mathbf{P}^*(m) = [P_1^*(m-1)(1 + \varepsilon_m) \quad \cdots \quad P_Q^*(m-1)(1 + \varepsilon_m) \quad \bar{P}_{\max}(m)]$.
4. Execute phase m of DCPC.
5. Test:
   (a) If $\Gamma_{Q+1}(m) = \Gamma_{Q+1}^{tar}$ and $P_{Q+1}(m) \leq P_{\max}$, then accept mobile $Q + 1$, stop IDAC, and continue with DCPC for mobiles $\{1, 2, \ldots, Q + 1\}$.
   (b) Otherwise go to step 2.

The main advantage of the IDAC is that it has zero AER as well as RER (in cases of perfect measurements and static channel). It is clear from the IDAC algorithm steps that we engage the new terminal in the network gradually. At the first step the new terminal will affect the other active terminals as small extra noise as:

$$P_i(k+1,m) = \min\left\{ P_{\max}, \Gamma_i^{tar}\left(\sum_{\substack{j=1 \\ j\neq i}}^{Q} \frac{G_{ij}}{G_{ii}} P_j(k,m) + \frac{N_i}{G_{ii}} + \underbrace{\frac{G_{i,Q+1}}{G_{ii}} P_{Q+1}(k,m)}_{\text{extra noise}} \right) \right\} \quad (7.7)$$

where k refers to the iteration number and m to the phase number. Because of this extra noise the stationary optimum power for the old active set will be higher, i.e.,

$$\mathbf{P}^*(m+1) > \mathbf{P}^*(m) \quad (7.8)$$

Hence, for every phase we increase the maximum allowed power for the new arrival terminal as far as $\max(\mathbf{P}^*(m)) < P_{\max}$, otherwise we reject the request.

The main concern about the IDAC algorithm is its convergence rate, that is, how many DCPC phases are required to reach a decision. Generally it suffers from slow convergence. The IDAC should be executed concurrently on uplink and downlink. It is concluded in [1] that the IDAC algorithm is very quick to make a decision if the mobile can be accepted. On the other hand, it is very slow when the new mobile request needs to be rejected.

7.5 Admission Control in UMTS

In this section we will discuss some practical procedures for admission control in UMTS networks. Why don't we discuss admission control for GSM? The admission control procedure for GSM is more straightforward than the UMTS case or other CDMA systems. GSM systems have hard capacity, for example, if we consider a cell with 10 carrier frequencies, then the capacity of the uplink and downlink is at most 10×8 (time slots) $= 80$ available channels (some channels are reserved for signaling purposes). The interference of other cells is usually avoided by good network planning (selecting a proper reuse factor). This is not the case in UMTS where all terminals share the same frequency band and time. In UMTS systems we have soft capacity, i.e., the system capacity depends on the interference structure as well as transmitter and receiver capabilities. If we utilize a better detection technique in CDMA systems so that we obtain 1 dB gain, this will be reflected directly on the system capacity. In GSM, this can be exploited by for example re-frequency planning which is not easy to do for established networks. This is one of the major benefits of using CDMA systems.

Generally, it is not possible to test the spectral radius for admission or load control as we showed in the theoretical analysis part. However, the AC needs to know some information about the current network load and available network resources in order to make the right decision to accept or to reject the new requests. The available network resources (such as power) are usually easy known by the AC. However, more sophisticated techniques are required to estimate how much the network is loaded and also to predict the future load if a certain service is accepted. The load estimation at the base station input (uplink) can be done by measuring the total received power as follows (all values are on average)

$$P_{i,total} = I_{i,total} + N_i \tag{7.9}$$

where $P_{i,total}$ is the total received power at base station i, $I_{i,total}$ is the total received power from all transmitters (mobile stations) where some of them are from the same cell and the others from other cells, and N_i is additive white noise at the receiver input. The main source of N_i is the receiver itself which can be specified by the receiver noise figure (N_F), another source of N_i is the external additive noise received by the base station antenna.

The uplink noise rise (NR) for base station i is defined as the ratio of the total received power to additive white noise power as follows [6]

$$NR_i = \frac{P_{i,total}}{N_i} = \frac{I_{i,total}}{N_i} + 1 \tag{7.10}$$

From Equation (7.10) we see that the noise rise gives an indication of power rise over the background noise according to the transmitting terminals.

The uplink (UL) load factor of base station i ($\eta_{i,UL}$) is defined as the total received power from all transmitters to the total received power at base station i, i.e.,

$$\eta_{i,UL} = \frac{I_{i,total}}{P_{i,total}} \tag{7.11}$$

We can see from Equation (7.11) that $\eta_{i,UL}$ gives an indication of the current load of the uplink. It can be related to the noise rise as:

$$\eta_{i,UL} = \frac{I_{i,total}}{P_{i,total}} = \frac{I_{i,total}}{I_{i,total} + N_i} = \frac{NR_i - 1}{NR_i} \tag{7.12}$$

It is possible to measure $P_{i,total}$ at the base station, furthermore N_i can be calculated using the N_F information of the base station subsystems. This means that it is possible to estimate the noise rise as well as the UL load factor. The load factor of the UL is an aggregation of all terminals' load impacts. If we define the load impact of mobile station j on base station i as ζ_{ij} then we can define the load factor as

$$\eta_{i,UL} = \sum_{j=1}^{Q} \zeta_{ij} + \hat{\zeta}_{i,oth} \tag{7.13}$$

where Q is the number of active terminals in the ith cell, and $\hat{\zeta}_{i,oth}$ is the load impact of other terminals from other cells. It is more convenient to model the other cell terminals' impact as a fraction from the own-cell terminals' impact as

$$\hat{\zeta}_{i,oth} = \varepsilon_r \sum_{j=1}^{Q} \zeta_{ij} \tag{7.14}$$

where $\varepsilon_r \geq 0$ is the fraction ratio. Now Equation (7.13) can be represented as

$$\eta_{i,UL} = (1 + \varepsilon_r) \sum_{j=1}^{Q} \zeta_{ij} \tag{7.15}$$

From Equation (7.12) we can express the load impact of terminal j as

$$\zeta_{ij} = \frac{\text{Pr}_j}{P_{i,total}} \qquad (7.16)$$

where Pr_j is the received signal power from terminal j.

The achieved $\left(\frac{E_b}{N_0}\right)$ at the base station is given by (see Chapter 8)

$$\left(\frac{E_b}{N_0}\right)_j = \frac{f_c}{R_j} \frac{\text{Pr}_j}{P_{i,total} - \text{Pr}_j} \qquad (7.17)$$

where f_c is the chip rate and R_j is the uplink data rate of terminal j. In Equation (7.17) we assume that the voice activity is one. From Equation (7.17), we can formulate the received power from terminal j as

$$\text{Pr}_j = \frac{P_{i,total}}{1 + \frac{f_c}{R_j(E_b/N_0)_j}} \qquad (7.18)$$

From Equations (7.12) and (7.18) we can express the load impact of terminal j as a function in the terminals' data rate as

$$\zeta_{ij} = \frac{1}{1 + \frac{f_c}{R_j(E_b/N_0)_j}} \qquad (7.19)$$

The UL load factor can be expressed using Equations (7.15) and (7.19) as follows:

$$\eta_{i,UL} = (1 + \varepsilon_r) \sum_{j=1}^{Q} \frac{1}{1 + \frac{f_c}{R_j(E_b/N_0)_j}} \qquad (7.20)$$

The expression in Equation (7.20) is a UL load factor estimation based on the terminal's data rates. If all terminals use similar QoS parameters (data rate, BER, etc.), then it is better to use Equation (7.12) to estimate the UL load factor. However, if the network supports different kinds of services with different data rates as well as different required performance it is better to use Equation (7.20). The UL load factor is an essential parameter in the UL AC decision as we will see later.

Example 7.3

The uplink of a certain base station in a UMTS network (with 5 MHz bandwidth) is loaded by 70%. Calculate:

(a) The UL load factor.
(b) The noise rise.
(c) If the composite noise figure (N_F) of the base station receiver is 2 dB, find the total received power from all terminals.
(d) If the terminals from other cells have 30% of the total interference, find the own-cell interference value.

Solution

(a) The UL load factor $\eta_{UL} = 0.7$

(b) From Equation (7.12), $NR = \frac{1}{1-\eta_{UL}} = \frac{1}{0.3} = 3.3$

(c) The background noise is given by $N_i = \kappa N_F T_0 B$, where κ is the *Boltzmann constant*, T_0 is the nominal equivalent noise temperature, and B is the system bandwidth, $N_i = 1.38 \times 10^{-23} \times 10^{0.2} \times 290 \times 5 \times 10^6 = 3.2 \times 10^{-14}$W. Using Equation (7.12), the total received power from all terminals, i.e., the total interference, is $I_{total} = N_i \frac{\eta_{UL}}{1-\eta_{UL}} = 7.4 \times 10^{-14}$W.

(d) $I_{own\_cell} = 0.7 \times 7.4 \times 10^{-14} = 5.18 \times 10^{-14}$W

Example 7.4

A certain UMTS cell supports four active terminals. The other-cells' impact is 35% from the own-cell interference. The chip rate is 3.84 Mchip/s. The supported data rates as well as the received signal quality of all terminals are shown in Table 7.1. Find the UL load factor of this cell.

Solution

You can find the UL load factor using Equation (7.20) as

$$\eta_{i,UL} = (1+0.35) \left[\frac{1}{1+\frac{3.84\times10^6}{64\times10^3\times10^{0.75}}} + 0.174 + 0.040 + 0.018 \right] = 0.429$$

The detailed calculation is shown above for the first terminal only.

Generally speaking, the DL load factor calculations are easier than UL. One reason is that the base station in the DL works in broadcasting mode, where it takes care of only its own resources. For example, if the total used power in the base station i is $\hat{P}_{i,total}$, and its maximum possible power is $\hat{P}_{i,max}$, then the DL load factor is simply given by

$$\eta_{i,DL} = \frac{\hat{P}_{i,total}}{\hat{P}_{i,max}} \tag{7.21}$$

Table 7.1 Terminals' parameters of Example 7.4

Terminal number	E_b/N_0	R_j [kb/s]
Terminal 1	7.5 dB	64
Terminal 2	8 dB	128
Terminal 3	7 dB	32
Terminal 4	6.5 dB	16

The DL load factor can also be given in terms of the downlink data rate. One formulation for the throughput-based load estimation is [6]

$$\eta_{i,DL} = \frac{\sum_{j=1}^{Q} \hat{R}_j}{\hat{R}_{i,\max}} \tag{7.22}$$

where Q is the total number of active terminals, $\hat{R}_j$ is the DL throughput for terminal j and $\hat{R}_{i,\max}$ is the maximum allowed throughput for base station i.

When a new request arrives at the radio network controller (RNC), the AC needs to estimate the load increase (in both UL and DL) if the new request is accepted. If the estimated new load is less than some maximum threshold in the UL and DL the new request is accepted otherwise it is rejected. Assuming a network with Q active terminals, we may estimate the UL load factor increase due to the $(Q + 1)^{\text{th}}$ request as

$$\eta_{i,UL}^{new} = \eta_{i,UL} + \frac{1}{1 + \frac{f_c}{R_{(Q+1)}(E_b/N_0)_{(Q+1)}}} \tag{7.23}$$

where the second term represents the load increase according to the new request requirements (data rate and the performance level). The AC should accept the new request (for uplink) if

$$\eta_{i,UL}^{new} \leq \eta_{UL,Threshold} \tag{7.24}$$

The UL load factor threshold $\eta_{UL,Threshold}$ is determined to specify the maximum allowed interference level at the base station. It is also possible to use the maximum total received power threshold or the maximum noise rise threshold to test the new request. All are related to each other as shown in Equation (7.12). The same procedure is also done for the DL. The AC will estimate (based on its channel gain as well as its requirements) the required power budget to handle the new request. The AC should accept the new request (for DL) if

$$\hat{P}_{i,total} + \hat{P}_{i,(Q+1)} \leq \hat{P}_{i,Threshold} \tag{7.25}$$

where $\hat{P}_{i,total}$ is the current DL total used power, $\hat{P}_{i,(Q+1)}$ is an estimation for the DL power required to support the new request, and $\hat{P}_{i,Threshold}$ is the DL power threshold. You can see that the AC usually uses certain threshold values instead of the maximum possible values. The reason for this is to avoid work closely to the maximum available resources. The channel gain of terminals as well as the network situation is time varying, so we should save some budget for these fluctuations in the network requirements. Furthermore, the operators usually leave some room for non-real-time applications, e.g., GPRS and HSDPA. Another important reason is that the interference level increases in an acute way when the load factors are close to the allowed maximum. This fact can be shown as follows. Equation (7.12) can be reformulated as

$$P_{i,total} = \frac{N_i}{1 - \eta_{i,UL}} \tag{7.26}$$

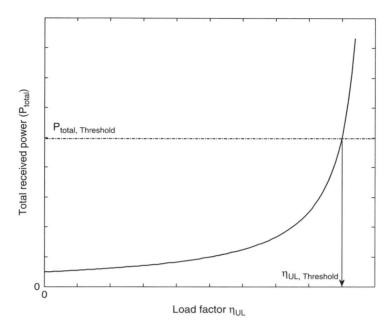

Figure 7.2   The total received power vs. load factor

Now let's check how the total received power at the base station will be changed with regard to the load factor:

$$\frac{dP_{i,total}}{d\eta_{i,UL}} = \frac{N_i}{\left(1 - \eta_{i,UL}\right)^2} = \frac{P_{i,total}}{\left(1 - \eta_{i,UL}\right)} \Rightarrow \Delta P_{i,total} \approx \frac{P_{i,total}}{\left(1 - \eta_{i,UL}\right)} \Delta \eta_{i,UL} \qquad (7.27)$$

From Equation (7.27), when the UL load factor is high, any small change in the load factor because of the channel fluctuations or new request acceptance will have a high impact on the total interference level. This relation can be demonstrated by Figure 7.2. In Figure 7.2 we can see that when the load factor is already high, any small increase in the load may cause congestion very quickly.

Example 7.5

In Example 7.4, if the UL load factor threshold is fixed at 0.75, compute how many new terminals the UL admission control will accept if all new terminals are the same as terminal 3.

Solution

We may solve the example by evaluating n in the following formula

$$0.429 + 0.040 n_{UL} = 0.75 \Rightarrow n_{UL} = 8 \text{ terminals}$$

Note that we assumed that the other-cell terminals' impact factor does not change. However, this is not completely true. When eight new terminals are accepted, the interference to other

Table 7.2   Downlink transmit power of Example 7.6

Terminal No.	$\hat{P}_j$ (W)
Terminal 1	0.40
Terminal 2	0.80
Terminal 3	0.30
Terminal 4	0.20

cells will increase, so that other-cells' terminals will increase their transmit power and this will increase the other-cells' terminals' impact factor. Anyhow, the solution does provide a rough estimate of the number of new terminals that can be accepted.

Example 7.6

As a continuation of Example 7.4, the transmit power in the DL for each terminal is shown in Table 7.2. The maximum DL power which the base station can handle is 10 W. Assume that 4 W are used for broadcasting cell information and signaling. The DL load factor threshold is 0.8.

(a) Calculate the DL load factor based on the transmit power.
(b) How many new terminals can the AC accept, if all terminals have the same parameters as terminal 3?

Solution

(a) From Equation (7.21) $n_{DL} = \frac{0.4 + 0.8 + 0.3 + 0.2}{6} = 0.28$.
(b) The total number of terminals n_{DL} can be calculated from $0.28 + \frac{0.3 n_{DL}}{6} = 0.8 \Rightarrow n_{DL} = \lfloor 10.4 \rfloor = 10$.

This is the number of new terminals that can be handled in the DL. But because the UL in the previous example can handle only eight terminals then it is not possible to accept all 10 terminals. The number of the terminals that can be supported in the cell is $n = \min(n_{UL}, n_{DL})$, where n_{UL} is the number of possible accepted terminals in the UL and n_{DL} is the number of possible accepted terminals in the DL.

7.6 Admission Control for Non-Real-Time Applications

For non-real-time applications such as HSDPA in UMTS, we may relax our admission procedure. Users of such kind of services can be classified into two sets: active and inactive sets. Active sets contain users who are served according to some asymptotically fair scheduling algorithm. Terminals in the inactive set can be moved to be active according to channel and network situations. The main difference between real- and non-real-time admission control is that the non-real-time application can be queued for a longer time until resources are available for services. In other words, it accepts much higher delays. However, even in non-real-time applications there will be some upper limit for the accepted delay and also there will be limits for the buffer capacity for the queued users' data. Several interesting algorithms are proposed in the literature to handle such kinds of problems.

7.7 Load Control (LC)

If the network is congested, the RRM must take fast steps to relax the network before many connections are dropped or the network becomes unstable. The steps that the RRM should take are:

1. Stop transmission of non-real-time data (in the uplink or downlink or both) and store data in the network buffers.
2. If the congestion is in the uplink, reduce the target uplink QoS for all terminals to their minimum required QoS.
3. If the congestion is in the downlink, reduce the target downlink QoS for all terminals to their minimum required QoS.
4. Start the handover process of some terminals to other networks or cells (if available).
5. Finally, if the above steps do not relax the congestion situation, the load control starts to drop some active (real-time) connections.

If all active terminals have the same priority, how do we select the correct terminal to be dropped? Usually the load control should drop the terminal which has the highest impact on the network. In this way we can minimize the number of dropped terminals.

Define $\mathbf{A}^{[k]}$ as a square sub-matrix of $\mathbf{A}$ with removed kth row and kth column. To minimize the number of dropped connections, the ith removed user should satisfy the following criteria

$$i = \arg \max_{k=1,\ldots,Q} G_a(\mathbf{A}^{[k]}) \tag{7.28}$$

where $\mathbf{A} = \mathbf{\Gamma H}$, and $G_a(\mathbf{A}^{[k]})$ is the achieved gain by removing the kth terminal, and it is given by

$$G_a(\mathbf{A}^{[k]}) = \left(\frac{1}{\rho(\mathbf{A}^{[k]})} - \frac{1}{\rho(\mathbf{A})} \right) \tag{7.29}$$

All connections need to be examined to select the optimum removed user. If dropping that link would relax the congestion then stop, otherwise repeat (7.28) with another dropping.

It is clear that the optimum removal algorithm has two disadvantages: the computational complexity and all mobiles' channel gain values are needed in order to compute the spectral radius. There are simpler algorithms to select the dropped (or removed) connection. One well-known method is the stepwise removal algorithm (SRA) introduced in [3], which is based on removing the link which has the maximum combined sum of its row and column values. The SRA is explained by the flowchart in Figure 7.3.

There is another modified version of the SRA algorithm called the stepwise maximum interference removal algorithm (SMIRA), which is proposed in [4]. The main difference between the SRA and the SMIRA is that the removed link is the one which has the maximum interference on other links. A simple, distributed and effective removal algorithm is proposed in [5]. It is applicable to the DCPC algorithm described in Chapter 5 and it is called the generalized DCPC (GDCPC) algorithm. When the network is congested, then there will be at least one terminal sent at maximum power $P_{\max}$ without achieving the minimum required QoS.

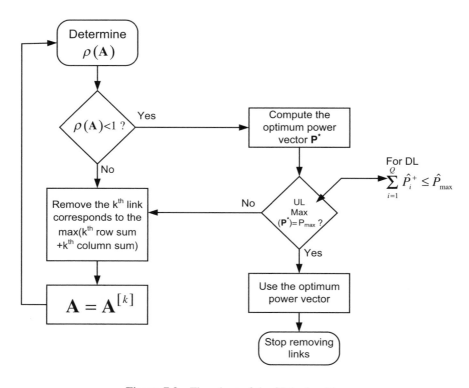

Figure 7.3 Flowchart of the SRA algorithm

In this case, this terminal will cause interference to all other terminals without successful connection. The GDCPC will then remove this terminal by setting its transmit power to zero. However, it can be powered up again when the network is relaxed (temporal removal). The convergence properties of the GDCPC are given in [5].

Exercise 7.3 A UMTS network consists of two cells. There are six active terminals in the network. The channel gains between every terminal and both base stations are given in Table 7.3. The table also shows the supported data rate for each terminal in both uplink and

Table 7.3 Terminals' parameters of Exercise 7.3

	Channel gain with base station 1 (db)	Channel gain with base station 2 (db)	Data rate (DL) (kb/s)	Data rate (UL) (kb/s)
Terminal 1	−110	−75	128	32
Terminal 2	−110	−120	64	16
Terminal 3	−90	−95	64	32
Terminal 4	−70	−95	256	64
Terminal 5	−130	−110	64	16
Terminal 6	−110	−95	384	64

downlink. The chip rate is 3.84 Mchip/s. Every terminal is assigned to the base station which has the better channel gain. The target E_b/N_0 is 8.5 dB for all terminals in both UL and DL. Assume the worst-case scenario for the orthogonality and voice activity factor. The maximum available power at the base stations is 5 W. The maximum allowed throughput of the cell is 2 Mb/s. The maximum transmit power of terminals is 23 dBm. Finally, the additive noise power in UL and DL is -90 dBm.

(a) Calculate the optimum transmit DL and UL power values.
(b) Is the system feasible (in UL and DL)? If not relax the network using the optimum removal algorithm.
(c) Calculate the UL and DL load factors.
(d) Calculate the UL noise rise.
(e) If the threshold value for both UL and DL load factors is 0.8, calculate the maximum possible acceptance of new terminals similar to terminal number 1 in Table 7.3.
(f) If terminal 2 requests a new service that requires increasing the UL data rate from 16 kb/s to 64 kb/s, will the AC accept this request?
(g) Explain the network behaviour if the channel gain of terminal 4 with base station 1 has decreased according to the shadowing effect $= -120$ dB.

References

[1] M. Andersin, Z. Rosberg and J. Zander, 'Soft and safe admission control in cellular networks', *IEEE/ACM Transactions on Networking*, **5**(2), 1997.
[2] N. Bambos, S. Chen and J. Pottie, 'Channel access algorithms with active link protection for wireless communication networks with power control', *IEEE/ACM Transactions on Networking*, **8**, 579–83, 2000.
[3] J. Zander, 'Performance of optimum transmitter power control in cellular radio systems', *IEEE Transactions on Vehicular Technology*, **41**(1), 1992.
[4] J. Lin, T. Lee and Y. Su, 'Power control algorithm for cellular radio systems', *Electronic Letters*, **30**(3), 1994.
[5] F. Berggren, R. Jantti and S-L. Kim, 'A generalized algorithm for constrained power control with capability of temporal removal', *IEEE Transactions on Vehicular Technology*, **50**(6), 2001.
[6] H. Holma and A. Toskala, *WCDMA for UMTS*, 3rd edition, John Wiley & Sons, Ltd, 2004.

8

Combining Different Radio Resources

The radio resource scheduler (RRS) is an essential part of the radio resource management (RRM). The main objective of the RRS is to schedule or quantify the available resources in order to achieve the target objectives for customers as well as operators. Different layers of the communication system (e.g., physical layer, MAC layer, network layer) may affect the values of the radio resources. Hence, in order to achieve the highest possible efficiency, we should utilize cross-layer optimization methods. Optimizing several resources at the same time is not an easy task, at least in practice. The problem becomes clear when we want to join algorithms from different layers. The communication between the different communication layers is very complex and governed by several protocols.

The configuration of RRS within the radio resource management is illustrated in Figure 8.1. The main role of the RRM is to assign resources to users according to the agreed QoS parameters which can be obtained through a negotiation phase between the terminals and the network. The QoS attributes are usually specified in terms of the bit error rate (BER), data rate, maximum delay, priority, and so on. As shown in Figure 8.1, the RRM mission starts by performing connection admission control (CAC). Since the decision is based on resource availability, CAC consults the RRS before accepting or rejecting the requested call/service. Upon call acceptance, the traffic classifier, another RRM component, categorizes the incoming traffic according to its QoS specification, which is typically included in each packet header [13]. Data flows are then directed to a corresponding queue according to its QoS field. Each QoS class is represented by at least one queue. Finally, the data dispatches from the buffers according to some priority logic after getting the assigned radio resources from the RRS, which relies on the channel conditions, the agreed QoS, and the network situation. Based on the above, it is evident that the RRS bears a great responsibility towards a successful RRM.

In this chapter we will introduce some important resources for scheduling and also some algorithms to combine those resources. Some important resources which can be controlled by

Systems Engineering in Wireless Communications   Heikki Koivo and Mohammed Elmusrati
© 2009 John Wiley & Sons, Ltd

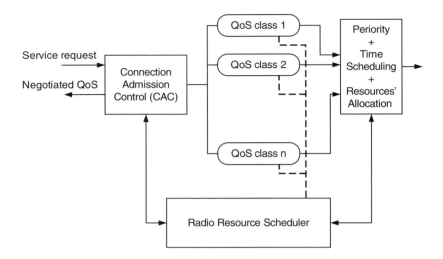

Figure 8.1   Block diagram of the radio resource manager

the RRS scheduler are:

- **Transmit power control:** in a multiuser environment, the power is adjusted to be just the right power value required to achieve the target objectives. Any extra transmit power will just increase the interference and reduce the system capacity and performance as well as decrease the lifetime of the mobile battery. The topic of transmit power control was covered in Chapters 5 and 6.
- **Data rate control:** data rate is a major item of the QoS requirements. Actually from an application point of view we are more concerned about the throughput than the data rate. However, they are closely related as will be shown when discussing error control. Data rate can be adapted by different means. On the carrier modulation side it can be adapted by the modulation level (BPSK, QPSK, 8PSK,16-QAM, etc.). In CDMA systems it can also be adapted by adjusting the processing gain (PG). Generally speaking, increasing data rate will degrade the system performance in terms of bit error rate (BER) or block error rate (BLER). A compromise should be made between data rate, required power, system capacity, and communication performance.
- **Error control:** the erroneous received symbols can be corrected either by requesting retransmission or by forward error correction. In both cases we need to add redundant bits to the information bits. Generally speaking, increasing the redundant bits (or reducing the coding rate) will increase the error control capabilities to improve the system performance by detecting and correcting more errors. With the same modulation level, this improvement will be at the expense of reducing the throughput, since more bits will be sent for the purpose of coding rather than information.
- **Transmission time control (scheduling):** for non-real-time applications, the scheduler has to find the optimum time to transmit its packets. When the time is not good, the coming packets will be stored in the buffer and wait. The optimum time to transmit is determined by many factors such as the channel quality, network situation, packet priorities, number of concurrent users, etc. For example, when the channel quality is excellent, we can send higher throughput within this time slot (by using proper power, modulation level, PG, coding rate,

and so on). This case can be clearly observed in downlink cellular systems. Because of the multiuser diversity there is a chance that at least one terminal will have excellent channel conditions for some time. The scheduler tries to exploit this channel diversity to maximize the downlink throughput. This is called an opportunistic scheduling algorithm. It is clear that this system could be unfair. Imagine if one user lives in an apartment which is in an excellent position to the base station so that the mobile of this user has an excellent channel condition most of the time. If this user uses his mobile for data communication (such as internet exploring), she/he will use most of the channel resources in the downlink, because the scheduler will mostly select her/him. For this reason, fair algorithms are proposed in literature such as the proportional fair algorithm.

- **Frequency sub-band control:** many modern wireless communication systems are based on orthogonal frequency division multiplexing (OFDM). The main idea of these systems is that the whole bandwidth is divided into much smaller sub-bands while preserving orthogonality between these bands using fast Fourier transform FFT and its inverse (IFFT). The main motivation of this band division is to mitigate the ISI problems associated with the wide-band transmission (frequency selective channels). These sub-bands may also be scheduled by the RRS to achieve the required objectives. When the multiuser communication environment uses OFDM, we call this multiple access as *orthogonal frequency division multiple access* (OFDMA). In such systems, the throughput which can be achieved by each user depends on several parameters such as the number of assigned sub-bands and the channel quality of those sub-bands. Since the total number of sub-bands is finite, the RRS will manage the locations and the number of sub-bands assigned for every terminal in order to maximize the system performance, capacity, and so on.

The RRS can use the above resources (and maybe more such as the spatial processing and the multiuser detection) to achieve different objectives like maximizing the number of simultaneous users, reducing the total transmitting power, and increasing the total throughput. The RRS is formulated mathematically as an optimization problem. The conventional way to formulate this optimization problem is by selecting one of the resources as a target to maximize or minimize and use the others as constraints. Other ways are based on genetic algorithms, neural networks, analytical multi-objective (MO) optimization, and the Kalman filter [4]. Here we address the problem of how to combine the above resources to achieve the target objectives. The next section discusses the interrelations between the radio resources.

8.1 Some Radio Resources Interrelations

The achieved throughput in wireless communication systems depends on different factors such as the available bandwidth, coding rate, modulation level, processing gain (for CDMA systems), and transmitted power. The previous parameters should be set (or optimized) based on the channel condition, required QoS and network situation. The SINR at the receiver is widely used to indicate the channel quality. The available bandwidth is usually given and it represents one main constraint in wireless communications (scarce resource!). The total available spectrum is divided between many different systems such as mobile networks, TV channels, FM and AM radio channels, military services, satellite, and GPS. Every wireless communication system has a specified and limited bandwidth. Almost all countries have very strict regulations for the spectrum usage. The remaining parameters can usually be adapted to overcome the fast and large dynamics of the wireless channels. In other words the achieved throughput at certain time slots depends on the channel quality at that time slot. For this reason we need an accurate estimator at the transmitter to estimate the channel quality at the time of transmission to accurately adapt its parameters. That is, high spectrum efficiency can be achieved if we have a feedback channel between the receiver and the transmitter. One example is the conventional closed-loop power control structure.

There is always a time difference between the channel estimation and resources allocation. This time delay can be critical if the channel time correlation is low such as in the case of high mobile speed. The main reason is that at very high mobile speed the channel properties can be completely different at the time of transmission than at the time of measurements. This means a low time-correlated channel. We may overcome this problem by making the time delays very short which can be achieved by decreasing the time slot duration. However, there are many limitations of this reduction. This problem is one of the main reasons for the performance degradation of the high speed packet access (HSPA) when the mobile is moving with very high speed. Several diversity techniques are used to mitigate such kinds of problems.

In this section we will explore the relations between the previous parameters without going into detailed analysis.

Regarding throughput, we have at least three different types of systems:

1. Fixed throughput systems where we need to achieve certain throughput flow to guarantee successful connection. One example is voice communication. These systems are usually sensitive to delay, i.e., real-time applications.
2. Adaptive throughput systems where we try to achieve as high average throughput as possible without a minimum limit, i.e., the minimum throughput limit can be zero. An example is internet exploring.
3. Elastic throughput system with minimum limit. In these systems we try to maximize the throughput as much as possible and there is a minimum throughput requirement. One example is video communication with MPEG4 coding.

Let's first discuss the limitations of using available resources. As you know when we increase the modulation level in phase modulation and fix the coding rate, we obtain more throughputs. Why cannot we just increase the modulation level until we obtain the maximum required throughput? In phase modulation systems when we increase the modulation level the distance between the transmitted symbols becomes closer and hence is more sensitive to noise and interference. In other words as we increase the modulation level, more errors in the detected symbols will occur. For example, in BPSK we transmit two symbols (every symbol represents one bit) which are $180°$ apart. The phase error should be at least $90°$ to observe errors in the received signal. In 8PSK, we transmit one of eight different symbols where every symbol represents three bits. The phase difference between two adjacent symbols (assuming regular distribution of symbols) in this case is $45°$. This means that a phase error of $22.5°$ is enough for error reception of the transmitted symbol. This explains why we should not use high modulation levels unless the SINR at the receiver is high enough.

Why should we just use a very high transmit power to increase the SINR at the receiver input? When we use high power in a multiuser environment, we will increase the interference to other terminals which means that their SINRs will be dropped. In this case other terminals will also try to increase their transmit power (to improve their SINRs), which makes more interference to our receiver, so that it may becomes even worse! There are many other reasons make the use of high transmit power unwise such as the battery life of our terminals. Hence, it is wise to use a minimum power value to achieve our target requirements (as throughput and error performance). Since we cannot increase the power as we want, what prevent us from using a very strong coding method that can correct all possible errors, so that we do not need to increase the power? Good question! The answer is that increasing the coding efficiency has some limitations as well. Generally increasing the ability of coding means that we increase the redundant bits which do not carry the actual information but some structure information about the transmit packet. For example if you use coding rate $= 1/2$, this means that half of the transmit bits in the packet are not the actual information but redundant bits! It is very important to use the proper coding method but at the same time it will not help to increase the throughput if the modulation level is higher than some value which depends on the channel quality. What if we serve only one or few terminals at a time and switch off the others during each time? Will that improve the average throughput? Yes, this is one very efficient way to maximize the total throughput through time scheduling. The idea is that for every time slot we select one (or few) terminal to be served and allocated all available resources. The served terminals are selected based on their instantaneous channel condition, their priorities, required QoS, and their past delivered services. There are several algorithms to manage fairness, total throughput, maximum latency and so on. However, this will only be efficient if the number of terminals is not too large and the channel quality can be good at least for one terminal at a time. Finally what about the processing gain (PG) of CDMA systems? The PG can be used as one parameter to adjust the transmit throughput on the systems. However, it is also limited by the channel equality. Reducing the PG will increase the throughput (with fixed coding rate), but at the same time it will increase the probability of errors. And vice versa, increasing the PG will decrease the throughput!

From this discussion we see that the problem of adjusting these parameters is not trivial. Actually, for a multiuser environment, even when we reduce the dimension of adjustable parameters to be only the power and rate, the problem is still difficult to optimize. As will be described later in this chapter, optimizing the power and rate for a multiuser environment is an NP-complete problem. Next we will explain the relations of each parameter on the system performance and also show the relations to each other.

In this section remember that our main objective is to achieve the required throughput (either fixed or the maximum possible value) at a predefined probability of error. For example, for voice communication in cellular systems it is required to achieve 22.8 kb/s at a BLER around 10^{-3} (higher BLER can be allowed in certain situations).

First the payload data (information bits) are packed into data packets. Assume that the length of payload data per packet is k bits. The length of this packet can be more than the payload length according to the coding and other bits added to each packet such as address information. The new packet length is n bits. We will consider only the increase in the packet length according to the coding bits. The channel coding is the redundancy added to the data to be able to detect and even correct some possible errors. The increase according to this redundancy is $(n - k)$ bits. The coding process can increase the system performance considerably at some loss of the throughput. Actually, many wireless systems such as the cellular system would not succeed without coding. There are two main types of coding:

1. *Error detection and retransmission:* the receiver utilizes the redundant bits (also called parity or check bits) to detect if any errors have occurred. In the case of erroneous reception

the receiver will request retransmission of the erroneous packet from the transmitter. There is no error correction process at the receiver.

2. *Forward error correction (FEC):* the code structure gives the possibility for the receiver to detect and even correct the errors. The number of detected and corrected errors depends on the coding rate as well as coder structure and type. The number of corrected errors is usually less than the number of detected errors. For example, we can define FEC code which is able to detect up to two errors within the packet and to correct one error. It is also possible to use joint coding system which corrects errors if possible and requests retransmission when the number of errors cannot be corrected.

In this brief introduction we will introduce only the FEC. Mainly there are two different kinds of coding: block code and convolution code. In block coding the $(n - k)$ redundant bits are distinct from the information bits; however, it depends on the information bits within the packet. Convolutional codes have a structure that extends over the entire transmitted bit stream, rather than being limited to codeword blocks [5]. If the SINR is very low, then the utilization of coding will not help to improve the BER. Actually it may reduce the performance to be worse than the system without coding. We can explain this behavior as in a word correction process in word-processing software. If the word-processing software just marks the wrongly typed words, then this is equivalent to error detection with retransmission (you should correct the word, maybe with the aid of the software). If you activate the automatic correction, this would be equivalent to FEC. With automatic correction, if you make a few letter errors (compared to the total number of letters in the word) there will be a big chance that the word processor corrects them successfully. But if you make a large number of errors, the word processor may suggest a different word than the correct one. This may give a different meaning of the context. In such a situation, it would be better to keep the word with errors than correct them to some meaningful words but *in the wrong way*! A large number of letter errors is equivalent to very low SINR. Example 8.1 explains the importance of coding as well as its effect on the throughput.

Example 8.1

Assume a wireless communication system uses a certain digital modulation method. Because of noise, channel fading, and the interference, the receiver makes errors in the estimation of the actual transmitted bits. The probability of bit error is 10^{-3}. The transmitter sends the information bits at fixed rate of 22 kb/s in the form of packets. Each packet contains 11 bits. Calculate the probability of block error at the receiver.

Now assume that we use block code by adding four bits to each block, i.e., we have (15,11) code. With this coding system the decoder at the receiver is able to detect and correct one error per packet. Calculate the probability of block error in this case. Also what is the throughput?

Solution

For the uncoded system, the probability of block error occurs if at least one of the 11 bits is received with error, i.e., $BLER = 1 - (1 - P_e)^{11}$, where $BLER$ is the probability of block error and P_e is the probability of bit error. From the above $BLER \approx 0.01$ which is generally a large error rate, on average every second we will have about 20 erroneous blocks of data!

By using coding methods we are now able to detect and correct one error out of the 15 bits. The probability of block error becomes $BLER = 1 - (1 - P_e)^{15} - 15P_e(1 - P_e)^{14}$. It is clear that the third term is the probability of one error occurring out of the 15 bits. By solving the above equation we obtain $BLER \approx 0.0001$! This means that the BLER has been reduced by about 100 times with this simple coding technique. What is the cost of this great improvement in the performance? The throughput will be slightly decreased, because the data rate is fixed and cannot be increased then the throughput will drop to $22 \times 10^3 \times \frac{11}{15} \approx 16 \, kb/s$. We can increase the throughput by increasing the code rate k/n, however, this will be at the expense of system complexity and delay. Generally speaking the more redundant bits used the more efficient will be the system in terms of error detection and correction, and the more reduction on the throughput. Therefore a compromise should be made between these two objectives. Usually when the channel quality is bad we use more redundant bits (in other word we reduce the coding rate) to increase the ability of the coding system to correct the high possible errors and vice versa.

What determines the bit error probability? There are different factors affecting the bit error probability such as noise, interference, intersymbol interference (ISI) caused by multipath in wireless channels, and channel fading because of the time-varying nature. In this analysis, we assume that we use receivers with good equalizers or Rake receivers so that the problem of ISI is solved. Moreover, the probability of total packet losses according to fading is assumed to be negligible because of interleaving bits. Hence, we consider only the noise and interference. The performance measure for digital signals is the energy per bit by the noise spectral density E_b/N_0. This parameter addresses the received bit quality compared to noise and interference. The probability of error depends directly on E_b/N_0 but with different mapping algorithms. The mapping algorithm depends on the channel type and the modulation method. For the additive white Gaussian noise channel and BPSK the relation between the probability of error and E_b/N_0 is given by:

$$P_B = Q\left(\sqrt{\frac{2E_b}{N_0}}\right), \tag{8.1}$$

where $Q(.)$ is called Q-function and it is given by:

$$Q(X) = \frac{1}{\sqrt{2\pi}} \int_X^\infty \exp\left(\frac{-u^2}{2}\right) du \tag{8.2}$$

It is also possible to send more than one bit per symbol without considerably affecting the available bandwidth when using PSK modulation. When we send k bits per symbol we call the system M-ary system where $M = 2^k$. As we increase the number of bits per symbol the probability of symbol error increases as indicated by the following mapping (for additive noise channel) [5]

$$P_E(M) \approx 2Q\left(\sqrt{\frac{2E_b \log_2(M)}{N_0}} \sin\left(\frac{\pi}{M}\right)\right) \tag{8.3}$$

where P_E is the probability of symbol error, and $E_S = E_b \log_2 (M) = kE_b$ is the energy per symbol. Observe that if E_S is fixed, then increasing the number of bits per symbol will reduce E_b.

Example 8.2

Using MATLAB® and Equation (8.3), draw the symbol probability versus M for $M = 4, 8, 16,$ 32, 64, 128, and 256, for two cases $E_S/N_0 = 10$ dB and $E_b/N_0 = 10$ dB.

Solution

It is straightforward to draw the relation as shown in Figure 8.2.

Of course, it is not possible to have a practical system working at such a high symbol error as shown in Figure 8.2. Therefore, high modulation level application is possible only in relatively high SINR channels. Try to repeat Example 8.2 with higher SINR values. The problem becomes more difficult if we consider the dynamical nature of the wireless channels. In this case E_b/N_0 becomes a random variable, and the above equations give only the instantaneous value of the probability of error. To obtain the average probability of error, we need to average over the probability density function of the E_b/N_0 parameter [3,6].

For a CDMA communication system the E_b/N_0 at receiver i is given by

$$\left(\frac{E_b}{N_0}\right)_i = PG_i \frac{P_i G_{ii}}{\sum\limits_{j \neq i}^{Q} P_j \theta_{ij} G_{ij} + N_i}, \tag{8.4}$$

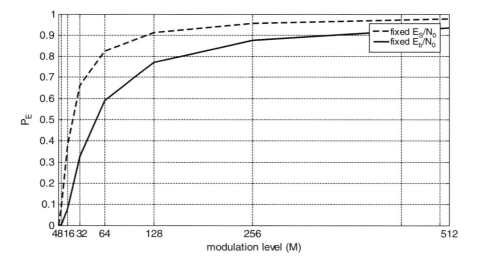

Figure 8.2   The probability of symbol error vs. modulation level in Example 8.2

where G_{ij} is the channel gain between transmitter j and receiver i, P_j is the transmit power of terminal j, θ_{ij} is the orthogonality factor ($0 \leq \theta_{ij} \leq 1$) between signals of terminals i, j at ith receiver, and N_i is the average noise power. The processing gain of user i, is defined as $PG_i = \frac{f_c}{f_{b,i}}$, where f_c is the chip rate and $f_{b,i}$ is the bit rate of user i. To obtain the target E_b/N_0 for user i we can adjust either the transmit power or the processing gain or both. Since the chip rate is usually fixed then we can change the bit rate to adjust the PG.

Usually the PG, modulation level, and coding are selected from a finite set. For example $PG \in \{2, 4, 8, \cdots, 256\}$, modulation level $\in \{BPSK, QPSK, 8PSK, 16QAM, \cdots\}$, and $Coding\_Rate \in \{\frac{1}{4}, \frac{1}{2}, \frac{3}{4}, \cdots\}$. Also the coding method can be changed.

For a general wireless communication system the bandwidth is usually given and taken as a constraint for the system. The performance of the system in terms of block error rate, throughput, maximum latency, and so on is determined by the designer and depends on the application.

Example 8.3

Assume a single-cell multiuser wireless communication system. The available bandwidth is 200 kHz. The maximum allowed block error rate is 10^{-4} where every block contains 11 bits of information. The required throughput is 10 kb/s. The system should be able to support five simultaneous users with the same requirements and connected to the same access point using the CDMA technique. The system is interference limited, i.e., we can ignore the additive noise value. Assume perfect power control is used and additive white noise channel approximation for the probability of error calculations. Check different possibilities to adjust the available parameters to achieve the objectives.

Solution

Let's first assume that we use a coding rate of 1, i.e., no coding at all! In this case the BER can be calculated from $BLER = 1 - (1 - P_b)^{11} = 10^{-4} \Rightarrow P_b = 9.1 \times 10^{-6}$, here we will need a very small bit error probability. If we assumed BPSK which gives the best performance in terms of probability of error, the required E_b/N_0 is

$$P_b = Q\left(\sqrt{\frac{2E_b}{N_0}}\right) \Rightarrow \frac{E_b}{N_0} = \frac{1}{2}[Q^{-1}(P_b)]^2 = \frac{1}{2}[Q^{-1}(9.1 \times 10^{-6})]^2 \approx 9.2$$

With perfect power control and ignoring the additive noise we can state that

$$\frac{E_b}{N_0} = PG \frac{P_i G_{ii}}{\displaystyle\sum_{j \neq i}^{Q} P_j \theta_{ij} G_{ij} + N_i} = PG \frac{1}{Q-1},$$

Remember that Q in the above equation is the number of terminals and has no relation with the *Q-function*. Since the required throughput is 10 kb/s then PG $= 200/10 = 20$. In this case, the maximum that E_b/N_0 we can achieve is

$$\frac{E_b}{N_0} = 20\frac{1}{5-1} = \frac{20}{4} = 5!$$

This value is less than the required E_b/N_0 to achieve the required probability of error! It means that with the first proposed parameters it is not possible to achieve the target. Actually we are too far from the target. We cannot support more than three terminals!

Let's change the coding rate using Hamming block code by adding four redundant bits to be able to correct one error in the packet. The packet size now is 15 bits. As we saw in Example 8.1 the bit error probability in this case is $P_b = 10^{-3}$, the required signal quality is now lower and can be found as before, $\frac{E_b}{N_0} = \frac{1}{2}[Q^{-1}(P_b)]^2 = \frac{1}{2}[Q^{-1}(1 \times 10^{-3})]^2 \approx 4.7$. With this new target, it is possible to support all users. However, the supported throughput per user is reduced to $10 kb/s \times \frac{11}{15} \cong 7.3 kb/s$. Moreover, to support the same throughput of 10 kb/s we need to increase the data rate to $R = 10\,kb/s \times \frac{15}{11} \approx 14\,kb/s$, i.e., the required processing gain becomes $PG = \lfloor \frac{200}{14} \rfloor = 14$. With this PG the achieved E_b/N_0 is reduced to 3.5 which is less than the target!

One option is to increase the modulation level to QPSK. We leave the reminder to the reader to complete the example.

8.2 Power and Rate Control

It is clear from the previous section that the signal quality at the receiver is the main player to determine the allocated resources as well as achieved objectives. In practice we usually have a certain set of predefined possible values of coding rate. Without loss of generality, for the rest of this chapter, we will consider only two degrees of freedom which are represented by the transmit power and data rate. The symbol rate is restricted by the available channel bandwidth. However, if the channel is free of noise and interference, we can send at any bit rate we desire (by increasing the number of bits/symbol). Unfortunately, this is not possible because of the inherent noise and interference. Shannon showed by his famous formula that the information rate is an increasing function in the SINR as

$$C = B \log_2(1 + SINR) \tag{8.5}$$

where B is the available bandwidth. Increasing the information rate is generally very desirable in data communication systems but it is restricted by the bandwidth and the received signal quality. The maximum available bandwidth B is usually given and cannot be changed.

Increasing the SINR can be achieved in two ways. The first way is by reducing the total interference and noise affected by that user. This depends on some characteristics of the noise and the interference. For example, if the structure of the interference from other users is known at the receiver then by applying one of the multiuser detection methods, that interference can be reduced. Also if the users are spatially distributed then the interference can be reduced by using adaptive beamforming with multi-antennas as will be described in Chapter 9. If the users concurrently use the channel (as in DS-CDMA) then the interference can be reduced by using power control techniques. From this discussion, we can see that some characteristics of the interference are assumed to be known or can be controlled. There are many sources of interference and noises that cannot be reduced such as thermal noise and interference from other cells.

The second way of increasing the SINR is simply by increasing the transmitted power. In a single user communication (point to point) or in broadcasting, this can be an acceptable solution and the main disadvantages are the possible cross/channel interference and the cost and the nonlinearities in the power amplifiers. But in a multiuser communication environment increasing the transmitted power means more co-channel and cross-channel interference problems.

Controlling the data rate as well as the transmitted power is an important topic in modern communication systems. The adaptive rate features are not critical for communication systems which are designed mainly for voice communication as in 1G and 2G cellular systems. In these systems the target SINR is specified, and the data rate is fixed (or changed within a very limited manner) and only the power is controlled as in IS-95. More recent communication systems (2.5G, 3G, 3.5G,4G) support multi-rate data communication because they are designed not only for voice communication but also for data and multimedia communication. An efficient combining algorithm for the power control and the rate control is required for these systems. The term 'efficient' here means how the transmitted power and data rate can be optimized to meet the required specifications.

There are many combining algorithms for the power and rate control proposed in the literature. The specifications of those algorithms are quite varied. Some algorithms suggest maximizing the throughput; others minimizing the packet delay or the total power consumption.

Although our analysis in this chapter can be applied for different communication schemes, we will frequently refer to the standard UMTS specifications.

The 3G mobile communication systems support multi-rate transmission. There are mainly two methods to achieve the multi-rate transmission: the multi-code (MC) scheme and the variable-spreading length (VSL) scheme. In UMTS the VSL scheme is called the orthogonal variable spreading factor (OVSF) scheme. In MC-CDMA system, all the data signals over the radio channel are transmitted at a basic rate, R_b. Any connection can only transmit at rates mR_b, referred to as m-rate, where m is a positive integer. When a terminal needs to transmit at m-rate, it converts its data stream, serial to parallel, into m basic-rate streams. Then each stream is spread by using different and orthogonal codes [7]. In VSL-CDMA system, the chip rate is fixed at a specified value (3.84 Mcps/s for UMTS) and the data rate can take different values. This means that the processing gain is variable. The processing gain can be defined as the number of chips per symbol. In UMTS, the processing gain (or the spreading factor) in the uplink can take one of the following values {4, 8, 16, 32, 64, 128, 256} [8]. The smallest spreading factor is equivalent to the channel bit rate 960 Kb/s and the largest spreading factor is equivalent to the channel bit rate 15Kb/s [8].

8.2.1 Optimal Centralized Power and Rate Control

In this section we will introduce an optimal methodology to find the rate and power values which achieve certain requirements. Since the computational cost of this method is very intensive, it is not possible to be implemented practically. In small dimensional problems it can be used for comparison purposes. Consider an uplink (the same as downlink) cellular cell with Q number of users. Each user has a set of m transmission rates $\mathbf{M} = \{r_1, r_2, \ldots, r_m\}$, $r_i \geq 0$, $i = 1, 2, \ldots, Q$, to choose from. The space of the achieved rates in the cell can be denoted as $\mathbf{N} = \{\mathbf{n}^1, \mathbf{n}^2, \ldots, \mathbf{n}^\kappa\}$, where $\mathbf{n}^j$ is the jth vector of allocated rates of users, $\mathbf{n}^j = \{n_1^j, n_2^j, \ldots, n_Q^j\}$, $n_i^j \in \mathbf{M} \; \forall i \in \{1, \ldots, Q\}$ and $j \in \{1, \ldots, \kappa\}$, where $\kappa \leq m^Q$. Each rate vector can be associated with a power vector which contains the transmitted power values required to achieve the rates. By defining the objective we can select the optimum power and rate vector. For example, if the objective is to maximize the total transmission rates, then the optimum rate vector is the maximum sum vector in the space N. For 20 users and eight data rates values, there are more than 10^{18} possible rate combinations that can be obtained in the set N (NP-complete problem). This number is only for one time slot. We can see the complexity of finding the optimal solution even when an efficient searching technique is used.

Example 8.4

To explain the optimal algorithm and how the solutions could be computed, assume a mobile communication system with two users. Each user can send at one of three available data rates. Assume a snapshot assumption with channel gain of $-80\,dB$ and $-95\,dB$, respectively. The data rates which each user can select from are $\{15, 60, 180\}$ Kb/s. The target SINR is fixed for both users at 7 dB. The additive noise has zero mean value with $-50\,dB_m$ variance. The modulation type is assumed to be VLS-CDMA with chip rate 3.84 M chip/s.

The space of the achieved rates of users contains $3^2 = 9$ pairs, namely $N = \{(15,15),(15,60),$ $(15,180),(60,15),(60,60),(60,180),(180,15),(180,60),(180,180)\}$ kb/s. The required transmitting power of users to meet these rates can be easily computed using a power control equation with different target SINR values (see Chapter 5). The space of transmitted power pairs needed to achieve the data rate space N is $Tp = \{(0.020,0.633),\ (0.021,2.529),\ (0.024,7.609),$ $(0.080,0.669),\ (0.085,2.687),(0.098,8.161),(0.241,0.768),(0.258,3.115),(0.307,9.711)\}$. Now we can select the optimum solution based on the required objectives. If there are no power constraints and the objective is to get the highest data rate then the ninth solution is the optimum one. If the maximum transmitted power is 1 W, and the objective is to get the highest total data rate, then the seventh solution is the optimum one. If there is no power constraint but the objectives are the minimum total power and the data rate of second user should be greater than 20 Kb/s, then the second solution is the optimum.

Theorem 8.1
The optimum rate allocation is NP-complete for both uplink and downlink.

Proof
We will prove the theorem by first formulating the optimization problem for both UL and DL. We then show that both of them have the form of the well-known knapsack problem [2]. Since the knapsack problem is known to be NP-complete [2], the optimum rate allocation is also NP-complete [1].

(a) For DL: find the optimum power vector $\mathbf{P} = [\hat{P}_1, \hat{P}_2, \cdots, \hat{P}_Q]'$ and rate vector $\hat{\mathbf{R}} = [\hat{R}_1, \hat{R}_2, \cdots, \hat{R}_Q]'$ which achieve the following objective and constraints:

$$\max \sum_{i=1}^{Q} \hat{R}_i \tag{8.6}$$

subject to:

$$\hat{\Gamma}_i = \frac{\hat{P}_i G_{ii}}{\sum_{\substack{j=1 \\ j \neq i}}^{Q} \hat{P}_j \hat{\theta}_{ij} \hat{G}_{ij} + \hat{N}_i} \geq \hat{\Gamma}_{\min,i} \qquad i = 1, \ldots, Q, \tag{8.7}$$

and

$$\sum_{i=1}^{Q} \hat{P}_i \leq \hat{P}_{\max}, \quad \hat{P}_i \geq 0, \, \hat{R}_i \in \{\hat{r}_1, \hat{r}_2, \ldots, \hat{r}_{m}\} \quad \forall i = 1, \cdots, Q \tag{8.8}$$

To include the case where the base station does not transmit to certain terminals we set $\hat{r}_1 = 0$.

(b) For UL: find the optimum power vector $\mathbf{P} = [P_1, P_2, \cdots, P_Q]'$ and rate vector $\mathbf{R} = [R_1, R_2, \cdots, R_Q]'$ which achieve the following objective and constraints:

$$\max \sum_{i=1}^{Q} R_i \tag{8.9}$$

subject to:

$$\Gamma_i = \frac{P_i G_{ii}}{\displaystyle\sum_{\substack{j=1 \\ j \neq i}}^{Q} P_j \theta_{ij} G_{ij} + N_i} \geq \Gamma_{\min,i}, \qquad i = 1, \ldots, Q, \tag{8.10}$$

and

$$P_{\min} \leq P_i \leq P_{\max}, \, P_{\min} \geq 0, \, R_i \in \{r_1, r_2, \cdots, r_m\} \, \forall i = 1, \cdots, Q \tag{8.11}$$

To include the case where certain terminals do not transmit we set $r_1 = 0$.

(c) The *multiple-choice knapsack problem* concerns choosing exactly one item j from each of k classes $Z_i, i = 1, 2, \cdots, k$ such that the profit sum is maximized. The *multiple-choice knapsack problem* is defined as

$$\max \sum_{i=1}^{k} \sum_{j \in Z_i} p_{ij} x_{ij} \tag{8.12}$$

subject to

$$\sum_{i=1}^{k} \sum_{j \in Z_i} w_{ij} x_{ij} \leq c, \tag{8.13}$$

$$\sum_{j \in Z_i} x_{ij} = 1, \quad i = 1, 2, \cdots, k,$$

$$x_{ij} \in \{0, 1\}, \quad i = 1, 2, \cdots, k, j \in Z_i$$

where p_{ij} is the profit of the jth element in the ith class where w_{ij} is its weight, $x_{ij} = 1$ states that item j was chosen from class i and c is the maximum allowed capacity. The constraint $\sum_{j \in Z_i} x_{ij} = 1, \quad i = 1, 2, \cdots, k$ ensures that exactly one item is chosen from each class.

Equation (8.6) (same as Equation (8.9)) can be reformulated as

$$\max \sum_{i=1}^{Q} \sum_{j=1}^{\hat{m}} \hat{R}_{ij} y_{ij} \tag{8.14}$$

where $\hat{R}_{ij} = r_j \in \{\hat{r}_1, \hat{r}_2, \cdots, \hat{r}_m\}$, $y_{ij} \in \{0,1\}$, and $\Sigma_{j=1}^{\hat{m}} y_{ij} = 1$. The transmit power depends on the selected data rate, as we described before, increasing the data rate will also increase the error rate, so we need to increase the transmit power to reduce the errors. The power values are calculated using Equation (8.7) to satisfy the minimum requirement of all active (selected) terminals. If we define $\hat{P}_{ij}$ as the transmit power in the DL for terminal i with data rate r_j. The constraint in Equation (8.8) can be reformulated as

$$\sum_{i=1}^{Q} \sum_{j=1}^{\hat{m}} \hat{P}_{ij} y_{ij} \leq P_{\max} \tag{8.15}$$

This states that the DL optimum rate allocation is a form of the *multiple-choice knapsack problem* which is well-known NP-complete. The UL proof is left to the reader.

Because of the complexity of finding the optimum rate allocation especially in large-scale networks, many heuristic methods have been proposed to optimize the rate allocation [2]. Assuming that the data rate is a continuous variable can considerably simplify the optimization problem. However, there is no guarantee that the continuity assumption of integer programming will lead to the optimum solution.

8.2.2 Centralized Minimum Total Transmitted Power (CMTTP) Algorithm

This algorithm is basically similar to the centralized power control algorithm described in Chapter 5. However, in the CMTTP, we will join the optimum power vector with other QoS requirements.

All the QoS requirements can be directly or indirectly mapped to the SINR as described before.

Generally we can formulate the relation as

$$\Gamma_i^{tar} = g(\delta_i^{tar}, R_i) \tag{8.16}$$

where $\delta_i^{tar} = \left(\frac{E_b}{N_0}\right)_i^{tar}$ is the target energy per bit to the noise density ratio for terminal i, and R_i is the data rate of terminal i. The mapping $g(\bullet)$ can be a linear or nonlinear function. It depends on the modulation type, channel model and the utilized multiple access method. For CDMA systems, assuming a fixed modulation level, i.e., δ_i^{tar} is not a function in the data rate, the relation between SINR and (R_i) can be given in linear form as:

$$\Gamma_i^{tar} = \frac{\delta_i^{tar}}{PG} = \frac{\delta_i^{tar}}{f_c} R_i \tag{8.17}$$

where PG is the processing gain which can be represented as the chip rate (f_c) to the data rate (R_i). It is clear that as the data rate increases the target SINR will increase linearly. If the modulation level changes then δ_i^{tar} will change as well, which makes Equation (8.17) nonlinear. Another example for the nonlinear relation between the SINR and the data rate can be observed in the Shannon channel capacity where the relation is given by

$$\Gamma_i^{tar} = 2^{R_i/f_c} - 1 \tag{8.18}$$

The above relation can be considered as the lower bound for the target SINR required to achieve a certain data rate (R_i).

The mathematical formulation of the CMTTP problem (in downlink) is as follows: *Find the power vector* $\hat{\mathbf{P}} = [\hat{P}_1, \ldots, \hat{P}_Q]'$ *and the rate vector* $\hat{\mathbf{R}} = [\hat{R}_1, \ldots, \hat{R}_Q]'$ *minimizing the cost function*

$$J(\hat{\mathbf{P}}) = \mathbf{1}'\hat{\mathbf{P}} = \sum_{i=1}^{Q} \hat{P}_i \tag{8.19}$$

subject to

$$\frac{\hat{P}_i \hat{G}_{ii}}{\sum_{\substack{j=1 \\ j \neq i}}^{Q} \hat{P}_j \hat{G}_{ij} + \hat{N}_i} \geq \hat{\Gamma}_{i,\min} = g_i(\hat{\delta}_{i,\min}, \hat{R}_i), \qquad \forall i = 1, \ldots, Q, \tag{8.20}$$

$$\sum_{i=1}^{Q} \hat{P}_i \leq \hat{P}_{\max}, \ R_i \geq \hat{r}_{i,\min} \in \{\hat{r}_1, \hat{r}_2, \cdots, \hat{r}_{\hat{m}}\}, \qquad \forall i = 1, \ldots, Q \tag{8.21}$$

where $\hat{\Gamma}_{i,\min}$ is the minimum required SINR, $\hat{r}_{i,\min}$ is the minimum required DL data rate for terminal i, and $\hat{\delta}_{i,\min}$ is the minimum required $\left(\frac{E_b}{N_0}\right)$ for terminal i. The problem presented in Equations (8.19)–(8.21) can be reduced to a system of linear equations. If the constraints presented in Equations (8.20) and (8.21) cannot be achieved then the problem is called infeasible. In this case either some users have to be dropped from the link or some of the constraints have to be relaxed. The optimal power vector is the one which can achieve all target QoS as follows

$$\frac{\hat{P}_i \hat{G}_{ii}}{\sum_{\substack{j=1 \\ j \neq i}}^{Q} \hat{P}_j \hat{G}_{ij} + \hat{N}_i} = \hat{\Gamma}_i^{tar} = g_i\left(\hat{\delta}_i^{tar}, \hat{r}_i^{tar}\right), \qquad \forall i = 1, \ldots, Q, \tag{8.22}$$

where $\hat{r}_i^{tar} \geq \hat{r}_{i,\min}$ is the target download data rate for terminal i.

From Equation (8.22) the optimum DL power for terminal i should achieve

$$\hat{P}_i = \hat{\Gamma}_i^{tar}\left[\sum_{\substack{j=1 \\ j \neq i}}^{Q} \frac{\hat{G}_{ij}}{\hat{G}_{ii}} \hat{P}_j + \frac{\hat{N}_i}{\hat{G}_{ii}}\right], \qquad \forall i = 1, \ldots, Q, \tag{8.23}$$

In matrix form, the optimum power vector is

$$\mathbf{P}^* = [\mathbf{I} - \hat{\mathbf{\Gamma}}^{tar}\hat{\mathbf{H}}]^{-1}\hat{\mathbf{\Gamma}}^{tar}\mathbf{u} \tag{8.24}$$

where

$$(\hat{\mathbf{H}})_{ij} = \begin{cases} 0 & i = j \\ \dfrac{\hat{G}_{ij}}{\hat{G}_{ii}} > 0 & i \neq j \end{cases} \tag{8.25}$$

$$(\mathbf{u})_i = \frac{\hat{N}_i}{\hat{G}_{ii}} \tag{8.26}$$

and

$$\hat{\mathbf{\Gamma}}^{tar} = diag\{\hat{\Gamma}_1^{tar} \quad \hat{\Gamma}_2^{tar} \quad \cdots \quad \hat{\Gamma}_Q^{tar}\} \tag{8.27}$$

In order to obtain a non-negative solution of Equation (8.24), the following condition must be held

$$\rho(\hat{\mathbf{\Gamma}}^{tar}\hat{\mathbf{H}}) < 1 \tag{8.28}$$

where $\rho(\mathbf{A})$ is the spectral radius of matrix $\mathbf{A}$ (see Chapter 5).

8.2.3 Maximum Throughput Power Control (MTPC) Algorithm

This algorithm has been suggested in [9]. The algorithm is based on the maximization of the total throughput in a cellular system. There is no need to generate all solutions in this method. Since the link gains and the interference of other users are required to calculate the transmitted power of each user, the MTPC algorithm is a centralized algorithm. The throughput of user i can be approximated when M-QAM modulation is used by

$$T_i = \Theta + \log_2(\Gamma_i) \tag{8.29}$$

where T_i is the throughput of user i, Θ is a constant, and Γ_i is the SINR of user i.

The total throughput T is given by

$$T = \sum_{i=1}^{Q} T_i = Q\Theta + \log_2\left(\prod_{i=1}^{Q} \Gamma_i\right) \tag{8.30}$$

where Q is the number of users.

Now the problem can be defined as follows: Given the link gains G_{ij} of the users, what is the power vector $\mathbf{P} = [P_1, P_2, \ldots, P_Q]'$ which maximizes the total throughput in Equation (8.29)? Since the first term in Equation (8.30) is constant and the logarithmic function is an increasing function, then maximizing the multiplicative term ($\prod_{i=1}^{Q} \Gamma_i$) will lead to maximizing the total throughput T. The problem considered in [9] is

$$\max_{\mathbf{P}} \left[\prod_{i=1}^{Q} \Gamma_i(\mathbf{P})\right] \quad \text{s.t.} \quad \mathbf{P} \in \Omega \tag{8.31}$$

where $\Omega = \{\mathbf{P} | P_{min} \leq P_i \leq P_{max}, i = 1, \ldots, Q\} \subset \Re^Q$.

The MTPC algorithm to solve (8.31) is given by

$$P_k(t+1) = \cfrac{1}{\displaystyle\sum_{r \neq k}^{Q} \cfrac{G_{rk}}{\left(\displaystyle\sum_{j \neq r}^{Q} G_{rj}P_j(t) + N\right)}}, \quad t = 0, 1, \ldots, \quad k = 1, \ldots, Q \tag{8.32}$$

$$P_{\min} \leq P_k(t+1) \leq P_{\max}$$

where G_{ij} is the channel gain between user j and base station i and N is an additive noise. Without loss of generality user i is assumed to be assigned to base station i. In [9], it has been shown that, starting from any initial vector $\mathbf{P}(0) \in \Omega$, the iteration specified by Equation (8.32) converges to a unique point $\mathbf{P}^* \in \Omega$, which achieves the global maximum.

8.2.4 Statistical Distributed Multi-rate Power Control (SDMPC) Algorithm

A distributed solution of the optimization problem given by Equations (8.19)–(8.21) is proposed for a single-cell case in [12]. It is assumed that every user has two states ON or OFF. The state ON refers to active state, i.e. the user sends data. The state OFF refers to idle state, where the transmitted power is zero. The transition probabilities of the ith user from idle to active state at any packet slot is v_i, and from active to idle state is ζ_i. The durations of the active and idle periods are geometrically distributed with a mean of $1/\zeta_i$ and $1/v_i$ (in packet slots), respectively. The optimization problem in Equations (8.19)–(8.21) is slightly modified to

Find

$$\min_{\mathbf{P}} J(\mathbf{P}(t)) = \sum_{i=1}^{Q} \beta_i(t) P_i(t) \tag{8.33}$$

subject to

$$\frac{R_s}{R_i} \frac{P_i G_{ki}}{\displaystyle\sum_{\substack{j=1 \\ j \neq i}}^{Q} P_j \beta_j(t) G_{kj} + N_i} \geq \delta_i^*, \quad \forall i = 1, \ldots, Q, \tag{8.34}$$

$$P_{\min} \leq P_i \leq P_{\max}, \quad R_i = r_i, \quad \forall i = 1, \ldots, Q, \tag{8.35}$$

One parameter has been added to the original optimization problem which is the indicator function $\beta_j(t)$. The indicator function is equal to one if the jth user is currently active, and zero otherwise. It is assumed in [12] that the random process $\hat{\beta}(t)$ has a Markovian property since geometric distribution is memoryless over the duration of traffic.

The centralized solution (if the system is feasible) is given by

$$P_i(t) = \frac{\beta_i(t) \gamma_i}{G_{ki}} \times \frac{N_i}{1 - \displaystyle\sum_{j=1}^{Q} \beta_j(t) \gamma_j} \tag{8.36}$$

where

$$\gamma_i = \frac{\delta_i^T}{\delta_i^T + R_s/R_i} \tag{8.37}$$

The main idea behind the SDMPC algorithm is to estimate the other users' information part. Therefore the term $(\sum_{j=1}^{Q} \beta_j(t)\gamma_j)$ is estimated. The Markovian property of the random process $\beta_j(t)$ has been exploited to obtain a good estimate of the other users' information part.

The SDMPC algorithm is given by

$$P_i(t) = \frac{\beta_i(t)\gamma_i}{G_{ki}} \times \frac{N_i}{1 - \hat{\beta}(t)} \tag{8.38}$$

where $\hat{\beta}(t)$ is the estimation of $\sum_{j=1}^{Q} \beta_j(t)\gamma_j$.

The estimated parameter $\hat{\beta}(t)$ has been derived in [12] for two cases: i. there is no 'collision' at t; and ii. a 'collision' occurs at t. There are at least three drawbacks to this algorithm:

1. In the cellular CDMA system a control channel is always active (when the mobile phone is ON).
2. In the SDMPC algorithm, the channel gain and the average power of the additive noise are assumed to be known. But in reality they should be estimated as well. Good estimation of the channel gain and the noise variance is usually difficult. In practice it is easier to estimate the SINR because it has a direct impact on BER [13].
3. It is assumed that the durations of active and idle periods are geometrically distributed, which can considered as an oversimplified assumption.

8.2.5 Lagrangian Multiplier Power Control (LRPC) Algorithm

As mentioned previously, the data rates which can be achieved belong to a set of integers. In the formulation of the optimization problem, to maximize the data rate we assume that the data rate is continuous. This assumption can be relaxed in the simulation by rounding the optimum data rate to the nearest floor of the data rate set. It can be proved that the solution of the optimization problem with continuity assumption is not necessarily the same as the solution of the actual discrete problem [14]. The advantage of the LRPC algorithm is that the optimization problem has been formulated without the continuity assumption of the data rates [15]. It has been assumed that each user has a set of m transmission rates $M = \{r_1, r_2, \ldots, r_m\}$ to choose from. Let the rates be ordered as, $r_1 < r_2 < \ldots r_m$. To properly receive messages at transmission rate r_k, mobile i is expected to attain $\Gamma_i(\mathbf{P}) \geq \Gamma_{i,k}^T$.

Define $\mathbf{Y} = [y_i^k]$ to be a [0 1] matrix such that, for every mobile i and rate r_k

$$y_i^k = \begin{cases} 1, & \text{if mobile i is transmitting with rate } r_k \\ 0, & \text{otherewise} \end{cases} \tag{8.39}$$

The combined rate and power control is formulated as the following optimization problem [15]

$$\mathbf{R} \triangleq \max_{Y,P} \sum_{i=1}^{Q} \sum_{k=1}^{m} r_k y_i^k \tag{8.40}$$

subject to the following constraints

$$\sum_{k=1}^{m} y_i^k \leq 1, \; y_i^k \in \{0,1\}, \; and \; 0 \leq P_i \leq P_{\max} \tag{8.41}$$

$$P_i + (1 - y_i^k) B_i^k \geq \frac{P_i \Gamma_{i,k}^T}{\Gamma_i(\mathbf{P})} \tag{8.42}$$

where B_i^k is an arbitrary large number satisfying

$$B_i^k \geq \max_P \frac{P_i \Gamma_{i,k}^T}{\Gamma_i(\mathbf{P})} \tag{8.43}$$

The above optimization problem is solved by using Lagrangian multiplier method. The main goal of the LRPC algorithm is to maximize the total throughput of the system. Although the LRPC improves the system throughput, its power consumption for supported users as well as the outage probability are rather high. So it is not recommended to be used in systems where the fairness is an important issue.

8.2.6 Selective Power Control (SPC) Algorithm

The SPC algorithm has been suggested in [5]. The SPC algorithm is a logical extension of the DCPC algorithm [16]. The main idea of the SPC algorithm is to adapt the target CIR of each user to utilize any available resources. The suggested SPC algorithm is given by

$$P_i(t+1) = \max_k \left\{ \frac{P_i(t) \Gamma_{i,k}^T}{\Gamma_i(\mathbf{P})} \times \chi \left(\frac{P_i(t) \Gamma_{i,k}^T}{\Gamma_i(\mathbf{P})} \leq P_{\max} \right) \right\}, \quad t = 0, 1, \ldots, i = 1, \ldots, Q \tag{8.44}$$

where $\chi(E)$ is the indicator function of the event E. Although the SPC algorithm improves the outage probability compared with LRPC algorithm, its outage is still high. The convergence speed of the SPC algorithm is slow [17].

Jäntti and Kim in [18] proposed an improved version of the SPC algorithm. It is called the selective power control with active link protection (SPC-ALP) algorithm [18]. The SPC-ALP algorithm has less outage probability and better performance than the SPC algorithm. The main idea of the SPC-ALP algorithm is to admit the new users into the network with at least the minimum data rate and also if possible allow old users to choose higher data rates. This is done by defining three different modes of operation for each user:

- Standard mode, where the user updates its power using the SPC algorithm. In this mode the rate cannot be increased but it could be decreased if needed. If there are more resources to be utilized by increasing the rate, the used mode is changed to the transition mode.

- Transition mode, where the user updates its power using the ALP algorithm. Also the rate is adapted to the maximum rate that can be supported.
- Passive mode, where the user stops its transmission.
- More details about the SPC-ALP algorithm can be found in [18].

8.3 Mathematical Formulation of the RRM Problem in the MO Framework

The application of the MO optimization method in RRM is introduced in this section. As stated in the introduction to this chapter, the QoS can be defined for a set of factors. In this section we will consider only the BER and the user data rate in the uplink. The objectives of the RRS can be defined as:

(a) Minimize the total transmitting power.
(b) Achieve the target SINR in order to achieve a certain BER level (depends on the application).
(c) Maximize the fairness between the users. In our definition, the system is fair as long as each user is supported by at least its minimum required QoS. In this sense, minimizing the outage probability leads to maximizing the fairness.
(d) Maximize the total transmitted data rate and achieve at least the minimum required data rate for each terminal.

It is clear that objective (a) is totally conflicting with objective (d) and partially conflicting with objective (b). Objective (c) is totally incompatible with objective (d). Objective (b) is partially contradictory to the objective (d).

In the literature the RRM problem is usually formulated as a single-objective (SO) optimization problem considering the others as constraints. Two very common formulations for solving the RRM problem in the literature are given. The first one is (e.g. MTPC and LRPC): *Find the rate vector* $\mathbf{R} = [R_1, \ldots, R_Q]'$ *and the power vector* $\mathbf{P} = [P_1, \ldots, P_Q]'$ *which maximize the following objective function*

$$\max \sum_{i=1}^{Q} \Upsilon(R_i) \tag{8.45}$$

subject to the constraints

$$\delta_i \geq \delta_{i,\min} \tag{8.46}$$

$$P_{\min} \leq P_i \leq P_{\max}, \mathrm{R}_{i,\min} \leq R_i \leq R_{i,\max}, \ i = 1, \ldots, Q, \tag{8.47}$$

where $\Upsilon(.)$ is a rate function, R_i is the data rate of user i, P_i is the transmitted power of user i, Q is the number of users, δ_i is the SINR for user i, $\delta_{i,\min}$ is the minimum allowed SINR of user i, $P_{\min}, P_{\max}$ are the minimum and maximum transmitted power of the mobile terminal, respectively, and $R_{i,\min}, R_{i,\max}$ are the minimum and maximum transmitted data rate of user i, respectively. The rate function is generally an increasing function of the user data rate R_i. In the literature, the rate function has been defined as the throughput [9], [10]. In [11], it has also been defined as a utility function, which is used to achieve certain QoS requirements. The allowed BER for user i is determined by the value of $\delta_{i,\min}$.

There is another SO optimization definition of the RRS problem in the literature. In this formulation, the total transmitted power is minimized (objective (a)) and the other objectives are defined as constraints. This formulation is widely used in the literature as e.g. in CMTTP and SDMPC: *Find the rate vector* $\mathbf{R} = [R_1, \ldots, R_Q]'$ *and the power vector* $\mathbf{P} = [P_1, \ldots, P_Q]'$ *which solves the following optimization problem for all* $i = 1, \ldots, Q$

$$\min \sum_{i=1}^{Q} P_i \tag{8.48}$$

subject to the constraints

$$\delta_i \geq \delta_{i,\min} \tag{8.49}$$

$$P_{\min} \leq P_i \leq P_{\max}, \tag{8.50}$$

$$R_{i,\min} \leq R_i \leq R_{i,\max} \tag{8.51}$$

We can see from the above two formulations (8.45)–(8.47) and (8.48)–(8.51) that the objectives (a)–(d) are optimized by a single objective and a number of constraints.

Solving the objectives (a)–(d) at the same time using the MO optimization technique leads to a more general solution than the conventional methods. Before we propose a general formulation of the RRS using MO optimization, let's make a quick review of MO optimization principles.

8.3.1 Multi-Objective Optimization

The MO optimization is a technique to find the best solution between different and usually conflicting objectives. In the MO optimization problem we have a vector of objective functions. Each objective function is a function in the decision (variable) vector. The mathematical formulation of the MO optimization problem is

$$\begin{aligned} &\min\{f_1(\mathbf{x}) \quad f_2(\mathbf{x}) \quad \cdots \quad f_m(\mathbf{x})\} \\ &\text{Subject to } \mathbf{x} \in \mathbf{S} \end{aligned} \tag{8.52}$$

where we have m (≥ 2) objective functions $f_i : \Re^n \rightarrow \Re$, $\mathbf{x}$ is the decision (variable) vector belonging to the (nonempty) feasible region (set) $\mathbf{S}$, which is a subset of the decision variable space $\Re^n$. The abbreviation *min* means that we want to minimize all the objectives jointly. Usually the objectives are at least partially conflicting and possibly incommensurable. This means that, in general, there is no single vector $\mathbf{x}$ that can minimize all the objectives simultaneously. Otherwise, there is no need to consider multiple objectives. Hence, the MO optimization technique is used to search for efficient (non-inferior) solutions that can best compromise between different objectives. Such solutions are called Pareto optimal solutions.

Definition 8.1   A decision vector $\mathbf{x}^* \in S$ is Pareto optimal, if there does not exist any other decision vector $\mathbf{x} \in S$ such that $f_i(\mathbf{x}) \leq f_i(\mathbf{x}^*)$ for all $i = 1,2,\ldots,m$ and $f_j(\mathbf{x}) < f_j(\mathbf{x}^*)$ for at least one index j [19]. The *Pareto optimal set* is a set of all possible (infinite number) Pareto optimal solutions. The condition of an optimal Pareto set is rather strict and many MO algorithms cannot guarantee to generate Pareto optimal solutions but only weak Pareto optimal solutions. Weak Pareto optimal solutions can be defined as follows.

Definition 8.2   A decision vector $\mathbf{x}^* \in S$ is weakly Pareto optimal if there does not exist another decision vector $\mathbf{x} \in S$ such that $f_i(\mathbf{x}) < f_i(\mathbf{x}^*)$ for all $i = 1,2,\ldots,m$.

The set of (weak) Pareto optimum solutions can be nonconvex and nonconnected [19]. Figure 8.3 shows the geometric interpretation of Pareto optimal and weakly Pareto optimal solutions. Note that all points on the line segment between points A and B are weakly Pareto optimal solutions. The reason is that it is possible to improve (reduce) one objective without affecting the other. All points on the curve between points B and C are Pareto optimal solutions. The reason is that there is no way to improve one objective without affecting the other. In this sense, the Pareto optimal set is the efficient solution. Every point in the set is optimum in some sense.

Two different MO optimization techniques are introduced in this section. The first method is called the *weighting method*. The weighting method transforms the MO problem posed in Equation (8.52) into a single objective as follows:

$$\min \sum_{i=1}^{m} \lambda_i f_i(\mathbf{x}) \tag{8.53}$$
$$\text{subject to } \mathbf{x} \in \mathbf{S}$$

where the tradeoff factors λ_i satisfy the following $\lambda_i \geq 0 \ \forall i = 1, \cdots, m$, and $\sum_{i=1}^{m} \lambda_i = 1$.

The weakly Pareto optimal set can be obtained by solving the optimization problem in Equation (8.53) for different tradeoff factor values. The second MO optimization technique has a special interest in the applications of MO optimization in RRS. It is the method of *weighted metrics*. If the global solutions of the objectives are known (or can be estimated) in advance,

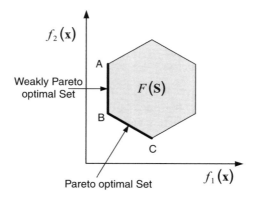

Figure 8.3   Pareto optimal sets

then problem (8.52) can be formulated as

$$\min \left(\sum_{i=1}^{m} \lambda_i |f_i(\mathbf{x}) - z_i^*|^p \right)^{1/p}$$

$$\text{subject to } \mathbf{x} \in \mathbf{S} \tag{8.54}$$

where $1 \leq p \leq \infty$, z_i^* is the optimum solution of objective i, and again the tradeoff factors satisfy $\lambda_i \geq 0 \, \forall i = 1, \cdots, m$ and $\sum_{i=1}^{m} \lambda_i = 1$.

It is clear that Equation (8.54) represents the minimization of the weighted p-norm distance. For $p = 2$ the weighted Euclidean distance is obtained. With $p = \infty$ problem (8.54) is called the weighted Tchebycheff problem.

8.3.2 General Multi-Objective Formulation of RRM

In this section we will present a general formulation of radio resources combining an analytical MO optimization. The detailed analysis and several novel algorithms in this area are presented in [20]. The field is very wide and many different algorithms and methods can be derived based on the MO optimization. One formulation of the RRS optimization problem can be defined as:

$$\min_{P_i, R_i} \left\{ \sum_{i=1}^{Q} P_i, \ - \sum_{i=1}^{Q} \Psi(R_i), O_P \right\}, \quad i = 1, \ldots, Q \tag{8.55}$$

subject to

$$P_{\min} \leq P_i \leq P_{\max}, R_{i,\min} \leq R_i \leq R_{i,\max} \tag{8.56}$$

where $\Psi(R_i)$ is utility function in the user data rate, O_P is the outage probability. The outage probability is defined as the probability that a user cannot achieve at least the minimum required QoS. We can see that the O_P can be formulated to reflect the fairness situation in the communication network. The minus sign associated with the sum of the rate utility function in Equation (8.55) refers to the maximization process of the aggregate utilities.

Defining the objectives and the constraints is the first step. Selecting the proper MO optimization method to solve the problem is the second step. Then the Pareto optimal set of solutions is generated, where every solution is an optimal in a different sense. Finally, the decision maker selects the optimum solution from the optimal set which best achieves the required specifications [20].

References

[1] M. Chatterjee, H. Lin and S. Das, 'Rate allocation and admission control for differentiated services in CDMA data networks', *IEEE Transactions on Mobile Computing*, **6**(2), 2007.

[2] D. Pisinger, *Algorithms for Knapsack Problems*, Department of Computer Science, University of Copenhagen, Denmark, 1995.

[3] H. Azad, A. Aghvami and W. Chambers, 'Multiservice, multirate pure CDMA for mobile communications,' *IEEE Trans. Vehicular Technology*, vol. **48**, no. **5**, 1999.

[4] M. Elmusrati, Radio resource scheduling and smart antennas for CDMA communication systems, Ph.D. Thesis, Helsinki University of Technology, Finland, 2004.

[5] B. Sklar, *Digital Communications: Fundamentals and Applications*, 2nd edition, Prentice Hall, 2001.

[6] M. Simon and M. Alouini, *Digital Communication over Fading Channels*, John Wiley & Sons, Ltd, 2004.

[7] D. Zhao,Radio resource management in cellular CDMA systems supporting heterogeneous services, Ph.D. thesis, Electrical and Computer Engineering, University of Waterloo, Ontario, Canada, 2002.

[8] H. Holma and A. Toskala, *WCDMA for UMTS*, John Wiley & Sons, Ltd, 2000.

[9] K. Chawla and X. Qiu, 'Throughput performance of adaptive modulation in cellular systems', *Proceedings of the IEEE International Conference on Universal Personal Communications*, pp. 945–50, Florence, Italy, October, 1998.

[10] S. Ulukus and L. Greenstein, 'Throughput maximization in CDMA uplinks using adaptive spreading and power control', *Proceedings of the IEEE International Symposium on Spread Spectrum Techniques and Applications*, pp. 565–9, New Jersey, USA, September, 2000.

[11] L. Song and N. Mandayam, 'Hierarchical SIR and rate control on the forward link for CDMA data users under delay and error constraints', *IEEE Journal of Selected Areas in Communications*, **19**, 1871–82, 2001.

[12] H. Morikawa, T. Kajiya, T. Aoyama and A. Campbell, 'Distributed power control for various QoS in a CDMA wireless system', *Proceedings of the IEEE PIMRC*, pp. 903–7, Helsinki, Finland, September, 1997.

[13] M. Moustafa, I. Habib, M. Naghshineh and M. Guizani, 'QoS-enabled broadband mobile access to wireline network', *IEEE Communications Magazine*, **40**, 50–6, 2002.

[14] J. Zander, 'Multirate resource management in wireless CDMA systems,' published slides on the internet at address: http://www.s3.kth.se/%7Ejensz/MultirateCDMA00.pdf.

[15] S. Kim, Z. Rosberg and J. Zander, 'Combined power control and transmission rate selection in cellular networks,' *Proceedings of the IEEE Vehicular Technology Conference*, pp. 1653–7, Amsterdam, Netherlands, September, 1999.

[16] S. Grandhi and J. Zander, 'Constrained power control in cellular radio systems', in *Proceedings of the IEEE Vehicular Technology Conference*, vol. 2, pp. 824–8, Piscataway, USA, June, 1994.

[17] F. Berggren, Power control, transmission rate control and scheduling in cellular radio systems, Licentiate thesis, Royal Institute of Technology, Stockholm, Sweden, 2001.

[18] R. Jäntti and S. Kim, 'Selective power control and active link protection for combined rate and power management', *Proceedings of the IEEE Vehicular Technology Conference*, pp. 1960–4, Tokyo, Japan, May, 2000.

[19] K. Miettinen, *Nonlinear Multiobjective Optimization*, Kluwer, 1998.

[20] M. Elmusrati, H. El-Sallabi and H. Koivo, 'Applications of multi-objective techniques in radio resource scheduling of cellular communication systems', *IEEE Transactions on Wireless Communications*, January, 2008.

9

Smart Antennas

The receiver and transmitter antennas are two of the most critical components in the design of wireless communication systems. A good design of the antenna can relax system requirements, improve overall system performance and greatly reduce the infrastructure costs [1]. Beam-forming (or adaptive/smart antennas) is one application of systems engineering in communications. It can be explained as a closed-loop adaptive control system to achieve predefined objectives. Figure 9.1 shows a simple block diagram for the beamforming system at the receiver site. The main objective of the beamforming is to adjust complex weights connected to every antenna branch to reshape (virtually) the receiver antennas' radiation pattern to achieve the required objective. One traditional objective is to maximize the received SNR. There are many algorithms to achieve this and other objectives as will be described later.

Exercise 9.1 Compare Figure 9.1 with the typical closed-loop control systems given in Chapter 2.

Adaptive beamforming or adaptive antennas is a more general name than smart antennas. Smart antennas are multiple antennas where the adaptation of the complex weights is according to certain optimization criteria. Some adaptive beamforming algorithms do not follow any optimization algorithms such as random beamforming which has certain applications such as increasing the fairness in opportunistic scheduling systems. Beamforming (without adaptive) can also be used to refer to the adaptive algorithms or to fixed beamforming such as sectoring. We will use these names interchangeably in the text. However, in our opinion, 'adaptive antennas' is not very accurate term because the antennas themselves are not adapting, but the weights connected to them. However, we will still use it in this text since there is general acceptance for the term. The performance enhancement of the adaptive antennas is due to the reduction in the interference power by attenuating the interference signals which have different directions of arrivals than the desired signal direction of arrival at the receiver antenna site. This can be done by exploiting the phase differences between the antenna terminals which depend on the direction of arrival of the signal. Such a process is known as spatial processing because the direction of arrival is related to the transmitter location.

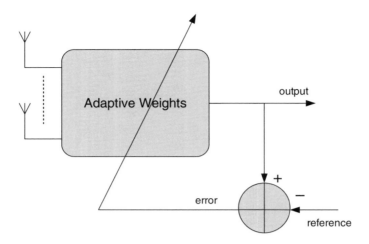

Figure 9.1   General adaptive beamforming structure

Generally, beamforming can be implemented at the transmitter or receiver sides. For example, if we assume a multiuser environment with access points (such as cellular system), beamforming can be implemented at the transmitter of the uplink (i.e., at the terminal side), the receiver of the uplink (i.e., at the access point side), at the transmitter of the downlink (i.e., at the access point side), and at the receiver of the downlink (i.e., at the terminal side). Hence, we have four situations for the beamforming. The gained benefits obtained by each depend on the application and the limitations. For example, in cellular networks, since many terminals communicate through the same access point, we will have high interference in the uplink. Utilizing beamforming at the receiver of the access point can enhance the performance and capacity of the uplink. Implementing beamforming in the receiver of the downlink, i.e., at the terminal or the handset, is not so attractive in cellular systems. There are some reasons for this such as: (i) the handset is small in size and limited in the processing power, therefore it is not so easy to implement many antenna terminals with beamforming capabilities; (ii) the number of strong interferers in the downlink in a macro-cell situation is limited with few other access points, so that the maximum achieved gain from beamforming should be carefully investigated. On other hand, using beamforming at the transmitter as well as the receiver sides in ad hoc networks may provide a very desirable influence on all system performance. Also, using beamforming in the transmitter of cognitive radio can be critical in some situations to avoid any harmful interference with primary users (see Chapter 10). Regardless of the applications, the beamforming concepts are similar. Hence, without loss of generality, we will consider in this chapter the beamforming in the uplink at the receiver side.

The system performance can be further improved by exploiting the delay spread of the received signals. The signal of each transmitter arrives at the receiver antenna in multipath form. Each path usually has its own delay and direction of arrival. Using the smart antenna alone means that we receive (ideally) only one path and ignore the others. Combining temporal processing and spatial processing can considerably enhance the overall system performance. However, it is important to emphasize that the smart antenna is not effective in the case of equally distributed power paths, i.e., the Rayleigh channel. In such a case it is much better to use MIMO techniques which can exploit the uncorrelated nature of the different paths. The MIMO topic will be introduced later. Smart antenna is very effective in line of sight propagation or a few strong multipaths.

The question now is, how can these multi-antennas determine (or estimate) the direction of arrival of signals? The answer is that there will be phase differences between the received signals at each antenna's output due to the trip delay of the electromagnetic wave between the antennas. This phase depends on the direction of arrival of the electromagnetic wave. Since we know that the electromagnetic waves travel with a speed close to the speed of light, and the difference between the antennas can be in terms of centimeters, how can we see the considerable phase difference between the signals in this case? This inquiry is handled in the following example.

Example 9.1

A WiFi receiver utilizes two antennas separated by 3 cm; if the signal comes from a direction perpendicular to the line joining the antennas, find the delay and the phase difference. The carrier is 2.4 GHz.

Solution

Assuming that the signal at the first antenna output has the following form: $x_1(t) = \text{Re}\{m(t)e^{j(2\pi f_0 t + \theta)}\}$. The signal at the second antenna's output will be delayed by τ, where $\tau = 3/(3 \times 10^{10}) = 1 \times 10^{-10}$ s. The signal at the second antenna will be $x_2(t) = \text{Re}\{m(t)e^{j(2\pi f_0(t-\tau)+\theta)}\} = \text{Re}\{m(t)e^{j(2\pi f_0 t - 2\pi f_0 \tau + \theta)}\} = \text{Re}\{m(t)e^{-j2\pi f_0 \tau}e^{j(2\pi f_0 t + \theta)}\}$. The phase difference is $2\pi f_0 \tau$ which is quite large due to the high carrier frequency $2\pi \times 2.4 \times 10^9 \times 1 \times 10^{-10} = 0.48\pi$.

Exercise 9.2   If the direction of arrival of the received signal is 0.2π, repeat Example 9.1.

9.1 Smart Antennas and Adaptation

Beamforming antenna systems can be classified as a fixed beamforming network, a switched beam system, and an adaptive antenna system or smart antennas. By using an adaptive antenna system, it is possible to achieve greater performance improvements than are attainable using a switched beam system or a fixed beamforming network.

A smart antenna system consists of a set of antenna elements distributed in a certain configuration. Each antenna terminal is connected through a complex weight as shown in Figure 9.2. By smart adaptation of these weights the radiation pattern of the antenna array can be adjusted in a proper way to minimize a certain error function or to maximize a certain reward function. This adaptation is performed using an adaptive algorithm. Many adaptive algorithms have been published in the literature [2,3].

The distance between the antenna elements is very small compared with the distance between the array and the transmitter antenna. Therefore it is convenient to ignore the differences in the amplitude between the received signals at each antenna terminal, but the differences in the phase cannot be ignored as shown in Example 9.1. Next, a brief summary of smart antenna models is given. In our analysis we have assumed one receiver (can be a base station) is dedicated to receive the signal from one transmitter (can be a mobile phone). There are Q transmitters transmitting at the same time from random locations. Our aim is to adjust the antenna's weights to reduce the effect of the $(Q-1)$ interferes at the receiver output. In the

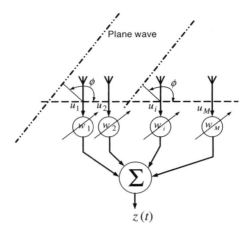

Figure 9.2   Antennas and weights configuration

sequel we refer to the transmitter as user or TX and the receiver as RX. Link i means the link between transmitter i and receiver i.

The phase difference $\Delta\Psi_m$ between the antenna element m and the reference element at the origin can be easily derived as

$$\Delta\Psi_m = \beta\Delta d_m = \beta(x_m \cos(\phi) \sin(\theta) + y_m \sin(\phi) \sin(\theta) + z_m \cos(\theta)), \quad m = 1, \ldots, M \tag{9.1}$$

where (θ, ϕ) are the elevation angle and the azimuth angle, respectively, $\beta = 2\pi/\lambda$ is the phase propagation factor, λ is the wavelength, and (x_m, y_m, z_m) is the Cartesian position of the antenna element m with respect to a reference element (assumed to be at the origin). When the receiver antenna is considerably higher than the transmitter antenna as in a macro-cell cellular system, it is very common to assume $\theta = \pi/2$. In this case Equation (9.1) becomes:

$$\Delta\Psi_m = \beta(x_m \cos(\phi) + y_m \sin(\phi)), \quad m = 1, \ldots, M \tag{9.2}$$

Exercise 9.3   With the help of Figure 9.2, derive Equation (9.1).

From Figure 9.2, the output signal $z(t)$ can be expressed as

$$z(t) = \sum_{k=1}^{M} w_k u_k(t), \tag{9.3}$$

or in a more compact vector form as

$$z(t) = \mathbf{w}^H \mathbf{u}(t), \tag{9.4}$$

where $\mathbf{w} = \begin{bmatrix} w_1 & \cdots & w_M \end{bmatrix}^H$ is the weight vector, the superscript H represents the Hermitian transpose, $\mathbf{u}(t) = [u_1(t) \cdots u_M(t)]'$ is the received signal vector, and $[\bullet]'$ means matrix transpose (without conjugating its elements). The phase differences between the signals at each antenna terminal ($\Delta\Psi_m$) depend on the direction of arrival (DoA) of each signal (θ, ϕ). It is convenient to take the first element as a reference element so that $\Delta\Psi_1 = 0$. We may define a vector of the relative phase differences with respect to the first element, which is called the *steering vector*, as

$$\mathbf{a}(\theta, \phi) = [1 \exp(-j\Delta\Psi_2(\theta, \phi)) \cdots \exp(-j\Delta\Psi_M(\theta, \phi))]' \tag{9.5}$$

The steering vector plays an essential role in the analysis of multi-antenna systems. It is a very strong and simple representation of the relation between the received signal at the antennas' output and the DoA of the signals.

Example 9.2

Find the general form of the steering vector for a three-receiving antenna system allocated on the *x*-axis as shown in Figure 9.3. The antenna elements are separated by a distance of 5 cm. The carrier frequency is 900 MHz.

Solution

From Equation (9.1), the phase difference for the *m*th element is given by

$$\Delta\Psi_m = \frac{2\pi}{\lambda} (x_m \cos(\phi) \sin(\theta) + y_m \sin(\phi) \sin(\theta) + z_m \cos(\theta)),$$

where

$$\lambda = \frac{3 \times 10^{10}}{900 \times 10^6} = 33.3 \ cm.$$

Let's assume that the first element is our reference, i.e., at (0,0,0), $\Delta\Psi_1 = 0$, for the second element at (5,0,0), $\Delta\Psi_2 = \frac{2\pi}{33.3} (5 \cos(\phi) \sin(\theta)) = 0.3\pi \cos(\phi) \sin(\theta)$, and finally $\Delta\Psi_3 = \frac{2\pi}{33.3}$ $(10 \cos(\phi) \sin(\theta)) = 0.6\pi \cos(\phi) \sin(\theta)$. From the above the steering vector becomes $\mathbf{a}(\theta, \phi) = [1 \exp(-j0.3\pi \cos(\phi) \sin(\theta)) \exp(-j0.6\pi \cos(\phi) \sin(\theta))]'$.

Example 9.3

Using the result of Example 9.2 compute the steering vector for the following two DoA signals:
(a) $(\theta, \phi) = (0.2\pi, 0.3\pi)$; (b) $(\theta, \phi) = (0.5\pi, 0.3\pi)$

Solution

(a) $\mathbf{a}(\theta, \phi) = [1 \exp(-j0.3\pi \cos(0.3\pi) \sin(0.2\pi)) \exp(-j0.6\pi \cos(0.3\pi) \sin(0.2\pi))]' =$
$[1 \quad 0.94 - j0.32 \quad 0.79 - j0.61]'$
(b) $\mathbf{a}(\theta, \phi) = [1 \quad 0.85 - j0.52 \quad 0.45 - j0.89]'$

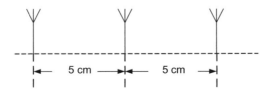

Figure 9.3 Antenna configuration of Example 9.2

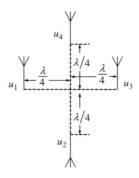

Figure 9.4   Antenna configuration of Example 9.3

Example 9.4

Find the steering vector of the antenna pattern shown in Figure 9.4 for carrier frequency of 1200 MHz.

Solution

First we compute λ such as: $\lambda = \frac{3 \times 10^{10}}{12 \times 10^8} = 25$ *cm*, Assuming that the first antenna element is our reference at (0,0,0), then the second antenna element u_2 location is (6.25,-6.25,0), u_3 at (12.5,0,0), and u_4 at (6.25,6.25,0). The steering vector can be determined by a direct substitution of the relative antenna locations in Equation (9.5) in the following manner: for $u_1 \Delta \Psi_1 = 0$, for u_2 $\Delta \Psi_2 = \frac{2\pi}{25} (6.25 \cos(\phi) \sin(\theta) - 6.25 \sin(\phi) \sin(\theta))$, for u_3 $\Delta \Psi_3 = \frac{2\pi}{25}$ $(12.5 \cos(\phi) \sin(\theta))$, and finally for u_4, $\Delta \Psi_4 = \frac{2\pi}{25} (6.25 \cos(\phi) \sin(\theta) + 6.25 \sin(\phi) \sin(\theta))$.

Exercise 9.4   Compute the steering vector for Example 9.4 for $(\theta, \phi) = (\pi/2, 0.3\pi)$.

The input signal at each antenna terminal is the convolution between the transmitted signal and the channel impulse response as follows:

$$u_{ij}(\tau, t) = s_i(t) * h_{ij}(\tau, t), \quad i = 1, \ldots, Q, j = 1, \ldots, M, \tag{9.6}$$

where the star indicates the convolution operation, $s_i(t)$ is the transmitted signal from transmitter i, and h_{ij} is the impulse response of the channel between transmitter i and antenna element j at the receiver, Q is the number of transmitters, and u_{ij} is the received signal from transmitter i at antenna terminal j of the receiver.

The wireless channel between the transmitter and the receiver can be represented as (see Chapter 1)

$$\mathbf{h}_i(\tau, t) = \sum_{l=1}^{B_i} \mathbf{a}_i(\theta_l, \phi_l) \alpha_{il}(t) \delta(t - \tau_l), \quad i = 1, \ldots, Q \tag{9.7}$$

where the subscript l represents the path number, $\mathbf{h}_i$ is the channel impulse response vector and τ_l is the time delay of the signal of TX i to the RX through path l. It is assumed also that there are B_i paths for the signal at RX i. The complex channel gain $\alpha_{il}(t)$ is given by

$$\alpha_{il}(t) = \sqrt{\rho_{il}} \exp(j(2\pi f_{il}t + \varphi_{il})), \quad i = 1, \ldots Q, \tag{9.8}$$

where ρ_{il} is the absolute channel gain given by

$$\rho_{il} \approx \frac{A_{il}}{d_{il}^{\eta_{il}}}, \quad i = 1, \ldots, Q, \tag{9.9}$$

Here A_{il} is the log-normal shadowing effect for path l of TX i, d_{il} is the distance between the RX and TX i through path l, η_{il} is the path-loss exponent for TX i through path l, f_{il} is the Doppler shift, and φ_{il} is the phase offset.

The received signal of TX i at antenna terminal j can be described as

$$\begin{aligned}
u_{ij}(t) &= s_i(t) * \sum_{l=1}^{B_i} \exp(-j\Delta\Psi_{jl})\sqrt{\rho_{il}} \exp(j(2\pi f_{il}t + \varphi_{il}))\delta(t - \tau_l) + n_j(t) \\
&= \sum_{l=1}^{B_i} \exp(-j\Delta\Psi_{jl})\sqrt{\rho_{il}} \exp(j(2\pi f_{il}t + \varphi_{il}))s_i(t - \tau_l) + n_j(t)
\end{aligned} \tag{9.10}$$

where $n_j(t)$ is an additive noise at antenna terminal j.

Equation (9.10) can be rewritten in a more compact form [14]:

$$\mathbf{u}_i(t) = \hat{\mathbf{a}}_i \boldsymbol{\alpha}_i(t) \mathbf{s}_i(t) + \mathbf{n}(t), \quad i = 1, \ldots, Q \tag{9.11}$$

where

$$\hat{\mathbf{a}}_i = \begin{bmatrix} \exp(-j\Delta\Psi_{11}) & \cdots & \exp(-j\Delta\Psi_{1B_i}) \\ \vdots & \cdots & \vdots \\ \exp(-j\Delta\Psi_{M1}) & \cdots & \exp(-j\Delta\Psi_{MB_i}) \end{bmatrix} \quad i = 1, \ldots, Q \tag{9.12}$$

is the multipath steering matrix,

$$\begin{aligned}
\boldsymbol{\alpha}_i(t) &= diag\left(\sqrt{\rho_{il}} \exp(j(2\pi f_{il}t + \varphi_{il}))\right) \\
&= \begin{bmatrix} \sqrt{\rho_{i1}}\exp(j(2\pi f_{i1}t + \varphi_{1l})) & 0 & 0 \\ \vdots & \ddots & \vdots \\ 0 & 0 & \sqrt{\rho_{iB_i}}\exp(j(2\pi f_{iB_i}t + \varphi_{iB_i})) \end{bmatrix}
\end{aligned} \tag{9.13}$$

and the transmitted signal vector $\mathbf{s}_i(t) = \begin{bmatrix} s_i(t - \tau_1) & \cdots & s_i(t - \tau_{B_i}) \end{bmatrix}'$, $i = 1, \ldots, Q$.

The total received signal from Equation (9.11) results in

$$\mathbf{u} = \sum_{i=1}^{Q} \mathbf{u}_i = \sum_{i=1}^{Q} \mathbf{a}_i \boldsymbol{\alpha}_i \mathbf{s}_i + \mathbf{n} \tag{9.14}$$

The output signal of receiver i (which is used to receive the signal from transmitter i is

$$z_i(t) = \mathbf{w}_i^H \mathbf{a}_i \boldsymbol{\alpha}_i(t) \mathbf{s}_i(t) + \mathbf{w}_i^H \sum_{\substack{k=1 \\ k \neq i}}^{Q} \mathbf{a}_k \boldsymbol{\alpha}_k(t) \mathbf{s}_k(t) + \mathbf{w}_i^H \mathbf{n}(t), \tag{9.15}$$

Now the beamforming problem can be stated as follows: What is the optimum weight vector $\mathbf{w}_i$ needed to enhance the performance of link i? Note that the additive white noise $\mathbf{n}(t)$ will not be considerably affected by adapting the antenna weights because usually it is not directive. For this reason we will drop the weight vector associated with the additive noise term in the coming analysis. The second term of Equation (9.15) represents the interference from other users. This term can be minimized by a proper selection of the weight vector.

Generally the weights should be adjusted to minimize (or reduce) the interference from other TXs, or equivalently to maximize the SINR at the RX output. From Equation (9.15) we can formulate the average output SINR (for independent process) to be

$$\Gamma_i(t) = \frac{E\left[|\mathbf{w}_i^H \mathbf{a}_i \boldsymbol{\alpha}_i(t) \mathbf{s}_i(t)|^2\right]}{E\left[\left\|\mathbf{w}_i^H \sum_{\substack{k=1 \\ k \neq i}}^{Q} \mathbf{a}_k \boldsymbol{\alpha}_k(t) \mathbf{s}_k(t) + \mathbf{n}(t)\right\|^2\right]} \tag{9.16}$$

where $E[.]$ is the expectation operator. For LoS case and with snapshot assumption (see Chapter 4), Equation (9.16) can be simplified as

$$\Gamma_i = \frac{|\mathbf{w}_i^H \mathbf{a}_i|^2 P_i G_{ii}}{\sum_{\substack{j=1 \\ j \neq i}}^{Q} |\mathbf{w}_i^H \mathbf{a}_j|^2 P_j G_{ij} + \delta_n^2} \tag{9.17}$$

where P_i is the transmit power of terminal i, G_{ij} is the channel gain between terminal j to receiver i.

Next we will present some commonly used algorithms for the weight computation. Without loss of generality, we will assume that every TX has one dominant path.

9.1.1 Conventional Beamformer

How should the antenna weights of the receiver be adjusted when only the DoA information of the required transmitter is known (i.e., the steering vector of the desired transmitter)? The antenna weights are selected to be the complex conjugates of the steering vector, i.e., for a single-path case the weight vector is selected as

$$\mathbf{w}_i^H \mathbf{a}_i = c, \quad i = 1, \ldots, Q, \tag{9.18}$$

where c is a positive real number. It is clear that that conventional beamforming makes coherent combining of signals received at antennas' terminals. This method does not take into consideration the interference of other users, however, it maximizes the output average SINR in additive white noise channels (i.e., there are no other directional transmitters) [3]. Another advantage of the conventional beamformer is its simplicity. On the other hand the method has many disadvantages, which makes it an unattractive choice to update the antenna weights. Some of these are:

1. It does not take into account the interference transmitters despite the fact that the main purpose of using smart antennas is to reduce the interference.
2. It needs to know the DoA of the desired user which is not always possible especially in a multipath environment.

The SINR at the array output can be derived by substituting Equation (9.18) into Equation (9.15) and setting the second term equal to zero. The average signal-to-noise power ratio where the noise is assumed to be uncorrelated becomes

$$SINR_i = \frac{c^2 \alpha_i^2 P_i}{\delta_N^2}, \quad i = 1, \ldots, Q \tag{9.19}$$

where P_i is the average transmitted signal power and δ_N^2 is the noise variance. Compared with a single antenna receiver, using the conventional beamformer gives some gain for SINR depending on the number of antenna elements.

Example 9.5

Using MATLAB® draw the radiation pattern of a beamforming antenna receiver utilizing the conventional beamforming method with two, four, and six antenna elements. The antennas are located on the x-axis and separated by $\lambda/2$. The DoA of the (line of sight) transmitter signal is $(\theta, \phi) = (\pi/2, 0.7\pi)$. The radiation pattern is performed according to the effects of the weighted sum of the antenna elements. We assume that every antenna element is isotropic, which is an ideal assumption. In practice every antenna element, e.g. dipole, has its own radiation pattern.

Solution

First we need to compute the steering vector of the received signal, and then we take its conjugate. The result is shown in Figure 9.5. It is clear that as the number of antennas increases, the beamwidth decreases and the relative gain increases. The resultant pattern is independent of the interferers' location and power. This is one disadvantage of the conventional beamforming method.

The radiation pattern that is shown here is the resultant after the weighting sum of the signals on the antenna terminals. It is not the physical radiation pattern of the antennas. In other words, every physical antenna has a radiation pattern depending on its configuration and construction. Exploiting the phase differences between the signals on the antenna terminals we can cancel some signals coming from certain DoAs and on the other side enhance other DoA signals.

Assume an array of parabola antennas with a front-side gain of 40 dB and at angle θ the gain is only 1 dB. By using the weighting sum of the signals we can cancel the reception from certain angles. If the desired signal comes from angle θ and the interference comes from the front side, then theoretically we may cancel the interference but we cannot enhance the antenna gain on the angle θ to be 40 dB. If the received signal is less than the sensitivity of the receiver, it could

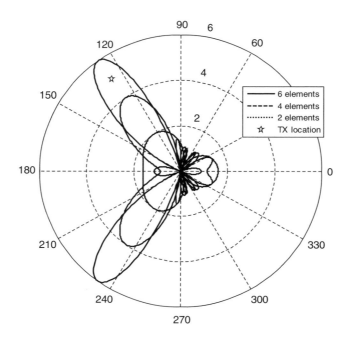

Figure 9.5   Radiation pattern result of Example 9.5

be better to use a physically rotated antenna to direct it in the direction of angle θ, so that we will have a physical gain of 40 dB in that direction. This maybe one reason why physically rotated radars are still in use although that we can use electronic scanning in the sky.

Exercise 9.5   Explain how the following smart antenna system with a delay line in Figure 9.6, could be used in conventional beamforming. What is the value of the delay T to obtain the conventional beamforming system?

9.1.2 Null-Steering Beamformer

This technique is more effective than the conventional beamformer in minimizing the signals of strong directional (intentional or unintentional) interferences. If we know the DoA of all interferers' signals, we can find the weights which null the interferers deterministically, at the condition that the number of interferers' signals is not larger than the number of antenna elements. Actually, even if it is larger, we can still reduce the interference effects considerably. If there are Q transmitters in the cell, and the weights are calculated for transmitter i, then the desired weight vector is the solution of the following system of linear equations (see Equation (9.15)):

$$\begin{aligned} \mathbf{w}_i^H \mathbf{a}_i &= 1, \quad \text{and} \\ \mathbf{w}_i^H \mathbf{a}_k &= 0, \quad \forall k \in \{1, 2, \dots, Q\} \quad \text{and} \quad k \neq i, \, i = 1, \dots, Q \end{aligned} \tag{9.20}$$

The above system of linear equations can be solved exactly if the number of transmitters Q is less than or equal to the number of antenna elements M. If the number of transmitters is larger

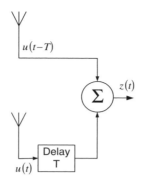

Figure 9.6 Conventional beamformer for two antenna elements

than the number of antenna elements we should use the minimum least-squares method to solve such an overdetermined linear system. Generally, the optimum weight vector can be determined by the following formula:

$$\mathbf{w}_i^H = D'(\mathbf{A}^H\mathbf{A})^{-1}\mathbf{A}^H, \quad i = 1, \dots, Q, \tag{9.21}$$

where $D = [0 \quad \cdots \quad 1 \quad 0 \quad \cdots \quad 0]'$; 1 is at the ith location, and $\mathbf{A} = [\mathbf{a}_1 \quad \cdots \quad \mathbf{a}_Q]$

Exercise 9.6   Prove that Equation (9.21) is the general solution for Equation (9.20).

The main advantage of the null-steering beamformer is its ability to null even strongly directed interferences (when their number is less than the number of antenna elements). The disadvantages are:

1. DoA of all transmitters should be known (or estimated) at the receiver.
2. The number of antenna elements should be comparable with the number of users. If the number of users is much more than the number of antenna elements, then the method becomes less effective and the weights approach zero.
3. It does not take the additive noise from other sources into consideration when calculating the optimum weights.

Example 9.6

Assume that the desired transmitter signal arrives at the receiver with DoA of $(\theta, \phi) = (\pi/2, 0.3\pi)$ and its transmitted power is 1 W. The distance between the transmitter and the receiver is given as 200 m. The jamming station tries to jam the communication link, transmitting with DoA of $(\theta, \phi) = (\pi/2, 0.7\pi)$ and with transmitted power of 20 W. The distance between the interferer and the receiver is 100 m. Calculate the received SINR if you use a single omnidirectional antenna at the receiver and assuming zero Doppler frequencies as well as zero phase shift. The path-loss exponent is assumed to be 4. The additive white noise at the receiver is −70 dBm and the carrier frequency is 1.8 GHz. The minimum SINR to decode the desired signal is 7 dB.

Solution

In this case we will not use the DoA information because we use only one antenna element. The average received signal power from the desired transmitter is $P_{rd} = \frac{P_{td}}{r^4} = \frac{1}{200^4} = 6.25 \times 10^{-4} \mu W$, and the received power from the jamming station is $P_{ri} = \frac{P_{ti}}{r^4} = \frac{20}{100^4} = 0.2 \mu W$. The SINR can be easily calculated as the average received power from the desired transmitter divided by the average received power from the interferer pulse the additive noise, i.e., $\Gamma = \frac{6.25 \times 10^{-4}}{0.2 + 10^{-4}} = 0.0031 \Rightarrow \Gamma_{dB} = -25 dB$. It is clear that the desired signal quality is very bad and much lower than the minimum required SINR. Hence, the signal cannot be decoded.

Example 9.7

In the previous example you want to improve the received SINR and to reduce the impacts of the jammed signal. At the receiver two antenna elements on the x-axis are installed and they are separated by $\lambda/2$. Calculate the optimum weight vector using the conventional and then the null steering beamformers. Calculate the resultant SINR in each case and compare it with the previous example.

Solution

The steering vector of the desired signal is $\mathbf{a}_d = [1 \ \exp(-j\pi \cos(0.3\pi))]'$, and for the interferer it is $\mathbf{a}_i = [1 \ \exp(-j\pi \cos(0.7\pi))]'$. The optimum weight using the conventional beamformer is $\mathbf{w}_c = \mathbf{a}_d^* = [1 \ \exp(j\pi \cos(0.3\pi))]'$. With the null-steering beamformer we are looking for the weight vector $\mathbf{w}_n = [w_1, w_2]^H$ such that $\mathbf{w}_n^H \mathbf{a}_d = 1$, and $\mathbf{w}_n^H \mathbf{a}_i = 0$, We can solve it directly using Equation (9.21), however, we will introduce it here in a step-by-step solution for explanation. The two equations are:

$$w_1 + \exp(-j\pi \cos(0.3\pi))w_2 = 1$$
$$w_1 + \exp(-j\pi \cos(0.7\pi))w_2 = 0$$

which can be solved easily to find that the optimum weight vector is $w_1 = 0.5 + j0.14$, $w_2 = j0.52$. Now with these weight values the signal of the interference signal will vanish completely (of course theoretically!). The resultant SINR can be computed using Equation (4.15). For a conventional beamformer the resultant SINR is

$$\Gamma = \frac{6.25 \times 10^{-4} \times 2}{0.2 \times 0.3 \times 10^{-4}} = 0.021 \Rightarrow \Gamma_{dB} = -15 dB$$

It is clear that the SINR has considerably improved (by order of 10) with the conventional beamformer. However, this is still not enough because of the high interference power. By calculating the SINR with the null-steering beamformer we obtain

$$\Gamma = \frac{6.25 \times 10^{-4} \times 1}{0.2 \times 0 + 10^{-4}} = 6.25 \Rightarrow \Gamma_{dB} \cong 8 dB$$

The SINR has greatly improved according to the complete cancellation of the interference signal. With two antenna elements we are able to completely cancel only one interference signal and at the same time maintain an accepted level for the desired signal. Example 9.8 demonstrates this point.

Example 9.8

Repeat Example 9.7 for another intentional jammed signal with DoA of $(\theta, \phi) = (\pi/2, 1.2\pi)$, its transmitted power is 10 W, and distance 200 m is added to the system. Comment on the results.

Solution

The steering vector of the added interferer is $\mathbf{a}_{i2} = [1 \ \exp(-j\pi \cos(1.2\pi))]'$ and its average received power value is $P_{ri2} = \frac{P_{t2}}{r^4} = \frac{10}{200^4} = 6.2 \times 10^{-3}\mu W$.

In this case we try to achieve the following system of linear equation with two weights:

$$w_1 + \exp(-j\pi \cos(0.3\pi))w_2 = 1$$
$$w_1 + \exp(-j\pi \cos(0.7\pi))w_2 = 0$$
$$w_1 + \exp(-j\pi \cos(1.2\pi))w_2 = 0$$

It is clear that generally there is no solution which can solve (exactly) the above overdetermined system of linear equations. However, it is possible to find the best solution according for example to the minimum square error. The solution in this case is computed using Equation (9.21) as $w_1 = 0.47 + j0.22$, $w_2 = 0.08 + j0.51$. The new weight may not cancel both interferers but will minimize their effects. To be able to cancel both interferers a third antenna element should be added to the system. The resultant SINR in this case can be calculated as:

$$\Gamma = \frac{6.25 \times 10^{-4} \times 0.87}{0.2 \times 0.03 + 6.2 \times 10^{-3} \times 0.04 + 10^{-4}} = 0.086 \Rightarrow \Gamma_{dB} \cong -10dB$$

It is clear that there is a large degradation in the performance when another interferer is added. In this example, because the first interferer is much stronger than the second one, it is better to adjust the weights to completely cancel it and ignore the second interferer. The SINR in this case can be computed as

$$\Gamma = \frac{6.25 \times 10^{-4} \times 1}{0.2 \times 0 + 6.2 \times 10^{-3} \times 0.13 + 10^{-4}} = 0.7 \Rightarrow \Gamma_{dB} \cong -1.5dB$$

which is still not good, but much better than the previous solution.

Exercise 9.7   Repeat Example 9.8 by adjusting the weight vector to cancel both interferers. Find the nontrivial solution then compute the resultant SINR.

Example 9.9

Show the antenna pattern resulting from the previous two examples. Show also the DoA of the desired signal as well as the interferers. The difference in the resultant radiation pattern between the null-steering and conventional beamformers is clear. The null-steering beamforming algorithm makes null or complete cancellation of the signals from the interferer DoA. However, the difference between Figures 9.7(b) and 9.7(c) is not visually clear. Hence the weight values are very close, but the difference in the performance is large as we showed in the previous examples.

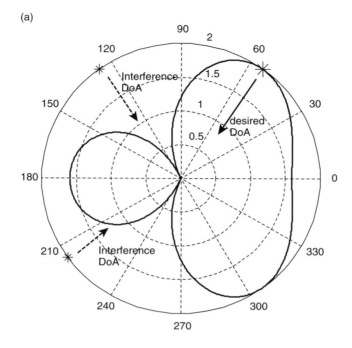

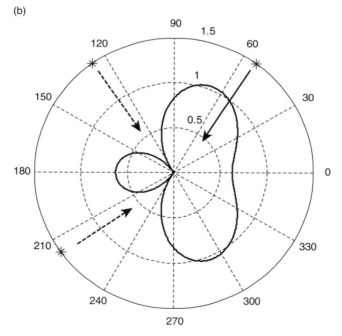

Figure 9.7 Radiation pattern results of Example 9.8

(c)

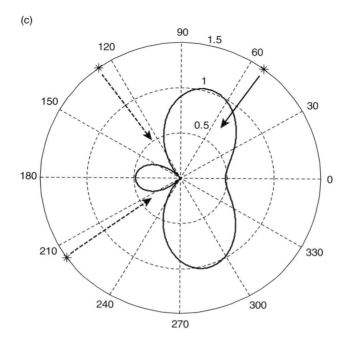

Figure 9.7 *(Continued)*

9.1.3 Minimum Variance Distortionless Response (MVDR) Beamformer

The MVDR beamformer overcomes the disadvantages of the null-steering beamformer. In this method only the steering vector of the desired user is needed. The concept of the MVDR beamformer is based on minimizing the average output array power while maintaining unity response in the looking direction. The problem can be described mathematically as follows:

$$\hat{\mathbf{w}}_i = \arg \min E[||z(t)|^2], \quad i = 1, \ldots, Q \tag{9.22}$$

subject to

$$\mathbf{w}_i^H \mathbf{a}_i = 1, \tag{9.23}$$

where $z(t)$ is given by Equation (9.4) and $E[.]$ is the expectation operator. The weights obtained by solving the optimization problem given in Equations (9.22) and (9.23) will minimize the total noise, including interferences and uncorrelated noise. So the MVDR beamformer maximizes the output SINR [3].

Substituting Equation (9.4) into Equation (9.22) the problem can be stated as follows: find the optimum weight vector $\mathbf{w}_i$ which minimizes the following objective function

$$\mathbf{w}_i^H R_{uu} \mathbf{w}_i, \quad i = 1, \ldots, Q \tag{9.24}$$

subject to

$$\mathbf{w}_i^H \mathbf{a}_i = 1. \tag{9.25}$$

The array correlation matrix of the received signal $\mathbf{u}$ can be computed by

$$\mathbf{R}_{uu} = E[\mathbf{u}\mathbf{u}^H]. \tag{9.26}$$

The MVDR problem can be solved using the Lagrange multiplier method as follows

$$C = \mathbf{w}_i^H \mathbf{R}_{uu} \mathbf{w}_i + \lambda_i(\mathbf{w}_i^H \mathbf{a}_i - 1)$$

$$\frac{\partial C}{\partial \mathbf{w}_i} = 2\mathbf{R}_{uu}\hat{\mathbf{w}}_i + \lambda_i \mathbf{a}_i = 0 \Rightarrow \lambda_i \mathbf{a}_i = -2\mathbf{R}_{uu}\hat{\mathbf{w}}_i \Rightarrow \hat{\mathbf{w}}_i = -\frac{1}{2}\lambda_1 \mathbf{R}_{uu}^{-1}\mathbf{a}_i$$

$$\frac{\partial C}{\partial \lambda_i} = \hat{\mathbf{w}}_i^H \mathbf{a}_i - 1 = 0 \Rightarrow -\frac{1}{2}\lambda_i \mathbf{a}_i^H \mathbf{R}_{uu}^{-1}\mathbf{a}_i = 1 \Rightarrow \lambda_i = \frac{-2}{\mathbf{a}_i^H \mathbf{R}_{uu}^{-1}\mathbf{a}_i}$$

$$\Rightarrow \hat{\mathbf{w}}_i = \frac{\mathbf{R}_{uu}^{-1}\mathbf{a}_i}{\mathbf{a}_i^H \mathbf{R}_{uu}^{-1}\mathbf{a}_i} \tag{9.27}$$

To find the optimum weights using the MVDR method, the DoA of the desired user is needed. Since in mobile communication systems, the users are moving, and the characteristics of the channel are time varying, an adaptive algorithm is needed to update the weights for the varying conditions. The sample matrix inversion (SMI) method can be used in adaptive beamforming algorithms. The weights are updated at every kth iteration using the K-sample correlation matrix:

$$\hat{\mathbf{R}}_K = \frac{1}{K}\sum_{j=1}^{K} \mathbf{u}(t_j)\mathbf{u}(t_j)^H \tag{9.28}$$

where $\hat{\mathbf{R}}_K$ is the unstructured maximum likelihood estimate of $\mathbf{R}_{uu}$. It converges to $\mathbf{R}_{uu}$ as $K \to \infty$ under the ergodic assumption. The SMI-based adaptive MVDR weights are given by [4]:

$$\hat{\mathbf{w}}_i = \frac{\hat{\mathbf{R}}_K^{-1}\mathbf{a}_i}{\mathbf{a}_i^H \hat{\mathbf{R}}_K^{-1}\mathbf{a}_i}, \quad i = 1,\ldots,Q, \tag{9.29}$$

which has the same form as in Equation (9.27).

Example 9.10

For a two-element linear smart antenna separated by $\lambda/2$, we have taken 10 samples at both antenna elements. The sampled signal is given by

$$u = \begin{bmatrix} 0.1 & 0.3 & -0.1 & 0.2 & 0.7 & 0.5 & -0.3 & 0.1 & 0.4 & 0.9 \\ 0.2 & 0.4 & -0.2 & 0.5 & 0.2 & -0.1 & 0.1 & 0.7 & -0.3 & 0.1 \end{bmatrix}$$

The DoA of the desired signal is estimated as $(\theta, \phi) = (\pi/2, 1.2\pi)$. Find the optimum weight vector.

Solution

$$\because \mathbf{a} = [1 \; \exp(-j\pi \cos(1.2\pi))]', \; \mathbf{R}_{uu} = \begin{bmatrix} 0.196 & 0.036 \\ 0.036 & 0.114 \end{bmatrix}$$

$$\Rightarrow \mathbf{w} = \frac{\mathbf{R}_{uu}^{-1}\mathbf{a}}{\mathbf{a}^H \mathbf{R}_{uu}^{-1}\mathbf{a}} = [0.389 - j0.055 \quad -0.535 + j0.299]'$$

Example 9.11

Using MATLAB, generate five users' locations with the following DoAs:

$$\left(\frac{\pi}{2}, \frac{4\pi}{3}\right), \; \left(\frac{\pi}{2}, \frac{2\pi}{5}\right), \; \left(\frac{\pi}{2}, \frac{\pi}{2}\right), \; \left(\frac{\pi}{2}, -\frac{3\pi}{7}\right), \; \text{and} \; \left(\frac{\pi}{2}, \pi\right)$$

All users communicate with the same access point which contains five receivers as shown in Figure 9.8. Compute the optimum weight vector for each user using conventional, null-steering and MVDR beamformers. Compare the SINR achieved with each one. Consider three linear antenna elements separated by $\lambda/2$. For the MVDR beamformer, generate a random vector with 1000 elements for each user, and then use it to construct the array correlation matrix. The distance of each user to the receiver is 120, 140, 200, 190, and 170 m. Assume zero Doppler as well as zero phase shifts and an exponent factor of 4. The additive noise is -80 dBm

Solution

First we will compute the optimum weight vectors for all users using the three discussed methods. Let's start with computing the steering vector for all receivers:

$$\mathbf{a}_1 = \left[1 \; \exp\left(-j\pi \cos\left(\frac{4\pi}{3}\right)\right) \; \exp\left(-j2\pi \cos\left(\frac{4\pi}{3}\right)\right)\right]',$$

$$\mathbf{a}_2 = \left[1 \; \exp\left(-j\pi \cos\left(\frac{2\pi}{5}\right)\right) \; \exp\left(-j2\pi \cos\left(\frac{5\pi}{5}\right)\right)\right]'$$

$$\mathbf{a}_3 = \left[1 \; \exp\left(-j\pi \cos\left(\frac{\pi}{5}\right)\right) \; \exp\left(-j2\pi \cos\left(\frac{\pi}{5}\right)\right)\right]'$$

$$\mathbf{a}_4 = \left[1 \; \exp\left(-j\pi \cos\left(\frac{3\pi}{7}\right)\right) \; \exp\left(-j2\pi \cos\left(\frac{3\pi}{7}\right)\right)\right]'$$

$$\mathbf{a}_5 = [1 \; \exp(-j\pi \cos(\pi)) \; \exp(-j2\pi \cos(\pi))]'$$

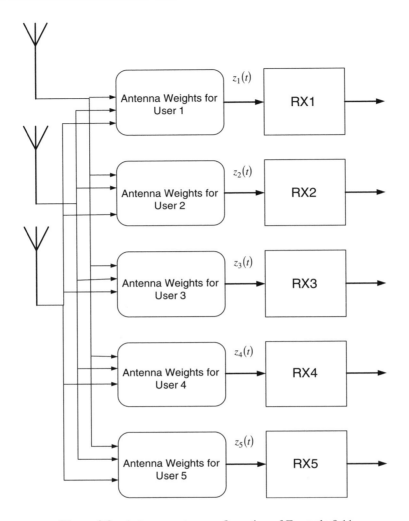

Figure 9.8 Antenna system configuration of Example 9.11

Using conventional beamforming the weight vector for every receiver is just the steering vector conjugate, i.e.,

$$\mathbf{w}_1 = \left[1 \ \exp\left(j\pi \cos\left(\frac{4\pi}{3} \right) \right) \ \exp\left(j2\pi \cos\left(\frac{4\pi}{3} \right) \right) \right]',$$

$$\mathbf{w}_2 = \left[1 \ \exp\left(j\pi \cos\left(\frac{2\pi}{5} \right) \right) \ \exp\left(j2\pi \cos\left(\frac{2\pi}{5} \right) \right) \right]'$$

$$\mathbf{w}_3 = \left[1 \ \exp\left(j\pi \cos\left(\frac{\pi}{5} \right) \right) \ \exp\left(j2\pi \cos\left(\frac{\pi}{5} \right) \right) \right]'$$

$$\mathbf{w}_4 = \left[1 \ \exp\left(j\pi \cos\left(\frac{3\pi}{7} \right) \right) \ \exp\left(j2\pi \cos\left(\frac{3\pi}{7} \right) \right) \right]'$$

$$\mathbf{w}_5 = \left[1 \ \exp(j\pi \cos(\pi)) \ \exp(j2\pi \cos(\pi)) \right]'$$

Using the null-steering beamforming algorithm, the optimum weight vectors are

$$\mathbf{w}_1^H = [1 \quad 0 \quad 0 \quad 0 \quad 0](\mathbf{A}^H\mathbf{A})^{-1}\mathbf{A}^H,$$

$$\because \mathbf{A} = \begin{bmatrix} 1 & 1 & 1 & 1 & 1 \\ \exp\left(-j\pi\cos\left(\frac{4\pi}{3}\right)\right) & \exp\left(-j\pi\cos\left(\frac{2\pi}{5}\right)\right) & \exp\left(-j\pi\cos\left(\frac{\pi}{5}\right)\right) & \exp\left(-j\pi\cos\left(\frac{3\pi}{7}\right)\right) & \exp(-j\pi\cos(\pi)) \\ \exp\left(-j2\pi\cos\left(\frac{4\pi}{3}\right)\right) & \exp\left(-j2\pi\cos\left(\frac{2\pi}{5}\right)\right) & \exp\left(-j2\pi\cos\left(\frac{\pi}{5}\right)\right) & \exp\left(-j2\pi\cos\left(\frac{3\pi}{7}\right)\right) & \exp(-j2\pi\cos(\pi)) \end{bmatrix},$$

$$\Rightarrow \mathbf{w}_1^H = [0.30 - j0.09 \ -j0.34 \ -0.3 - j0.09],$$

$$\mathbf{w}_2^H = [0\,1\,0\,0\,0](\mathbf{A}^H\mathbf{A})^{-1}\mathbf{A}^H = [0.15 + j0.02\ 0.10 + j0.14\ -0.03 + j0.15],$$

$$\mathbf{w}_3^H = [0\,0\,1\,0\,0](\mathbf{A}^H\mathbf{A})^{-1}\mathbf{A}^H = [0.22 + j0.01\ -0.09 + j0.06\ 0.07 - j0.20],$$

$$\mathbf{w}_4^H = [0\,0\,0\,1\,0](\mathbf{A}^H\mathbf{A})^{-1}\mathbf{A}^H = [0.16 + j0.06\ 0.16 + j0.13\ 0.08 + j0.14],$$

$$\mathbf{w}_5^H = [0\,0\,0\,0\,1](\mathbf{A}^H\mathbf{A})^{-1}\mathbf{A}^H = [0.18 + j0.00\ -0.16 - j0.00\ 0.18 - j0.00],$$

Finally, the weight vectors using the MVDR beamforming algorithm can be found after constructing the correlation matrix of the received signal.

$$\mathbf{w}_1^H = [0.29 - j0.13 \quad 0.01 - j0.35 \quad -0.36 - j0.11],$$

$$\mathbf{w}_2^H = [0.29 + j0.04 \quad 0.23 + j0.34 \quad -0.06 + j0.29]$$

$$\mathbf{w}_3^H = [0.37 - j0.09 \quad -0.21 + j0.14 \quad 0.22 - j0.32],$$

$$\mathbf{w}_4^H = [0.41 + j0.24 \quad 0.14 + j0.13 \quad 0.31 + j0.36]$$

$$\mathbf{w}_5^H = [0.38 + j0.12 \quad -0.24 - j0.01 \quad 0.38 - j0.12]$$

We can calculate the SINR for all terminals using the above three methods by using Equation (9.16).

Without beamforming the achieved SINR for the first terminal (with omnidirectional antenna) is:

$$\Gamma_1 = \frac{P_{r1}}{\sum_{k=2}^{5} P_{rk} + \delta_n^2}$$

where P_{rk} is the received power from terminal k and δ_n^2 is the average noise power. From the above

$$\Gamma_1 = \frac{1/120^4}{1/140^4 + 1/200^4 + 1/190^4 + 1/170^4 + 10^{-11}} = 0.93 \Rightarrow \Gamma_{1\_dB} = -0.33dB$$

We can calculate all other SINR in the same manner:

$$\Gamma_{2\_dB} = -4.55dB, \Gamma_{3\_dB} = -11.77dB, \Gamma_{4\_dB} = -10.82dB, \Gamma_{5\_dB} = -8.68dB$$

Using conventional beamforming method the achieved SINR for all terminals are

$$\Gamma_{1\_dB} = 9.22dB, \Gamma_{2\_dB} = 2.07dB, \Gamma_{3\_dB} = -0.53dB, \Gamma_{4\_dB} = -6.53dB, \Gamma_{5\_dB} = 1.83dB$$

The improvement in the SINR is very clear after using conventional beamforming. For example the SINR of the first terminal has been improved by about 10 dB, i.e., the SINR has been increased about 10 times.

Using the null-steering beamforming we obtain the following SINR values

$$\Gamma_{1\_dB} = 23.15dB, \Gamma_{2\_dB} = 3.31dB, \Gamma_{3\_dB} = -3.31dB, \Gamma_{4\_dB} = -3.77dB, \Gamma_{5\_dB} = 1.33dB$$

Finally the achieved SINR values when using the MVDR beamforming algorithm is

$$\Gamma_{1\_dB} = 25.50dB, \Gamma_{2\_dB} = 5.21dB, \Gamma_{3\_dB} = -2.71dB, \Gamma_{4\_dB} = -7.44dB, \Gamma_{5\_dB} = 2.33dB$$

The improvement in the SINR when using null-steering beamforming compared to conventional beamforming is clear especially for the first terminal where the SINR has been increased more than 20 times. However, the third terminal performance has been reduced by about 3 dB. The average results of using MVDR beamforming is better than the other algorithms. The main reason for this is the concept of minimizing the average output power except at the looking direction. This leads to minimizing the effects of the additive noise as well.

The previous MVDR algorithm was derived for a single path of the desired signal. If we have several significant paths of the desired signal, how can we derive the optimum weight vector? Next we show our derivation for this general case, and we call it the general MVDR (GMVDR) algorithm.

The problem can be described mathematically as follows: find the optimum weight vector $\mathbf{w}_i$ which achieves the following optimization problem:

$$\min \mathrm{E}\left[||z_i|^2\right], \qquad i = 1, 2, \ldots, Q, \tag{9.30}$$

subject to

$$\begin{aligned}
\mathbf{w}_i^H \mathbf{a}_{i,1} &= 1 \\
\mathbf{w}_i^H \mathbf{a}_{i,2} &= 1 \\
&\vdots \\
\mathbf{w}_i^H \mathbf{a}_{i,M_i} &= 1
\end{aligned} \tag{9.31}$$

where $\mathbf{a}_{i,k}$ is the steering vector of path k for user i. If we assume that the first element is the reference element then the steering vector can be defined as

$$\mathbf{a}_{i,k} = [1 \, \exp(-j\Delta\Psi_{2ik}) \cdots \exp(-j\Delta\Psi_{Nik})]', \quad i = 1, \ldots, Q \tag{9.32}$$

In the normal MVDR algorithm, we have only one equality constraint to represent the user's DoA.

Using the Lagrange multiplier method, the total cost function becomes

$$C_i = E[\mathbf{w}_i^H \mathbf{u} \mathbf{u}^H \mathbf{w}_i] + \sum_{k=1}^{M_i} \gamma_k (\mathbf{w}_i^H \mathbf{a}_{i,k} - 1) \tag{9.33}$$

where γ_k is the kth Lagrange multiplier factor.

Necessary conditions for minimization are

$$\frac{\partial C_i}{\partial \mathbf{w}_i} = 2\mathbf{w}_i^H \mathbf{R}_{uu} + \sum_{k=1}^{M_i} \gamma_k \mathbf{a}_{i,k}^H = 0, \quad i = 1, \dots, Q, \tag{9.34}$$

$$\frac{\partial C_i}{\partial y_j} = \mathbf{w}_i^H \mathbf{a}_{i,j}^H - 1 = 0, \quad j = 1, 2, \dots, M_i, \tag{9.35}$$

where $\mathbf{R}_{uu} = E[\mathbf{u}\mathbf{u}^H]$.

From Equation (9.34) the optimum weight vector is obtained as

$$\mathbf{w}_i^H = -\frac{1}{2} \sum_{k=1}^{M_i} \gamma_k \mathbf{a}_{i,k}^H \mathbf{R}_{uu}^{-1}, \quad i = 1, \dots, Q \tag{9.36}$$

Now substituting Equation (9.36) in Equation (9.35) for all j results in

$$\mathbf{a}_{i,1}^H \mathbf{R}_{uu}^{-1} \mathbf{a}_{i,1} \gamma_1 + \mathbf{a}_{i,2}^H \mathbf{R}_{uu}^{-1} \mathbf{a}_{i,1} \gamma_2 + \cdots + \mathbf{a}_{i,M_i}^H \mathbf{R}_{uu}^{-1} \mathbf{a}_{i,1} \gamma_{M_i} = -2,$$

$$\mathbf{a}_{i,1}^H \mathbf{R}_{uu}^{-1} \mathbf{a}_{i,2} \gamma_1 + \mathbf{a}_{i,2}^H \mathbf{R}_{uu}^{-1} \mathbf{a}_{i,2} \gamma_2 + \cdots + \mathbf{a}_{i,M_i}^H \mathbf{R}_{uu}^{-1} \mathbf{a}_{i,2} \gamma_{M_i} = -2,$$

$$\vdots \tag{9.37}$$

$$\mathbf{a}_{i,1}^H \mathbf{R}_{uu}^{-1} \mathbf{a}_{i,1} \gamma_1 + \mathbf{a}_{i,2}^H \mathbf{R}_{uu}^{-1} \mathbf{a}_{i,1} \gamma_2 + \cdots + \mathbf{a}_{i,M_i}^H \mathbf{R}_{uu}^{-1} \mathbf{a}_{i,1} \gamma_{M_i} = -2.$$

The optimum Lagrange multiplier factors can be obtained by solving the system of linear equations given in Equation (9.37).

In the matrix form, Equation (9.37) can be represented as

$$\mathbf{A}_i \Gamma_i = -2\mathbf{1}, \quad i = 1, \dots, Q \tag{9.38}$$

where

$$\mathbf{A}_i = \begin{bmatrix} \mathbf{a}_{i,1}^H \mathbf{R}_{uu}^{-1} \mathbf{a}_{i,1} & \mathbf{a}_{i,2}^H \mathbf{R}_{uu}^{-1} \mathbf{a}_{i,1} & \cdots & \mathbf{a}_{i,M_i}^H \mathbf{R}_{uu}^{-1} \mathbf{a}_{i,1} \\ \mathbf{a}_{i,1}^H \mathbf{R}_{uu}^{-1} \mathbf{a}_{i,2} & \mathbf{a}_{i,2}^H \mathbf{R}_{uu}^{-1} \mathbf{a}_{i,2} & \cdots & \mathbf{a}_{i,M_i}^H \mathbf{R}_{uu}^{-1} \mathbf{a}_{i,1} \\ \vdots & & & \vdots \\ \mathbf{a}_{i,1}^H \mathbf{R}_{uu}^{-1} \mathbf{a}_{i,M_i} & \cdots & \cdots & \mathbf{a}_{i,M_i}^H \mathbf{R}_{uu}^{-1} \mathbf{a}_{i,M_i} \end{bmatrix} \tag{9.39}$$

$$\Gamma_i = [\gamma_1 \gamma_2 \cdots \gamma_{M_i}]' \tag{9.40}$$

$\mathbf{1} = [11 \cdots 1]'$ is M × 1 vector of ones

From Equation (9.38), the optimum Lagrange factors become

$$\Gamma_i = -2\mathbf{A}_i^{-1}\mathbf{1} \tag{9.41}$$

Equation (9.36) can be rewritten as

$$\mathbf{w}_i^H = -\frac{1}{2}\Gamma'\hat{\mathbf{a}}_i\mathbf{R}_{uu}^{-1}, \qquad i = 1, \ldots, Q \tag{9.42}$$

where

$$\hat{\mathbf{a}}_i = \begin{bmatrix} \mathbf{a}_{i,1}^H \\ \vdots \\ \mathbf{a}_{i,M_i}^H \end{bmatrix} \tag{9.43}$$

Substituting Equation (9.41) into Equation (9.42) gives

$$\mathbf{w}_i^H = \mathbf{1}'\mathbf{A}_i^{-1}\hat{\mathbf{a}}_i\mathbf{R}_{uu}^{-1}, \qquad i = 1, \ldots, Q \tag{9.44}$$

This result can be compared with the conventional (single path) MVDR algorithm given by Equation (9.27) as:

$$\hat{\mathbf{w}}_i^H = \frac{\mathbf{a}_i^H\mathbf{R}_{uu}^{-1}}{\mathbf{a}_i^H\mathbf{R}_{uu}^{-1}\mathbf{a}_i}, \qquad i = 1, \ldots, Q. \tag{9.45}$$

It is clear that algorithm (9.45) is a special case (one path) of Equation (9.44).

From Equation (9.39) it is clear that $\mathbf{A}_i$ can be decomposed as

$$\mathbf{A}_i = \hat{\mathbf{a}}_i\mathbf{R}_{uu}^{-1}\hat{\mathbf{a}}_i^H, \qquad i = 1, \ldots, Q \tag{9.46}$$

so that in order to get a non-singular matrix, the number of antenna elements should be greater than or equal to the number of Rake fingers (or required number of paths). Generally, beamforming is not efficient in multipath channels as mentioned in the introduction. Hence, the GMVDR algorithm may be effective when the number of paths is not large, for example 3–5 strong paths per user is a reasonable target.

9.1.4 Minimum Mean Square Error (MMSE) Beamformer

In all previous discussed algorithms we need to estimate the DoA at least for the desired signal. This requirement may complicate the weight adaptation algorithms. It is possible to adapt the antennas' weights without direct estimation of the DoA by using some pilot signal. The pilot signal is a specific signal transmitted periodically from the transmitter and it is known at the receiver. It is used to optimize the equalizer to mitigate the multipath fading problems. It can also be used to adapt the antenna weights using different algorithms. In wireless communication systems, we try to utilize the available bandwidth to transmit the required information and signaling signals so that using a pilot signal should be very limited. This is one disadvantage of using a pilot signal to adapt the antenna weights.

One of the methods which uses a reference signal is the MMSE beamformer. The MMSE is based on finding the optimum weights to minimize the mean square error

$$\hat{\mathbf{w}}_i = \arg \min E\left[\left|\mathbf{w}_i^H \mathbf{u}(t) - d_i(t)\right|^2\right], \quad i = 1, \ldots, Q, \tag{9.47}$$

where $d_i(t)$ is the pilot signal (or training sequence) for user i at time t. The optimum weights can be obtained by setting the gradient of the cost function with respect to $\mathbf{w}_i$ equal to zero such as

Let us define $C = E\left[\left|\mathbf{w}_i^H \mathbf{u}(t) - d_i(t)\right|^2\right] = E[(\mathbf{w}_i^H \mathbf{u}(t) - d_i(t))(\mathbf{w}_i^H \mathbf{u}(t) - d_i(t))^H],$

$$\because \frac{\partial C}{\partial \mathbf{w}_i} = 0 = 2E\left[\mathbf{u}(t)(\mathbf{w}_i^H \mathbf{u}(t) - d_i(t))^H\right] \Rightarrow E\left[\mathbf{u}(t)\mathbf{u}^H(t)\right]\mathbf{w}_i - E\left[\mathbf{u}(t)d_i^*(t)\right] = 0$$

Let us drop the time symbol for simplicity and define:

$$\mathbf{P}_i = E[\mathbf{u}d_i^*] \text{ and } \mathbf{R}_{uu} = E[\mathbf{u}\mathbf{u}^H],$$

From the above the optimum weight vector can be computed as:

$$\hat{\mathbf{w}}_i = \mathbf{R}_{uu}^{-1}\mathbf{P}_i, \quad i = 1, \ldots, Q \tag{9.48}$$

This is called the Wiener–Hopf equations [5].

A recursive form of Equation (9.48) is given in [3,5] as:

$$\mathbf{w}_i(t+1) = \mathbf{w}_i(t) - \mu(E[\mathbf{u}(t)\mathbf{u}^H(t)]\mathbf{w}_i(t) - E[\mathbf{u}(t)d_i^*(t)]) \tag{9.49}$$

where constant μ is a positive scalar (gradient step size) that controls the convergence characteristic of the algorithm, that is, how fast and how close the estimated weights approach the optimal weights.

If we assume that the signals are ergodic, then the adaptive algorithm can be approximated as

$$\mathbf{w}_i(t+1) = \mathbf{w}_i(t) - \mu\mathbf{u}(t)e_i(t), \quad t = 0, 1, \ldots, \quad i = 1, \ldots, Q \tag{9.50}$$

$$e_i(t) = \mathbf{w}_i^H(t)\mathbf{u}(t) - d_i^*(t). \tag{9.51}$$

Here $e_i(t)$ is the instantaneous error between the array output and the desired response. The main disadvantage of the adaptive LMS algorithm and its different versions is its slow convergence speed [5]. This is an essential problem in mobile communication due to the nature of fast-varying channel characteristics. The convergence of the LMS algorithm depends on the eigenvalue distribution of the correlation matrix [5].

Example 9.12

The pilot signal from a certain TX is given by [1 1 0 0 1 1 0 1 0 1], the received signal on the terminal of two antennas separated by $\lambda/2$ is

$$u = \begin{bmatrix} 0.1+j0.2 \ 0.3-j0.1 \ -0.1+j0.2 \ 0.2-j1.1 \ 0.7-j0.3 \ 0.5+j0.1 \ -0.3-j0.2 \ 0.1+j1.2 \ 0.4+j0.5 \ 0.9-j0.1 \\ 0.2+j1.2 \ 0.4-j0.5 \ -0.2+j0.1 \ 0.5-j0.4 \ 0.2+j0.1 \ -0.1+j1.2 \ -0.1+j1.5 \ 0.7-j1.2 \ -0.3-j0.2 \ 0.1-j0.8 \end{bmatrix}$$

Starting from the zero weight vector, find the final value of the weights.

Solution

We can solve this example directly by implementing Equations (9.50) and (9.51) using any programming language. The result is $\mathbf{w} = \begin{bmatrix} 0.42 + j0.26 \\ 0.64 - j0.16 \end{bmatrix}$

Exercise 9.8   Repeat Example 9.11 with MMSE using Equations (9.48) and (9.50). Compare the results with other techniques.

9.1.5 Recursive Least Squares (RLS) Algorithm

The RLS algorithm is more efficient (in many situations) than the LMS algorithm. In [4], the RLS algorithm has been proposed for weight adaptation in the uplink of a CDMA mobile communication system.

The RLS algorithm minimizes the cumulative square error [3]

$$\mathbf{w}_n = \arg\min\left\{ \sum_{t=0}^{n} \mu_t \left\| e(t) \right\|^2 \right\}, \tag{9.52}$$

where the error $e(t)$ is the difference between the reference signal and the actual array output. The weight updating algorithm is [5]:

$$\mathbf{w}(t) = \mathbf{w}(t-1) + \mathbf{K}(t)e(t). \tag{9.53}$$

Here $\mathbf{K}(t)$ is the update gain and it is given by

$$\mathbf{K}(t) = \frac{\mathbf{A}(t-1)\mathbf{u}(t)}{\frac{1}{\mu_t} + \mathbf{u}^H(t)\mathbf{A}(t-1)\mathbf{u}(t)}, \tag{9.54}$$

where $\mathbf{A}(t) = \mathbf{R}_{uu}^{-1}(t)$. and it is solved recursively as

$$\mathbf{A}(t) = \mathbf{A}(t-1) - \frac{\mathbf{A}(t-1)\mathbf{u}(t)\mathbf{u}^H(t)\mathbf{A}(t-1)}{\frac{1}{\mu_t} + \mathbf{u}^H(t)\mathbf{A}(t-1)\mathbf{u}(t)}. \tag{9.55}$$

μ_t is a real scalar, it is called the forgetting factor. There are different updating algorithms for the forgetting factor such as:

$$\mu_t = \alpha\mu_{t-1} + (1 - \alpha) \quad 0 < \alpha < 1 \tag{9.56}$$

so that the old samples are de-emphasized.

Exercise 9.9   Using MATLAB, write the weight adaptation code using the RLS algorithm then repeat Example 9.11.

Example 9.13

To compare between some of the discussed algorithms, compute the average SINR and also plot the radiation pattern of the first terminal for the following simulation environment:

- number of terminals: 6
- number of antenna elements: 4 (linear array)

- utilized beamforming algorithms: null-steering, RLS, and MVDR
- average additive noise power: $-80\,\text{dB}$
- DoA of the terminals (in degrees): 308, 304, 86, 286, 111, and 251

Solution

The resultant average SINRs (over all terminals) in dB are:

- null-steering $= 9\,\text{dB}$
- RLS $= 19\,\text{dB}$
- MVDR $= 17\,\text{dB}$

Figures 9.9(a), (b), and (c) show the radiation pattern of the first terminal array for null-steering, RLS, and MVDR beamforming, respectively.

9.1.6 Subspace Methods for Beamforming

In the subspace technique the structure of the signals at the antenna array output is exploited for beamforming applications [7–9] and [10]. If there are Q users in the cell then the sum of received signals at an M antenna element array (see Equation (9.14)) is

$$\mathbf{u} = \sum_{i=1}^{Q} \mathbf{u}_i = \sum_{i=1}^{Q} \mathbf{a}_i \alpha_i \mathbf{s}_i + \mathbf{n}. \tag{9.57}$$

Since each user i has B_i different paths, which are assumed to come from different directions of arrivals (DoA), Equation (9.57) can be written in a more compact form using matrix vector notation:

$$\mathbf{u} = \mathbf{AS} + \mathbf{n}, \tag{9.58}$$

where $\mathbf{A} = [\hat{\mathbf{a}}_1 \alpha_1 \ \vdots \ \hat{\mathbf{a}}_2 \alpha_2 \ \vdots \ \cdots \ \vdots \ \hat{\mathbf{a}}_Q \alpha_Q]$ is an $M \times D$ dimensional matrix, $\hat{\mathbf{a}}_i$ is an $M \times B_i$ matrix (see Equation (9.43)), $\mathbf{S} = [\mathbf{s}_1 \vdots \mathbf{s}_2 \vdots \cdots \vdots \mathbf{s}_Q]'$ an $B_i \times T$ matrix, $\mathrm{D} = \Sigma_{i=1}^{Q} \mathbf{B}_i$ and $T =$ number of samples.

For simplicity we will consider only one single path for each user. In this case the dimension of matrix $\mathbf{A}$ will be $M \times Q$ and the dimension of matrix S will be $Q \times T$, where Q is the number of users or in a more general term, the number of uncorrelated signals.

Using the data model of Equation (9.58), the input correlation matrix $\mathbf{R}_{uu}$ can be expressed as

$$\mathbf{R}_{uu} = E[\mathbf{u}\mathbf{u}^H] = \mathbf{A}E[\mathbf{SS}^H]\mathbf{A}^H + E[\mathbf{nn}^H] \tag{9.59}$$

or

$$\mathbf{R}_{uu} = \mathbf{AR}_{ss}\mathbf{A}^H + \delta_n^2 \mathbf{I}, \tag{9.60}$$

where $\mathbf{R}_{ss} = \mathrm{E}[\mathbf{SS}^H]$ is the signal correlation matrix.

The matrix $\mathbf{R}_{uu}$ can be decomposed, for example by singular value decomposition, to obtain

$$\mathbf{R}_{uu} = \mathbf{w}\Sigma V' \tag{9.61}$$

where $\mathbf{W}$ and $\mathbf{V}$ are $M \times M$ orthogonal matrices, and $\Sigma = diag(\sigma_1, \sigma_2 \cdots, \sigma_M)$ is a diagonal matrix with $\sigma_i \geq 0$. The nonnegative numbers $\{\sigma_i\}$ are called the singular values of $\mathbf{R}_{uu}$ and

(a)
Null-Steering Beamforming

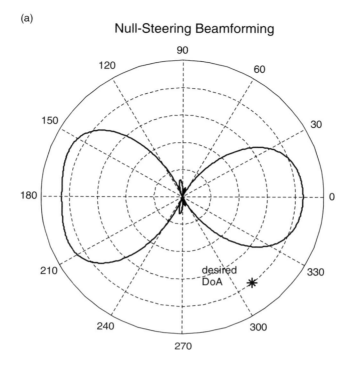

(b)
RLS Beamforming

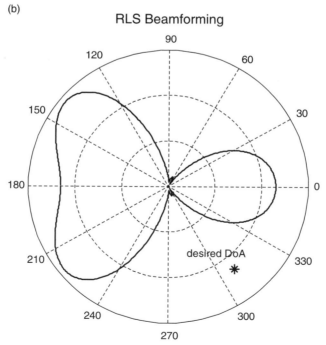

Figure 9.9 Radiation pattern result of Example 9.13. (a) null-steering; (b) RLS; (c) MVDR

$\sigma_1 \geq \sigma_2 \geq \cdots \geq \sigma_M$. If $\mathbf{R}_{uu}$ has rank r then $\mathbf{R}_{uu}$ has exactly r strictly positive singular values so that $\sigma_r > 0$ and $\sigma_{r+1} = \sigma_{r+2} = \cdots = \sigma_M = 0$.

Assume that all incident signals are not highly correlated and their number is less than the number of antenna elements, i.e., $M > D$. Then by examining the singular values of $\mathbf{R}_{uu}$ we will find D singular values with considerable values. The other $(M - D)$ singular values have very small values. These small values represent the variance of the background noise.

The received signal space can be decomposed into two subspaces. The first subspace that is spanned by the eigenvectors associated with the first D eigenvalues is called the *signal subspace*. The second subspace that is spanned by the eigenvectors associated with the last $(M - D)$ eigenvalues is called the *noise subspace*. It has been proven that the noise subspace is orthogonal to the steering vectors [2]. This fact can be exploited in the estimation of the DoA of the signals. When the DoA of the interference signals is estimated it can be cancelled by making the weight vector orthogonal to the interference subspace [10].

Many different algorithms exploit the eigenstructure of the array-correlation matrix of the received signal [3,7] and [10]. The main disadvantage of beamformers that are based on the eigen decomposition (ED) methods is that the number of users is limited by the number of antenna elements. Therefore they are not suitable for commercial CDMA applications, which need to support a large number of users [35]. There are some techniques to overcome this limitation. In [7] the fact that in CDMA systems the desired user power is much larger than that of each interference power due to the processing gain of the CDMA demodulation has been exploited. The authors used the eigenvector corresponding to the largest eigenvalue as the optimum weight of the array. The main advantage of this method is that it can be used with any number of users, if the SINR of the desired user is high enough [7].

Example 9.14

Construct the following simulation environment:

- number of users: 3
- number of antenna elements: 4
- additive noise power: $-80\,$dBm
- distance of terminals from receiver side: 20, 25, and 30 m
- DoA of the signals (in degrees): 35, 48, and 118
- average transmitted power of terminals: 10 W
- number of samples: 1000

Compute the correlation matrix $\mathbf{R}_{uu}$. Compute the SVD of the correlation matrix (use command *svd* in MATLAB). Show that the noise subspace (represented by the eigenvector associated with noise singular values) is orthogonal to the steering vectors.

Solution

The correlation matrix is

$$
\mathbf{R}_{uu} = \begin{bmatrix}
0.480 & -0.366+j0.246 & 0.121-j0.419 & 0.106+j0.419 \\
-0.366+j0.246 & 0.470 & -0.359+j0.254 & 0.121-j0.421 \\
0.121-j0.419 & -0.359+j0.254 & 0.476 & -0.373+j0.250 \\
0.106+j0.419 & 0.121-j0.421 & -0.373+j0.250 & 0.492
\end{bmatrix} \times 10^{-6}
$$

Applying the SVD for the correlation matrix $\{[\mathbf{W},\Sigma,\mathbf{V}']=\mathrm{svd}(\mathbf{R}_{uu})\}$, we obtain the following diagonal matrix

$$\Sigma = \begin{bmatrix} 0.178 & 0 & 0 & 0 \\ 0 & 0.007 & 0 & 0 \\ 0 & 0 & 0.005 & 0 \\ 0 & 0 & 0 & 0.0000 \end{bmatrix} \times 10^{-5}$$

We can see that we have three non-zero (or large enough) singular values. This indicates that we have three signals (anyhow we know that!). Now the eigenvector associated with the fourth eigenvalue represents the noise subspace. It is the fourth vector on the V matrix, and it is equal to $V_{ns}^{H} = [-0.427 - 0.535 + j0.174 - 0.525 - j0.204 - 0.428 - j0.025]$. Multiplying the noise subspace by the steering vector gives close to zero values indicating that it is orthogonal to all DoA signals. Figure 9.10 shows the radiation pattern using weights equal to the eigenvectors associated with the noise subspace and the DoA of the terminals.

9.1.7 Adaptive Beamforming using the Kalman Filter

The constrained optimization problem presented in Equations (9.24) and (9.25) can be solved using the Kalman filtering approach [11]. Equation (9.22) can be rewritten as follows:

$$\min E[|0\text{-}z(t)|^2], \tag{9.62}$$

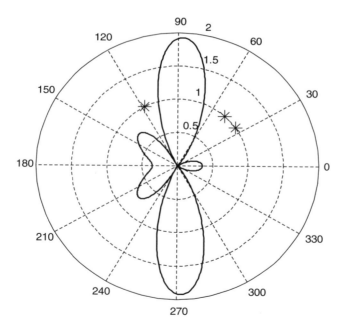

Figure 9.10   Radiation pattern using eigenvectors of noise subspace

subject to

$$\mathbf{w}_i^H \mathbf{a}_i = 1 \tag{9.63}$$

To incorporate Kalman filtering, the measurement equation can be written as

$$\begin{bmatrix} 0 \\ 1 \end{bmatrix} = \begin{bmatrix} \mathbf{u}^H(t) \\ \mathbf{a}_i^H \end{bmatrix} \mathbf{w}_i(t) + \begin{bmatrix} v_1(t) \\ v_2(t) \end{bmatrix}. \tag{9.64}$$

Here $v_1(t)$ is the residual error and $v_2(t)$ is the constraint error. These errors are assumed to be zero mean Gaussian and independent random variables. In matrix form Equation (9.64) becomes

$$\mathbf{Y} = \mathbf{B}^H(t)\mathbf{w}_i(t) + \mathbf{V}(t), \tag{9.65}$$

where $\mathbf{Y} = \begin{bmatrix} 0 \\ 1 \end{bmatrix}$, $\mathbf{B}^H(t) = \begin{bmatrix} \mathbf{u}^H(t) \\ \mathbf{a}_i^H \end{bmatrix}$ and $\mathbf{V}(t) = \begin{bmatrix} v_1(t) \\ v_2(t) \end{bmatrix}$.

Further, the correlation matrix of $\mathbf{V}(t)$ is

$$\mathbf{Q} = \begin{bmatrix} \delta_{v_1}^2 & 0 \\ 0 & \delta_{v_2}^2 \end{bmatrix} \tag{9.66}$$

The state-space model of the constrained Kalman algorithm may be written as

$$\mathbf{w}_i(t) = \mathbf{w}_i(t-1) \tag{9.67}$$

Now, we may use the Kalman filter to solve Equations (9.65) and (9.67) to minimize the residual error in the mean-square sense while maintaining a distortionless response along the looking direction. The discrete Kalman filter can be written as [6]:

$$\hat{\mathbf{w}}_i(t) = \hat{\mathbf{w}}_i(t-1) + \bar{\mathbf{K}}(t-1)[\mathbf{Y} - \mathbf{B}^H(t-1)\hat{\mathbf{w}}_i(t-1)]. \tag{9.68}$$

The Kalman gain $\bar{\mathbf{K}}(t)$ is given by

$$\bar{\mathbf{K}}(t) = \mathbf{G}(t-1)\mathbf{B}(t)[\mathbf{B}^H(t)\mathbf{G}(t-1)\mathbf{B}(t) + \mathbf{Q}], \tag{9.69}$$

where the filtered weight-error correlation matrix $\mathbf{G}(t)$ is

$$\mathbf{G}(t) = [\mathbf{I} - \bar{\mathbf{K}}(t)\mathbf{B}^H(t)]\mathbf{G}(t-1). \tag{9.70}$$

It has been proven that the constrained Kalman-type array processor can converge to the minimum-variance distortionless-response (MVDR) beamformer [11].

Exercise 9.10   Prove that the constrained Kalman-type array processor converges to the same MVDR beamforming algorithm.

9.1.8 Blind Beamforming

Blind beamforming refers to a class of adaptive algorithms which do not need specified signals (training or pilot) for weight adaptation. Nevertheless, blind algorithms utilize some of the inherent features in signals (either band-pass or baseband) to indicate the received signal quality. This will enhance the efficiency because there is no need to transmit training data, moreover there is no need for the DoA information. Blind algorithms have many applications in digital communications and generally in most of the signal processing areas. We will just briefly introduce them in this chapter. In digital communications, information is represented by a finite set of symbols. For example in the binary system we have only two symbols $S \in \{s_0, s_1\}$ which is 0 and 1 in digital representation. In a 4-ary communication system, data is represented as a finite set of four symbols such as $S \in \{s_0, s_1, s_2, s_3\}$, and so on. The flow of information is sent over the sequence of these finite symbols. A simplified model of the received signal is $r = \alpha x_i + y + n$, where α is the channel effects, x_i is the transmitted band-pass signal which also has finite forms, i.e., $s_i \rightarrow x_i$, y is the cochannel interferences, and n is the additive noise. Assume that we use a perfect equalizer so that we can remove the effects of channel fading, then we can represent the received signal as $r = x_i + y + n$. After beamforming the received signal is $z = \mathbf{w}^H \mathbf{x}_i + \mathbf{V}$, where $\mathbf{x}_i$ is the vector of the received signal (desired) at each antenna element, and $\mathbf{V}$ is the interference and noise at each antenna element (as described previously). We are looking to maximize the desired signal portion of the received signal and since the desired signal belongs to a finite set of possibilities, we use an estimation technique (such as maximum likelihood) to estimate the transmitted symbol (s_i) from the received signal. Next we regenerate x_i (the reverse process is possible as well), then we adapt the antenna weights in order to minimize the following error function

$$e = E\left[||\mathbf{w}^H \mathbf{x}_i - r||\right]. \tag{9.71}$$

Therefore we have utilized the fact that digital systems use a finite set of possibilities to construct an error function which can be used to adapt the antenna weights. Furthermore, this error function can be used for other purposes such as equalizers and multiuser detection. It is a nice method because we do not need to estimate the DoA or to send a specific pilot or training signal. Figure 9.11 shows a simple M-ary digital receiver utilizing blind beamforming. The dashed lines refer to the possibility of computing the error signal at baseband as well. We assume M-ary digital transmission. The correlator receiver computes the correlation between the received signal with all possible prototypes and selects the one which has the highest match with the received signal. The detected symbol is used again as reference to compute the error signal. What do you think is the main problem with this technique? The success of this method is based on the probability of correct detection, because if we detect the symbol incorrectly we will be incorrect in our weight adaptation. In other words, this blind techniques work efficiently if we have a high enough SINR ratio at the receiver input. This raises another question: if the SINR is already high, why do we need beamforming? One reason is that we will be able to reduce the transmitted power of users. This leads to higher system capacity, longer handset battery-life, lower interference to other systems, and other benefits. The joining algorithm between beamforming and power control will be discussed later.

One interesting application of blind beamforming is the least-squares despread respread multitarget array (LS-DRMTA). The LS-DRMTA has been proposed mainly for CDMA communication systems [2,12]. The idea of this algorithm is based on respreading the received data bits. The respread signal is compared with the received signal (before the despreading process), and the difference is used as an error signal. This error is minimized by adjusting the antenna weights. Figure 9.12 shows the block diagram of the LS-DRMTA for user i.

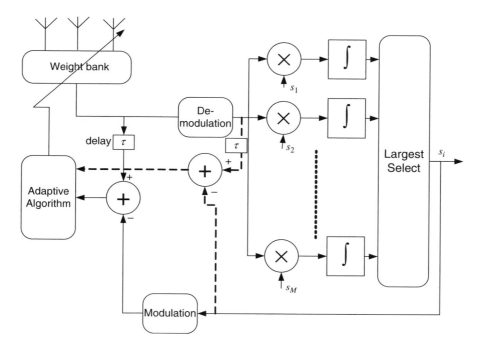

Figure 9.11   Blind beamforming using transmitted symbols

The respread signal is given by

$$r_i(t) = b_{in}C_i(t - \tau_i) \quad (n-1)T_b \leq t < nT_b, \quad i = 1, \ldots, Q, \tag{9.72}$$

where $C_i(t)$ is the spreading code for user i and b_{in} is the nth received data for user i.

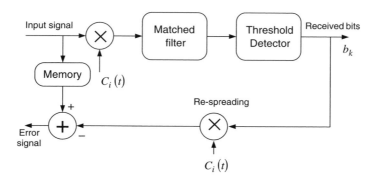

Figure 9.12   LS-DRMTA block diagram for user i

The LS-DRMTA is used to minimize an error function by adjusting the weight vector $\mathbf{w}_i$. The error function is given by

$$F(\mathbf{w}_i) = \sum_{k=1}^{K} ||y_i(t) - r_i(t)||^2 = \sum_{k=1}^{K} \left| \mathbf{w}_i^H \mathbf{x}(t) - r_i(t) \right|^2, \qquad (9.73)$$

where K is the data block size and is set equal to the number of samples in a one-bit period in the LS-DRMTA. There are several advantages of using the LS-DRMTA algorithm such as the DoA information is not needed and there is no need to use a pilot signal to adapt the antenna weights. It is clear from the algorithm that the received data bits should be decoded correctly in order to be able to adapt the antenna weights correctly. This is possible when the SINR is high enough at the receiver input even before the weight adjusting.

Exercise 9.11   Suggest a blind beamforming algorithm for a BPSK transmitter. Derive the weight adaptation algorithm based on the least mean square algorithm.

Exercise 9.12   Assume that we want to design a beamforming algorithm to track a mobile transmitter based on its carrier frequency. There are no other users in the area using this frequency. Derive a blind beamforming algorithm based on the detection of the carrier periodicity. Write the adaptation using MATLAB. Show the tractability of such an algorithm within a certain dynamic scenario with Q users (with different carrier frequencies) and an additive white noise channel. Hint: one interesting method is to use cyclostationary analysis given in Chapter 10.

9.2 Spatial-Temporal Processing

The capacity and the performance of cellular communication systems can be greatly enhanced by exploiting any known characteristics of the communication link. The natural spatial distribution of the users and the access delay distribution of the signal paths are two important characteristics which can be exploited. Using the adaptive antenna array we may enhance the reception from certain direction of arrivals (DoA) and attenuate others, as has been shown in the previous sections. Usually the signal of interest (SoI) arrives at the receiver's antennas as multipath components, where each component has its own DoA as well as access delay. For wideband signals these multipaths could be uncorrelated. Using CDMA will minimize the correlation between paths when the delay is larger than the chip duration. Exploiting the other paths will improve the communication link through time diversity. Beamforming and time diversity can be achieved jointly by using two different methods. The first is to use a general wideband array as shown in Figure 9.13 [2]. If the length of each tapped delay line is long enough to capture the delayed multipath components, then the wideband array can capture power in components which arrive with different delays and recombine them [2]. The other method is to use a Rake receiver with the adaptive antenna array. The Rake receiver is capable of receiving multiple signal paths and adding them coherently using multiple fingers. Each Rake finger is time locked to a different delay to capture the multipath components arriving with different path delays. Hence, we obtain better performance through time diversity.

Combining the Rake receiver with the adaptive antenna array means we can exploit spatial as well as time distributions of the signals. Figure 9.14 shows the conventional way to combine a Rake receiver with an adaptive antenna [2]. Any adaptive algorithm can be used to compute the optimum weight vector for each significant path which is captured by the Rake fingers. The

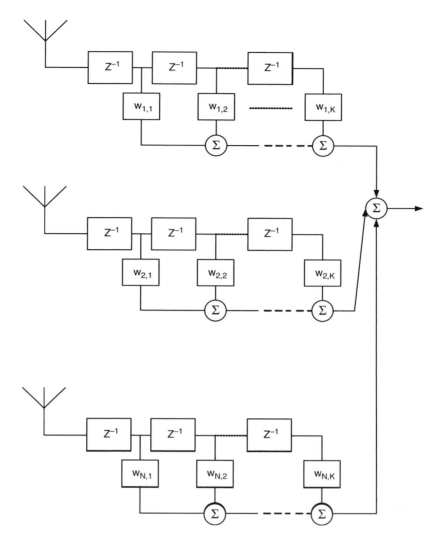

Figure 9.13 Wideband adaptive antenna system

main problem of spatial/temporal processing is the high computational cost. For example if the number of antennas is six and the number of Rake fingers is six, then 36 weights should be adapted each time. More efficient methods (from a computational point of view) could be used such as that proposed in Equation (9.44). There are other techniques to reduce the computational complexities of the receivers [13].

9.3 Joining Radio Resources with Beamforming

It can be concluded that the main task of beamforming is to reduce the directional interferences, and hence enhance the SINR. The main target of any wireless communication system is to

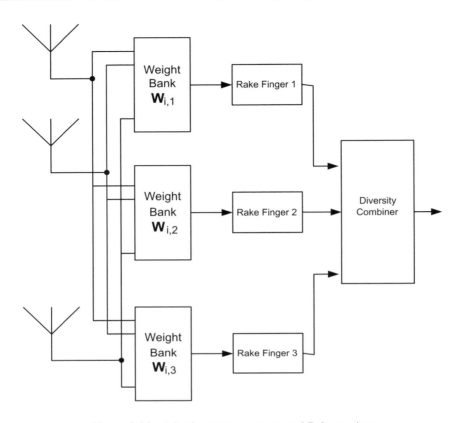

Figure 9.14   Adaptive antenna system and Rake receiver

achieve certain QoS parameters for the communication link. Such parameters include maximum allowed bit error rate (BER), minimum required throughput, latency constraints, probability of packet losses, probability of outage and probability of blocking. These requirements can be easily achieved if we have only a single link such as microwave links or broadcasting services. On the other hand, it is a very challenging solution to achieve in a multiuser environment with limited radio resources, for example, a multiuser communication system within a certain city that should be available for 100,000 users, with probability of blocking not more than 0.01, and within a strictly limited bandwidth. There could be other requirements as well as other constraints such as maximum allowed transmission power.

However, the most scarce radio resource is the bandwidth, because most of the useful spectrum is already allocated and licensed. How can we optimize our available resources among all users to achieve the best performance and target QoS? The main three radio resources over which we have control are the transmitted power, data rate, and time of transmission (scheduling). Integrating other resources will greatly enhance the performance of our system and relax the design parameters. Beamforming is a strong candidate for enhancing the performance. Assume for example a communication link where the transmitter sends with its maximum allowed power to achieve the target SINR at its receiver. Now, we integrate beamforming at the receiver side, with a good weight adaptation, we are able to considerably reduce the interference. Hence, the SINR is increased significantly. Now we are happy because we get free improvement which can be exploited in different ways such as:

1. Reducing the transmission power to come back to the target SINR. This reduces the interference on other terminals, so that they may reduce their transmission power as well. Hence, higher capacity (new users can start) and better influence over all the network.
2. We may keep the transmission power high, but now we can exploit the higher SINR by increasing the throughput by using higher modulation levels and lower coding rate (this topic is discussed in Chapter 8).

Actually most of the QoS parameters are directly affected by the value of the SINR. Without loss of generality, we will consider only the concepts of joining power control with beamforming. The results can be easily generalized as discussed in Chapter 8.

Combining the beamforming and power control is an efficient technique to increase the channel capacity and to improve the QoS. In practice, the smart antenna weights are updated using for example one of the adaptation techniques described previously such as LMS, RLS, or MVDR. The power control is updated using for example one of the distributed power control algorithms described in Chapters 5 and 6 such as DCPC, MODPC, or B-BPC. The beamforming weights can be updated on a symbol-by-symbol basis or even slower, while the transmission powers are adjusted on a slot-by-slot basis to mitigate the fast fading problems. The joining algorithm of power control and smart antenna should minimize the transmission power and at the same time achieve the target QoS performance. Convergence to an acceptable performance should be fast enough to handle the dynamical behavior of the mobile communication systems. The joining algorithm is the solution of a multivariable optimization problem.

The joining of smart antenna with power control can exploit the spatial distribution of users, which is sometimes referred to as multiuser diversity. A smart antenna can minimize the interference between users with different DoA signals, while the power control can mitigate the near-far problem at the same time. Using the Rake receiver after the smart antennas' weights can enhance the instantaneous SINR by coherently combining strong signal paths in a frequency selective channel scenario.

It has been shown that when using RLS or LMS beamforming the smart antenna weights are dependent on the received signal power 'to construct the correlation matrix'. Moreover, power control adjusts the transmitted power according to the received SINR. And the SINR depends on the smart antenna weights, as shown in Equation (9.16). Therefore we can see the direct effect of power control and beamforming on each other. By optimal or even sub-optimal joining of the two interference management approaches, we can greatly improve the capacity and the QoS of the mobile communication system [17]. Theoretically we can combine any available number of interference management techniques. For example in [16] three interference management approaches were combined: transmit power control, multiuser detection, and beamforming. Mathematically we may describe the joining power control and beamforming algorithms for terminal i as follows.

Find the weight matrix $\mathbf{W}$ and power vector $\mathbf{P}$ which achieve the following objective

$$\min \sum_{i=1}^{Q} P_i \qquad (9.74)$$

subject to

$$\Gamma_i = \frac{E\left[\left|\mathbf{w}_i^H \mathbf{a}_i \boldsymbol{\alpha}_i \mathbf{s}_i\right|^2\right]}{E\left[\left\|\mathbf{w}_i^H \sum_{\substack{k=1 \\ k \neq i}}^{Q} \mathbf{a}_k \boldsymbol{\alpha}_k \mathbf{s}_k + \mathbf{n}\right\|^2\right]} \geq \Gamma_i' \quad \forall i = 1, 2, \cdots, Q \qquad (9.75)$$

where the weight matrix $\mathbf{W} = [\,\mathbf{w}_1 \quad \mathbf{w}_2 \quad \cdots \quad \mathbf{w}_Q\,]$, the power vector $\mathbf{P} = [\,P_1 \quad P_2 \quad \cdots \quad P_Q]'$, and Γ_i' is the target required SINR for terminal i.

The above optimization problem can be solved in distributed manner, for example by combining Equations (9.50) and (9.51), and the DPC algorithm in Chapter 5. More detailed information can be found in [17]. Other joining algorithms have been proposed in literature, one example uses Kalman filters [14].

For cellular systems, the power control and beamforming can be performed for both the uplink and the downlink. Let's consider uplink power control and an uplink smart antenna installed at the receiver in the base station (BS). Assume that the power control update rate equals the antenna weight update rate. The joining algorithm of power control and beamforming can be summarized for user i by the following steps:

1. The mobile station (MS) transmits with an initial transmit power based on the open-loop control algorithm.
2. The BS starts with an initial weight vector for the smart antenna open-loop control algorithm.
3. The BS measures (estimate) the SINR value and informs the MS (either directly or indirectly).
4. Based on the SINR, the MS updates the transmit power using a power control algorithm such as the DPC. It is clear that this process represent closed-loop control.
5. The receiver at BS updates the weight vector by using one of the known algorithms, for example MMSE or MVDR.
6. Go to step 3 and so on.

Note that step 5 does not have to follow step 4, both can be done simultaneously or with different update rates. When the antenna weights are updated in the right direction toward the optimum, the interference signals from different DoAs will be reduced. This leads to enhancing the SINR at the receiver input. Thus, less transmit power will be needed to achieve the target SINR.

Next we will investigate the theoretical effects of beamforming on power control. The influence of beamforming on the maximum achievable SINR can be examined from the normalized channel matrix (see Chapter 5). The normalized channel matrix H when using beamforming can be derived using Equation (9.17) such as

$$\mathbf{H} = \begin{bmatrix} 0 & \dfrac{\left|\mathbf{w}_1^H \mathbf{a}_2\right|^2 G_{12}}{\left|\mathbf{w}_1^H \mathbf{a}_1\right|^2 G_{11}} & \cdots & \dfrac{\left|\mathbf{w}_1^H \mathbf{a}_Q\right|^2 G_{1Q}}{\left|\mathbf{w}_1^H \mathbf{a}_1\right|^2 G_{11}} \\[2mm] \dfrac{\left|\mathbf{w}_2^H \mathbf{a}_1\right|^2 G_{21}}{\left|\mathbf{w}_2^H \mathbf{a}_2\right|^2 G_{22}} & 0 & & \dfrac{\left|\mathbf{w}_2^H \mathbf{a}_1\right|^2 G_{2Q}}{\left|\mathbf{w}_2^H \mathbf{a}_2\right|^2 G_{22}} \\[2mm] \vdots & & \ddots & \\[2mm] \dfrac{\left|\mathbf{w}_Q^H \mathbf{a}_1\right|^2 G_{Q1}}{\left|\mathbf{w}_Q^H \mathbf{a}_Q\right|^2 G_{QQ}} & \dfrac{\left|\mathbf{w}_Q^H \mathbf{a}_2\right|^2 G_{Q2}}{\left|\mathbf{w}_Q^H \mathbf{a}_Q\right|^2 G_{QQ}} & & 0 \end{bmatrix} \qquad (9.76)$$

It is clear that by optimizing the antenna weights we can reduce the above channel matrix. Consequently, this leads to a reduction in the value of the spectral radius of the matrix. The maximum achievable SINR is related to the spectral radius as:

$$\gamma^* = \frac{1}{\rho(\mathbf{H})} \qquad (9.77)$$

where γ^* is the maximum achievable SINR and $\rho(\mathbf{H})$ is the channel spectral radius. Hence, optimizing the beamforming weights in a multiuser environment will enhance the theoretical system capacity.

Example 9.15

Assume three CDMA transmitters (with LoS) at the following locations compared to the access point: $(x_1, y_1) = (30, 70)m$, $(x_2, y_2) = (-20, 66)m$, and $(x_3, y_3) = (60, -60)m$. Calculate the optimum transmit power from each transmitter to achieve $E_b/N_0 = 15\,dB$, where $E_b/N_0 = PG \times SINR$, $PG = 64$ is the processing gain. The additive noise is $-70\,dBm$. The exponent factor of the channel is 4. Repeat the calculations for two-antenna element beamforming with the null-steering algorithm.

Solution

First we calculate the Euclidean distance between the terminals and the access point such that $d \cong [76, 69, 85]m$. The optimum power values which can achieve the required SINR can be calculated directly using Equation (5.19) in Chapter 5 as $\mathbf{P} = \begin{bmatrix} 0.14 \\ 0.09 \\ 0.22 \end{bmatrix} W$.

For the antenna weights calculations we need to compute the steering vector. It can be easily calculated using Equation (9.5) as $\mathbf{a}_1 = \begin{bmatrix} 1 \\ 0.33 - j0.95 \end{bmatrix}$, $\mathbf{a}_2 = \begin{bmatrix} 1 \\ 0.61 - j0.79 \end{bmatrix}$, and $\mathbf{a}_3 = \begin{bmatrix} 1 \\ -0.61 - j0.80 \end{bmatrix}$. The optimum weight vectors using null-steering beamforming can be calculated using Equation (9.21) as $\mathbf{w}_1 = \begin{bmatrix} 0.25 + j0 \\ 0.08 - j0.24 \end{bmatrix}$, $\mathbf{w}_2 = \begin{bmatrix} 0.44 + j11 \\ 0.19 - j0.42 \end{bmatrix}$, and

$\mathbf{w}_3 = \begin{bmatrix} 0.31 + j11 \\ -0.27 - j0.18 \end{bmatrix}$. We can calculate the optimal power values by using the centralized power control equation with the $\mathbf{H}$ matrix given in Equation (9.76). In this example it can be computed

as $\mathbf{H} = \begin{bmatrix} 0 & 0.225 & 0.779 \\ 0.071 & 0 & 0.055 \\ 0.517 & 0.115 & 0 \end{bmatrix}$. Substituting in Equation (5.19) in Chapter 5 where

$\mathbf{u} = \begin{bmatrix} \dfrac{10^{1.5} \times 10^{-10}}{64 \left| \mathbf{w}_1^H \mathbf{a}_1 \right| G_{11}} & \dfrac{10^{-8.5}}{64 \left| \mathbf{w}_2^H \mathbf{a}_2 \right| G_{22}} & \dfrac{10^{-8.5}}{64 \left| \mathbf{w}_3^H \mathbf{a}_3 \right| G_{33}} \end{bmatrix}$ we obtain the following optimum power vector:

$\mathbf{P} = \begin{bmatrix} 0.011 \\ 0.002 \\ 0.010 \end{bmatrix}$ W. It is clear that there is great reduction in the required power values to achieve

the target SINR. Note that in this example we used two antennas for three terminals, i.e., the effects of the interferers have not been completely removed. However, we still have considerable gain of using beamforming.

Exercise 9.13   Repeat the previous example with three antenna elements.

9.4 Multiple-Input Multiple-Output (MIMO) Antennas

The MIMO structure refers generally when multiple antennas are used at the transmitters as well as at the receivers. There are three main techniques of utilizing this structure: (i) MIMO beamforming; (ii) MIMO channel (or spatial multiplexing); and (iii) MIMO diversity. The general structure of MIMO antennas is shown in Figure 9.15.

In MIMO beamforming we use the same concepts of weight adaptation as explained earlier to optimize the virtual antenna pattern for both transmitter and receiver. A feedback channel is needed to adjust the weight vector of the transmitter. For LoS channels and time division duplex (TDD), the conjugate of the weight vector computed at the receiver could be used at the

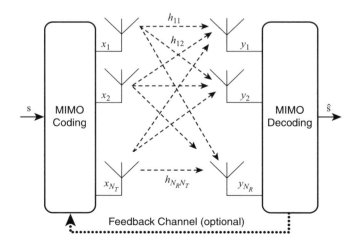

Figure 9.15   MIMO antenna structure

transmitter. MIMO beamforming may enhance the link performance if the channel is LoS. Otherwise, if the channel is Rayleigh or the number of users is very large compared to the number of antenna elements and the users are uniformly distributed, the achieved performance will be very limited.

It was observed (theoretically and practically) that it is possible to have uncorrelated parallel channels when using the MIMO structure in Rayleigh channels. The number of parallel channels is at the most equal to the lower number of antennas at the transmitter and receiver sides. If the number of antennas at the transmitter and receiver is notated as N_T and N_R respectively, then the number of parallel channels is $N_C \leq \min(N_T, N_R)$. This greatly increases the channel capacity which can be in the best case N_C times the capacity achieved by single antennas. The capacity here can be obtained by multiplexing and pre-filtering. For highly dynamic channels and high data rate we will have a high probability of outage according to the high packet losses due to fading. The reliability in this case can be enhanced through diversity. Using the MIMO structure we can transmit the same symbol with different coding over all transmit antennas and this will greatly minimize the deep fading effects. Which MIMO scheme is the best? Actually all MIMO schemes are compatible and selection depends on the application and the channel situation. We have already introduced several beamforming algorithms, therefore it is good now to present a basic analysis of MIMO channels and to show how the feedback channel could enhance the performance even in a Rayleigh channel type. From Figure 9.15, we may formulate the received signal at each receiving antenna element as:

$$
\begin{aligned}
y_1 &= h_{11}x_1 + h_{12}x_2 + \cdots + h_{1N_T}x_{N_T} + n_1 \\
y_2 &= h_{21}x_1 + h_{22}x_2 + \cdots + h_{2N_T}x_n + n_2, \\
&\;\;\vdots \\
y_{N_R} &= h_{N_R1}x_1 + h_{N_R2}x_2 + \cdots + h_{N_RN_T}x_{N_T} + n_{N_R},
\end{aligned}
\tag{9.78}
$$

The original relation should be convolution, i.e., $y_i(t) = \sum_{j=1}^{N_T} h_{ij}(t,\tau)*x_j(t) + n_i$, where $y_i(t)$ is the received signal at antenna element i, $x_j(t)$ is the transmitted signal through antenna element j, '*' is the convolution operation, $h_{ij}(t,\tau)$ is the channel impulse response between transmit antenna j and receive antenna i, and τ represents the time-varying nature of the channel. However, this relation can be reduced to a simple multiplication between the transmitted signal with a complex number representing the channel for flat fading and quasi-static channels (see Equation 1.7). In matrix form Equation (9.78) can be formulated as:

$$
\mathbf{y} = \mathbf{H}\mathbf{x} + \mathbf{n}
\tag{9.79}
$$

The capacity of such a MIMO communication system is given by

$$
C = BE_{\mathbf{H}}\left[\log_2\left(\det\left[\mathbf{I}_{N_R} + \frac{1}{N_0B}\mathbf{H}\mathbf{R}_{XX}\mathbf{H}^H\right]\right)\right]
\tag{9.80}
$$

where $E_{\mathbf{H}}[.]$ is the expectation operator and it is used because the channel matrix $\mathbf{H}$ is usually a random process, det[.] is the determinant operation for matrices, $\mathbf{R}_{XX} = E[\mathbf{x}\mathbf{x}^H]$, $\mathbf{x}^H$ is the conjugate transpose of transmit vector x, and $\mathbf{I}_{NR}$ is the identity matrix with size N_R, B is the bandwidth, and N_0 is the noise spectral density.

If the signals over all transmit antennas are orthogonal then:

$$\mathbf{R}_{XX} = diag(P_1, P_2, \cdots, P_{N_T}), \sum_{k=1}^{N_T} P_k = P_{\max} \tag{9.81}$$

where P_{max} is the maximum power available at the transmitter. How should the transmitter divide the available power among the transmit antennas? We will answer this interesting question later, but if the transmitter does not know anything about the channel, it may divide the power equally among transmit antennas, i.e.,

$$\mathbf{R}_{XX} = \frac{P_{\max}}{N_T} \mathbf{I}_{N_T} \tag{9.82}$$

The capacity equation (9.80) can be considerably simplified by two operations: singular value decomposing (SVD) and Sylvester's determinant theorem. First, with SVD, $\mathbf{H}^H\mathbf{H} = \mathbf{Q}\Lambda\mathbf{Q}^H$, where $\mathbf{Q}$ is a unitary matrix and Λ is a diagonal matrix with the squares of the singular values of the channel matrix. Sylvester's determinant theorem states that for any matrices $\mathbf{A}^{n \times m}$ and $\mathbf{B}^{m \times n}$ then $\det[\mathbf{I}_n + \mathbf{AB}] = \det[\mathbf{I}_m + \mathbf{BA}]$. Hence, the capacity equation becomes (assuming $N_T \leq N_R$):

$$C = BE_{\mathbf{H}}\left[\log_2\left(\det\left[\mathbf{I}_{N_R} + \frac{P_{\max}}{N_T N_0 B}\mathbf{HH}^*\right]\right)\right] = BE_{\mathbf{H}}\left[\log_2\left(\det\left[\mathbf{I}_{N_T} + \frac{P_{\max}}{N_T N_0 B}\mathbf{H}^*\mathbf{H}\right]\right)\right]$$

$$\Rightarrow C = BE_{\mathbf{H}}\left[\log_2\left(\det\left[\mathbf{I}_{N_T} + \frac{P_{\max}}{N_T N_0 B}\mathbf{Q}\Lambda\mathbf{Q}^*\right]\right)\right] = BE_{\mathbf{H}}\left[\log_2\left(\det\left[\mathbf{I}_{N_T} + \frac{P_{\max}}{N_T N_0 B}\mathbf{Q}^*\mathbf{Q}\Lambda\right]\right)\right]$$

$$\Rightarrow C = BE_{\Lambda}\left[\log_2\left(\det\left[\mathbf{I}_{N_T} + \frac{P_{\max}}{N_T N_0 B}\Lambda\right]\right)\right] = BE_{\sigma_k}\left[\log_2\left(\prod_{k=1}^{N_T}\left(1 + \frac{P_{\max}\sigma_k^2}{N_T N_0 B}\right)\right)\right]$$

where σ_k is the singular value of the channel matrix $\mathbf{H}$. The above capacity can be finally expressed as:

$$C = B\sum_{k=1}^{N_T} E_{\sigma_k}\left[\log_2\left(1 + \frac{P_{\max}\sigma_k^2}{N_T N_0 B}\right)\right] \tag{9.83}$$

Actually the number of non-zero singular matrix σ_k represents the rank of channel matrix $\mathbf{H}$ and at the same time the number of parallel channels. General representation of Equation (9.83) is given by:

$$C = B\sum_{k=1}^{\min(N_T, N_R)} E_{\sigma_k}\left[\log_2\left(1 + \frac{P_{\max}\sigma_k^2}{N_T N_0 B}\right)\right], \tag{9.84}$$

The capacity equation can be further enhanced if we optimize the allocated power per transmit antenna element. The optimization formula becomes:

$$C = BE_{\rho}\left[\max_{P_k:P_k \geq 0, \sum P_k \leq P_{\max}} \sum_{k=1}^{\min(N_T, N_R)} \log_2\left(1 + \frac{P_k}{N_0 B}\rho_k^2\right)\right], \tag{9.85}$$

Solving the above optimization for fixed singular values leads to the well-known water-filling power allocation algorithm given by

$$P_k = \left(\mu - \frac{N_0 B}{\rho_k^2} \right)^+, \quad k = 1, 2, \cdots, \min(N_T, N_R), \tag{9.86}$$

where μ is the water-fill level which is adjusted to achieve the following power constraint

$$\sum_{k=1}^{\min(N_R, N_T)} P_k \leq P_{\max}, \tag{9.87}$$

However, it is clear that we need feedback information about the channel in order to enhance the capacity through the water-filling algorithm. It should be noted that the channel information available at the transmitter is useless if the channel has very low temporal correlation. The reason is that when the receiver makes a channel estimation and then send this information (or part of it) back to the transmitter, the process takes some time. Therefore the information received by the transmitter represents (relatively) old measurements. Hence, if the channel variation is faster than the channel information update rate, then this information is outdated. The transmitter can still use outdated information to estimate or predict the channel, however, if the time correlation of the channel is very low (as for very fast moving mobiles), then the estimation/prediction would be too poor. In such situations, it is better to use diversity rather than multiplexing.

Exercise 9.14 Write a report with MATLAB® simulations about the maximum achieved capacity for a static and dynamic MIMO channel for the following cases of the available channel information at the transmitter:

(a) Full feedback knowledge: when **H** is known at the transmitter.
(b) Partial feedback I: only the singular values are known at the transmitter.
(c) Partial feedback II: only SINR at each antenna element is known at the transmitter (full and quantized).
(d) Partial feedback III: only the rank number is known at the transmitter

References

[1] C. Balanis, 'Antenna theory: a review', *Proceedings of the IEEE*, **80**, 7–23, 1992.
[2] J. Liberti and T. Rappaport, *Smart Antennas for Wireless Communications*, Prentice-Hall, 1999.
[3] L. Godara, 'Application of antenna array to mobile communications, part II: beam-forming and direction of arrival considerations', *Proceedings of the IEEE*, **85**, 1195–245, 1997.
[4] K. Bell, Y. Ephraim and H. Van Trees, 'A Bayesian approach to robust adaptive beamforming', *IEEE Transactions on Signal Processing*, **48**, 386–98, 2000.
[5] J. Candy, *Signal Processing – The Modern Approach*, McGraw-Hill, 1987.
[6] M. Grewal and A. Andrews, *Kalman Filtering*, Prentice-Hall, 1993.
[7] S. Choi and D. Yun, 'Design of an adaptive array for tracking the source of maximum power and its application to CDMA mobile communication', *IEEE Transactions on Antennas Propagation*, **45**, 1393–404, 1997.
[8] W. Youn and C. UN, 'Eigenstructure method for robust array processing', *IEE Electronic Letters*, **26**, 678–80, 1990.

[9] S. Kwon, I. Oh and S. Choi, 'Adaptive beamforming from the generalized eigenvalue problem with a linear complexity for a wideband CDMA channel', *Proceedings of the IEEE Vehicular Technology Conference*, pp. 1890–4, Amsterdam, Netherlands, September, 1999.

[10] A. Haimovich and Y. Bar-Ness, 'An eigenanalysis interference canceller', *IEEE Transactions on Signal Processing*, **39**, 76–8, 1991.

[11] Y. Chen and C. Chiang, 'Adaptive beamforming using the constrained Kalman filter', *IEEE Trans. Antennas Propagat.*, vol. **41**, pp. 1576–1580, Nov. 1993.

[12] Z. Rong, T. Rappaport, P. Petrus and J. Reed, 'Simulation of multitarget adaptive array algorithms for wireless CDMA systems', *Proceedings of the IEEE Vehicular Technology Conference*, pp. 1–5, Phoenix, USA, May, 1997.

[13] J. Choi, 'A receiver of simple structure for antenna array CDMA systems', *IEEE Transactions on Vehicular Technology*, **48**, 1332–40, 1999.

[14] M. Elmusrati, Radio resource scheduling and smart antennas in cellular CDMA communication systems, Ph.D. Thesis, Control Engineering Laboratory, Helsinki University of Technology, Finland, 2004.

[15] Y. Liang, F. Chin and A. Kot, 'Adaptive beamforming and power control for DS-CDMA mobile radio communications', *Proceedings of the IEEE International Conference on Communications*, pp. 1441–5, Helsinki, Finland, June 2001.

[16] A. Yener, R. Yates and S. Ulukus, 'Interference management for CDMA systems through power control, multiuser detection, and beamforming', *IEEE Transactions on Communications*, **49**, 1227–39, 2001.

[17] R. Farrokh, K. Liu and L. Tassiulas, 'Transmit beamforming and power control for cellular wireless systems', *IEEE Journal on Selected Areas in Communications*, **16**(8), 1998.

10

Cognitive Radios and Networks

Wireless media can easily be used in wireless channels to transmit and receive our signals. However, it is not free media except for very small and narrow bands. The main reason for this is the limited bandwidth available for a huge number of applications. Several different reasons are behind this limitation of the bandwidth. To transmit information signals wirelessly we need to define several things such as:

- *Signal bandwidth:* this parameter is very important because it specifies the bandwidth which will be occupied by the signal. This bandwidth is related (not necessarily equal) to the original data bandwidth. The occupied bandwidth can be controlled by the utilized modulation level, coding rate, and so on. Usually, more transmitted information per unit time will need more bandwidth. However, the allowed bandwidth is usually given as a strict constraint. There are international and national regulations stating the available bandwidth at each carrier frequency. These regulations are defined to avoid possible interference between different services. Some of these services are: radio broadcasting (LW, MW, SW, FM); TV broadcasting (analog and digital); cellular mobile (GSM, UMTS, IS95); WiFi (IEEE 802.11a/b/g/n); WiMAX; GPS; Bluetooth; Zigbee; satellite communication (C, K, Ku, etc.) bands; radars; navigations; military bands; and many more!
- *Carrier frequency:* electromagnetic wave propagation behavior depends on the signal frequency. Hence, to achieve the required transmission conditions and also to realize the multi-usage of the wireless channel, it is crucial to modulate the information signal with the sinusoidal signal (called the carrier) at the target carrier frequency. The main purpose of the carrier signal is to shift the information signal to the required band. Carrier frequency must be much larger than the information bandwidth. This represents another constraint. For example, if our baseband signal bandwidth is 5 MHz, then the applied carrier frequency should be for example >50 MHz. Moreover, we cannot use any carrier frequency we wish because of hardware limitations and biological reasons. For example, it is not feasible (yet) to implement personal commercial transmitters and receivers working at 100 GHz. Furthermore, most probably you will not be eager to stick such a device close to your ear!

Systems Engineering in Wireless Communications   Heikki Koivo and Mohammed Elmusrati
© 2009 John Wiley & Sons, Ltd

- *Modulation type:* the modulation with a carrier can be done by using carrier amplitude (amplitude modulation), frequency (frequency modulation), phase (phase modulation), and hybrid, for example amplitude–phase modulation as in QAM.
- *Multiple-access type:* defines the way that multiple users, operators and services could use the same channel simultaneously. There are several well-known techniques for multiple access such as: frequency division multiple access (FDMA); time division multiple access (TDMA); code division multiple access (CDMA); spatial division multiple access (SDMA); carrier sense multiple access (CSMA); and demand assigned multiple access (DAMA). Some descriptions about multiple access techniques were given in Chapter 1.
- *License costs:* licensing new bands for commercial use is usually extremely expensive.

From the above points and limitations we can realize that wireless available bandwidth is very scarce. Almost all bands under 3 GHz have already been allocated (or licensed) for many different services, and hence cannot be used (except for a small number of bands left free!). Usually it is illegal to use bands without permission. All countries have stern laws for wireless bandwidth usage. It is interesting to know that even when some bandwidth is licensed for a certain service, there are still strict rules about the maximum transmission power (or more accurately effective power) that can be used. As wireless communication services increase, how can we initiate new communication systems? New systems will need new bands, but most of the technically feasible bands are already allocated or licensed. One attractive solution for this problem is the recycling of the licensed bands. This solution is defined as the secondary usage of the band by cognitive radios (CR). The original owner or licensed user of the band is called the primary user. Cognitive radios are secondary users in this context. Cognitive radio can be expressed as a green technology, because it avoids the usage of new bands with extremely high frequencies.

The main and curial condition allowing cognitive radios to work is to avoid any harmful interference with primary users or the original licensed users of the bands. For example, cellular mobile operators will never permit secondary users to interfere with their system in any aspect. Actually, no one can blame them as they have already paid huge license costs. Fortunately, the FCC has realized through spatial and time scanning that most of the licensed bands are underutilized. For some licensed bands, the utilization can be as low as 5% or even less [1, 2]. This means that for 95% of the time or area this band could be used by secondary users without making new license requirements, given that the primary user is fully protected whenever and wherever to use its licensed band. There are three main situations where the secondary users can use available bands:

1. *Free bands (ISM):* there are a few free bands (i.e., unlicensed bands) left intentionally to be used freely for Industrial, Scientific and Medical (ISM) applications. Some famous bands in this category are 5.725–5.875 GHz, 2.4–2.5 GHz, 902–928 MHz (not all countries), 433.05–434.79 MHz (not all countries), and 40.66–40.70 MHz. Secondary users can use those bands freely. However, the interference level is usually high because of many services (such as Bluetooth, WiFi, cordless phones, amateur radios, and TVs). Moreover, different secondary users may compete with each other to use the free bands.
2. *Spectrum holes or white space:* spectrum holes refer mainly to licensed bands that are not in use for certain time periods. In this case secondary users may exploit this chance to send and receive over that band. However, they must stop immediately as soon as the primary

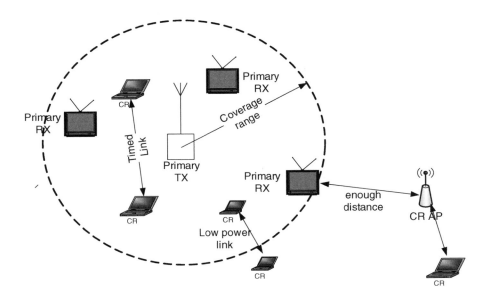

Figure 10.1 Primary and secondary users

user starts to use its band. Spectrum holes can also be found in certain geographical areas where the primary user's signal disappears (i.e., it becomes close or less than the noise floor). In such situations, where it is impossible for the primary receiver to use its band, the secondary user may assume this band as white band and use it. For example, assume that the primary broadcasting station uses a certain transmission power to cover an area of maximum radius 30 km. Secondary users located at distances ≫30 km could use the same band with a very strict power budget in order not to harmfully interfere with the primary receiver. Figure 10.1 illustrates some of these situations. Assuming a TV broadcasting service as the primary user, where this TV channel is switched off for a certain period daily (e.g., 1 am–6 am), then secondary users can utilize this idle time for their transmission (called a timed link).

3. *Gray spaces:* this situation is observed when the secondary user is within range of the primary receiver, but the SINR at the input of the primary receiver is much higher than the required target. In this case the secondary user may use this band also with small transmission power in order that no harmful interference would be observed by the primary receiver as shown in Figure 10.1. This could be explained as the energy gap that secondary users could fill as shown in Figure 10.2. Actually, it may not be legal to use the band without permission in the presence of the primary signal even without harmful interference. But what makes the primary user eager to give such permission? One proposal is that the primary user may lease its band when it is not in use or when there is no harmful interference to secondary users. This can be done via an auction between secondary users. Several economical models have been proposed to handle this situation using game theory as a tool. However, cognitive radios could act selfishly and may also be used fraudently and or even 'steal' the spectrum. Several interesting analysis and design algorithms based on game theory are available in the literature, e.g., [3, 4].

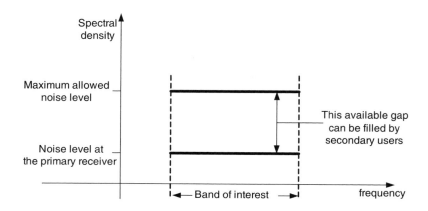

Figure 10.2   Energy gaps can be exploited by secondary users

The ability of cognitive radios to sense, assist and learn the spectrum situation and then adapt their radio parameters to the optimum makes them attractive to many applications. Some of the possible applications are:

- *Internet providers:* cognitive radio has been proposed to provide data communication links within TV licensed bands. There are several conditions for this usage in order to protect the primary users. There is already an ongoing IEEE standard for this application (IEEE 802.22). Other bands (than TV bands) are also possible at least from a theoretical point of view.
- *Military applications:* we are almost certain that there are intensive secret research activities in this field of application. It sounds attractive for the military to have communication links which can skip and even maneuver the jamming signals of enemies!
- *Personal area networks (PAN):* Bluetooth is a successful PAN for some moderate rate applications. However, it is very slow if you want to download 4 GB video from your camera to the computer. One proposal is to use cognitive radio transceivers which can work at many bands with very limited maximum power transmission. The cognitive radio devices will check the free bands and use them. For example surveillance radars which work at about 3 GHz, with about 200 MHz bandwidth, seem a possible option. One feature of radars is that the transmitter and receiver are allocated to the same place, hence, if cognitive radio does not detect the radars' signal, it means that it can use its band.

There are so many proposed applications for cognitive radios. However, there are so many challenges to making these proposals optimistic. In the rest of this chapter we will address some of those challenges.

10.1 Concepts of Cognitive Radios

From the previous introduction, we may define cognitive radios as transceiver devices that are able to perform at least the following three operations:

1. Watching the spectrum either by periodically sensing it to detect if there are primary user signals, or by using databases containing information about the temporal and spatial availability of the spectrum holes. The ability of assisting and evaluating the environment and learning from it, and moreover the possibility to cooperate with other cognitive radios, provide the cognition feature of such transceivers.
2. Fast adaptation of its radio transmission parameters according to the channel situation such as power, rate, carrier frequency, bandwidth and modulation type.
3. Use the available bandwidth efficiently through channel identification and feedback channel information between the receiver and the transmitter.

Why have cognitive radios not appeared before? There are several reasons, but at least two of them are especially important. The first is the recent accelerated needs for new wireless systems. The spectrum has become more crowded during the last decades with many new wireless standards. The second reason is the recent advances in VLSI technologies especially fast ADC which is the core of software-defined radios (SDR). Software-defined radio is a very flexible and generic radio platform [5].

Consequently software-defined radios are the core of cognitive radios and are required to achieve operation of the full radio parameters adaptation. One general structure of cognitive radio is shown in Figure 10.3, where cognitive tasks refer to the applied smart algorithms (evaluation, predication, cooperation, utility maximization, risk minimization, etc.). Several methods can be applied in the cognition engine such as game theory, neural networks, fuzzy algorithms and expert systems [6].

Is it true that such radio devices are cognitive? The answer depends on how we define cognition. Humans have at least five types of sensing inputs and one powerful processing unit (the brain) which perform many tasks such as learning, reasoning, and awareness. There is a lot of debate about the exact definition and limitation of each of these capabilities. However, there is a general consensus that the main source of cognition is environment sensing followed by processing power. Cognitive radios are able to sense the environment and then process data

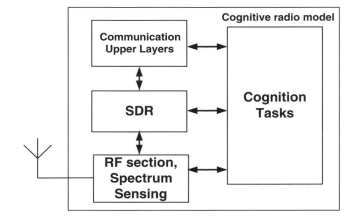

Figure 10.3 General structure of cognitive radio

and take actions. Therefore, by general analogy with human capability they are called cognitive devices; although they are still a long way off from even a fraction of insects' capabilities.

10.2 Spectrum Attention of Cognitive Radios

It is clear from previous discussions that spectrum attention is critical for the success of cognitive radio systems. Currently, spectrum sensing is one of the main technical problems for cognitive radios. Cognitive radios must be careful not to make any harmful interference with the primary receivers. In broadcasting systems (such as radio, GPS and TV), the receivers do not send signals back to the transmitters. Hence, how can cognitive radio sense if there are any close receivers? They may sense the transmitter signal, but we are mainly concerned with the primary receivers, not the transmitters. One proposal is to sense the local signals generated inside the receivers such as the receivers' local oscillators. This is only possible if the cognitive radio is close enough to the receiver. Therefore, this is not a practical solution. Another proposal is to integrate a transmitter in all receivers which send 'Hello' signal over a predefined frequency band. Hence, cognitive radios are able to detect any close primary receivers. This is not practical at least for systems in the near future. There are, however, two practical methods for spectrum attention adopted by the IEEE802.22 cognitive radio standard [7]:

- Using a database which contains information about channel organization and scheduling. This solution is valid for fixed cognitive radios since it depends on the location. Moreover, it can be applied to pre-specified primary users such as TV channels (the main target of IEEE802.22). It is necessary for mobile cognitive radios to know their location in order to use the available database. In this regard, it is suggested to integrate the GPS receiver within the cognitive radio. However, this increases the cost, complexity, and is a problem when there is not enough GPS coverage (such as indoors).
- Building knowledge about the spectrum through direct spectrum sensing. This solution seems to be the most promising approach for current and future cognitive radios. We will discuss this in more detail and show its largest challenges.

10.3 Direct Spectrum Sensing

Here each cognitive radio builds its own knowledge about the spectrum situation and whether there are spectrum holes in certain bands. There are two major classes: energy-based and feature-based detection methods. In energy-based detection methods, we just sense the power density level of the scanned spectrum. It is possible to perform this kind of measurement in time or in frequency domains as we will explain later. The problem is that there is always a noise power floor at the input of the cognitive radio. Often the cognitive radio is required to detect the presence of a primary user signal even if its average power is less than the noise average power (negative SINR in dB scale). One of the requirements of the IEEE802.22 standard is that the cognitive radio is able to detect the presence of primary users' signal at a minimum SINR of -22 dB! Another problem of energy-based detection is that it is not possible to determine the type of the detected signal – whether it is from a primary user or other secondary users. On the other hand, energy-based detection methods are relatively easy to implement and need less processing power.

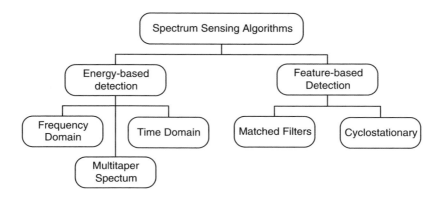

Figure 10.4 General classifications of spectrum-sensing algorithms

Many of the energy-based detection disadvantages can be overcome with feature-based detection methods. The idea of feature-based detection methods is to search for one of the known primary signal features (such as carrier frequency, training sequence) within the specified spectrum. Figure 10.4 shows a general classification of direct spectrum-sensing methods, however, there are several other techniques which are not shown [8].

Before we describe briefly the spectrum-sensing methods shown in Figure 10.4, let's first explain why spectrum sensing is a big challenge. The reason is that it is not possible to have an unbiased and consistent estimation for the spectrum. Practically, we need to estimate the spectrum based on finite time observation measurements. However, as you remember from Fourier transform, the frequency domain presentation needs to integrate over infinite time. So that when we take only finite observation time, the transform will be for the original time-domain signal multiplied by the rectangular window as explained in the following two equations:

$$X(f) = \int_{-\infty}^{\infty} x(t)e^{-j2\pi ft}dt \tag{10.1}$$

$$\hat{X}(f) = \int_{-T}^{T} x(t)e^{-j2\pi ft}dt = \int_{-\infty}^{\infty} (w(t)x(t))e^{-j2\pi ft}dt \tag{10.2}$$

where $w(t)$ can be rectangular function. However, it is better in terms of estimation variance to use more smooth windows such as the Kaiser window. For rectangular windows, $w(t)$ is given by:

$$w(t) = \begin{cases} 1, & -T \leq t \leq T \\ 0, & elsewhere \end{cases} \tag{10.3}$$

As we know, multiplication in the time domain is transformed as convolution in the frequency domain, this leads to:

$$\hat{X}(f) = W(f) * X(f) \tag{10.4}$$

So to estimate the original $x(f)$ we will need to deconvolve the measurement with the sinc function for rectangular windows (the sinc function is the Fourier transform of the rectangular time-domain signal). There are many problems associated with the deconvolution process which makes the perfect construction of the original signal in noisy measurements impossible! However, deconvolution is not the only problem of spectrum estimation. There are many others. One of which is the discrete nature of the measurements. Cognitive radios are digital devices with limited sampling rate and memory capacity. The discretization of time-domain signals gives periodic spectrum with maximum frequency of $f_s/2$, where f_s is the sampling rate. Taking discrete Fourier transforms (DFT) of these time samples defines the frequency at only discrete values. This means that this set of frequency samples represents period replication of the measured time-domain signal. This leads to leakage problems in the spectrum. Example 10.1 illustrates these concepts.

Example 10.1

Using MATLAB® generate a sinusoidal signal with frequency 1 kHz and sampling rate 10 kHz. Draw the absolute value of FFT (using the *fftshift* function) in two cases:

  i. from 0 to 2 ms window;
 ii. from 0 to 2.5 ms window.

Solution

A straightforward code to perform the first part is given below:

```
Tf = 2e-3; % Time window 0<=t<=Tf
fs = 10e3; % sampling frequency
Ts = 1/fs; % sampling duration
t = 0:Ts:Tf-Ts; % simulation time
N = length(t); % simulation sample length
Tf = 2e-3; % Time window 0<=t<=Tf
fs = 10e3; % sampling frequency
Ts = 1/fs; % sampling duration
t = 0:Ts:Tf-Ts; % simulation time
plot(f,fftshift(abs(X))) %plot the spectrum
```

The resultant spectrum is shown in Figure 10.5. It is exactly as we expected, that is the magnitude of the spectrum of the sinusoidal signal is two lines at the wave frequency.

If we increase the simulation period to 2.5 ms as required in the second part of the example, we obtain the spectrum shown in Figure 10.6.

This second result is strange and does not fulfill our expectations. It seems strange because when we increased the simulation time we obtained a worse estimation. Why is this? Let's keep this 2.5 ms window and increase the sampling rate from 10 kHz up to 100 kHz. The result is shown in Figure 10.7. No improvement! This is called spectrum leakage. At the 2 ms period we have two complete periods of the sinusoidal signal. When we take the FFT, we have discrete frequency which represents the periodic spectrum of the original signal. The replication of

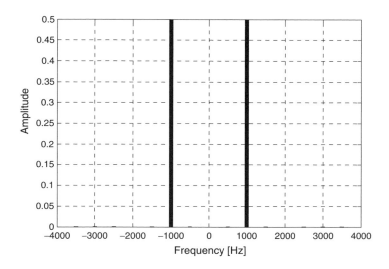

Figure 10.5 The spectrum of sinusoidal signal when $T = 2$ ms

the complete periods generates the original infinite-length sinusoidal signal. On the other hand when we increase the time duration to 2.5 ms, the FFT process was done over 2.5 periods, and the replication of this signal is not a perfect sinusoidal signal. Increasing the sampling frequency will not reduce this leakage problem. Hence, even for this trivial case, we cannot obtain perfect estimation of the spectrum unless we know exactly the frequency of the sinusoidal signal in order to guarantee full-period windows. In reality we deal with non-periodic, random,

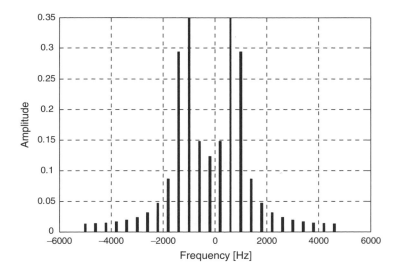

Figure 10.6 The spectrum of sinusoidal signal when $T = 2.5$ ms

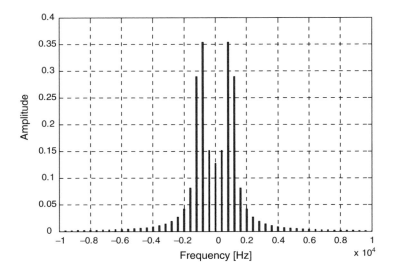

Figure 10.7   The spectrum of sinusoidal signal when $T = 2.5$ ms and higher sampling rate

and noisy signals, which make the spectrum estimation problem much more complicated. One practical method to mitigate this leakage problem is to use smooth windowing for the time-domain signal. This prevents the sharp discontinuity of the signal at signal ends.

In spectrum sensing, the signals are usually random with any possible distribution. Hence, it is more convenient to visualize the spectrum using the power spectral density (PSD) function, which is given as the Fourier transform of the autocorrelation function as

$$S(f) = \Im\{R_{xx}(\tau)\} = \int_{-\infty}^{\infty} R_{xx}(\tau)e^{-j2\pi f\tau}d\tau, \tag{10.5}$$

This expression assumes a stationary random process $x(t)$. If it is not stationary the problem is more complicated and the PSD is time dependent. The autocorrelation function $R_{xx}(\tau)$ is generally given as $E[x(t)x(t-\tau)]$, where $E[.]$ is the expectation operator. Since generally we do not know the probability density functions of the sensed signals, the expectation will be based on time averaging. In this case we have two equivalent forms of the estimated PSD (the *periodogram*). The first form is given by

$$\hat{S}(f) = \frac{1}{2T}\left|\hat{X}(f)\right|^2 = \frac{1}{2T}\left|\int_{-T}^{T} x(t)e^{-j2\pi ft}dt\right|^2 \tag{10.6}$$

Here, we have considered only the finite measurement time effects. We can also use DFT summation instead of the integration. However, since we have already seen the impact of sampling on the spectrum estimation, we will concentrate here on the natural inconsistency in the spectrum estimation even in continuous cases. The effects of time windowing can be better described in the second formulation of the periodogram which is shown in Equation (10.5), but for finite time measurement, i.e., the autocorrelation estimation becomes [9]:

$$
R_T(\tau) = \begin{cases} \dfrac{1}{2T - |\tau|} \displaystyle\int\limits_{-T + \frac{|\tau|}{2}}^{T - \frac{|\tau|}{2}} x\left(t + \dfrac{\tau}{2}\right) x\left(t - \dfrac{\tau}{2}\right) dt, & \text{for } |\tau| < 2T \\[2em] 0, & \text{for } |\tau| > 2T \end{cases} \tag{10.7}
$$

It is clear that as $|\tau| \to 2T$ the error becomes very large because we average over a small period. If we want to estimate the autocorrelation of stationary signal at a certain correlation time τ_0, we can apply Equation (10.7) and select $T \gg \tau_0$. This does not work for the spectrum estimation, because we should integrate over all correlation time τ as:

$$
\hat{S}(f) = \int\limits_{-2T}^{2T} R_T(\tau) e^{-j2\pi f \tau} d\tau, \tag{10.8}
$$

This means that whatever the length of the measurement time, we will have estimation variance as long as the measurement time duration is finite. The estimation variance could be reduced at the cost of increasing the estimation biasing. A compromise can be made between the biasing and the variance, for example to minimize the mean square errors. The topic is interesting, however, the detailed analysis is beyond the scope of this chapter.

After this quick introduction about the inherent problems of spectrum estimation, let's go back to Figure 10.4 which introduced the main concept of every spectrum-sensing method.

10.3.1 Energy-Based Detection

This is the most common method of spectrum sensing. It is simply based on the detection of any transmitters within a certain range. This method has several advantages such as low cost and complexity. Furthermore, it is a general technique and can be used with any system without prior knowledge about the primary users, and it has a fast sensing time if the primary signal power is high (i.e., high SINR). However, it suffers from many disadvantages such as: it is not possible to distinguish between primary usage of the spectrum or other secondary users; it has a large sensing time at low SINR; and there is a large probability of detection errors at low SINR [10]. There are two main types of detection errors. First, when the detection output indicates incorrectly the presence of primary user, this is called a false alarm. The second type is false detection, which is observed when the detector cannot detect the primary user signal and indicates free spectrum incorrectly. Some of the energy-based detection problems can be considerably mitigated with some system enhancement such as using multi-antennas at the cognitive radio and also through spectrum information exchange with other cognitive radios, i.e., using cooperative networks. We will start with time-domain sensing as shown in Figure 10.8.

Time-domain sensing is based on time sampling of the required band, which is determined by the sharp band-pass filter (BPF). The sampled signal is quantized using an analog to digital

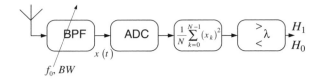

Figure 10.8   Energy-based spectrum sensing in time domain

converter (ADC). The following operations are performed in digital form using a digital processer. Assuming that the number of samples is N, we next compute the average power content as shown in Figure 10.8. Finally, we compare the result with a certain threshold λ. If the result larger than the threshold, we make the decision that the spectrum is occupied (i.e., a primary signal exists, Hypothesis H_1), otherwise the spectrum is white (Hypothesis H_0), i.e.,

$$
x_k = \begin{cases} n_k, & H_0 \ (white) \\ s_k + n_k, & H_1 \ (occupied) \end{cases}
\tag{10.9}
$$

where n_k is sampled noise signal, s_k is sampled primary user signal, and x_k is the sampled received signal. It can be assumed that the received signal x_k samples are identical independent Gaussian distribution samples. We may also assume that the average values of both primary and noise signals are zero. Moreover, the variances of the primary and noise signals are σ_s^2 and σ_n^2 respectively. Our decision is based on the test statistic:

$$
u = \frac{1}{N} \sum_{k=0}^{N-1} (x_k)^2
\tag{10.10}
$$

It is clear that the test statistic u is a random variable with a chi-square probability density function, which has the following form:

$$
f_U(u) = \frac{N(Nu)^{\frac{N}{2}-1}}{\sigma^N 2^{N/2} \Gamma\left(\frac{N}{2}\right)} e^{-Nu/2\sigma^2}, \quad u \geq 0
\tag{10.11}
$$

where the value of σ depends on the sampled data such as:

$$
\sigma^2 = \begin{cases} \sigma_n^2, & \text{under } H_0 \\ \sigma_n^2 + \sigma_s^2, & \text{under } H_1 \end{cases}
\tag{10.12}
$$

It is more common to approximate the distribution of u as a normal distribution. This approximation is valid for large enough samples N which is satisfied with practice. In this case, we need to compute the mean and variance of u from Equation (10.10) which can be easily found as σ_x^2 and $2\sigma_x^4/N$ respectively. With respect to the sampled signal situation, the distribution of u can be formulated as:

$$
f_U(u) = \begin{cases} G(\sigma_n^2, 2\sigma_n^4/N), & \text{under } H_0 \\ G(\sigma_i^2, 2\sigma_i^4/N), & \text{under } H_1 \end{cases}
\tag{10.13}
$$

where $G(a, b)$ is normal distribution with mean (a) and variance (b), and $\sigma_i^2 = \sigma_s^2 + \sigma_n^2$. If $SNR = \sigma_s^2/\sigma_n^2$, we can state that $\sigma_i^2 = (1 + SNR)\sigma_n^2$.

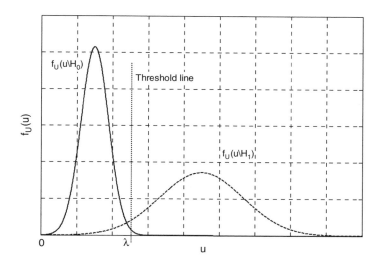

Figure 10.9   Distributions of measurement statistic u with two hypothesis

The distributions of u under both hypotheses are shown in Figure 10.9.

From Figure 10.9, in order to reduce the probability of false alarm we need to increase the threshold value λ, however, this will reduce the probability of primary user detection.

On the other hand, in order to increase the probability of primary user detection, we need to reduce the threshold value λ, however, this will increase the probability of false alarm! We need to compromise between these two objectives. It is pretty clear that the probabilities of error detection can be represented in forms of the Q-function. For example, the probability of false alarm is given by [11]

$$P_{\text{false}} = P_r(u > \lambda | H_0) = Q\left(\frac{\lambda - \sigma_n^2}{\sigma_n^2/\sqrt{N/2}}\right) \tag{10.14}$$

Exercise 10.1   Find the probability of primary user detection based on a normal approximation.

Exercise 10.2   Determine the expression of the value of the threshold which guarantees that we detect the primary user signal with probability of 0.99. Based on this value, find the probability of false alarm.

Hint: The relation between the threshold value and the probability of detection can be determined directly from the algorithm derived in Exercise 10.1. It is given by:

$$\lambda_{\text{det}} = \sigma_n^2(1 + SNR)\left(1 + \frac{Q^{-1}(P_{\text{det}})}{\sqrt{N/2}}\right) \tag{10.15}$$

Exercise 10.3   Discuss the impact of the number of samples (N) on the probability detection and the threshold value given in Equations (10.14) and (10.15), respectively.

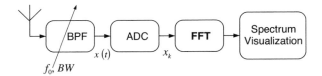

Figure 10.10   Energy-based detection method in frequency domain

Exercise 10.4   Address the relation between the SNR and the number of required samples for a certain probability of detection.

It can be shown from the result of Exercise 10.4 that the required number of samples to achieve a certain probability of primary detection can be very large for low SNR. This is one of the problems of energy-based detection methods as described previously. Moreover, it assumes an accurate knowledge of the average power of the background noise, which is not always true. In many situations, this average noise power is time varying (non-stationary), where we will need to estimate this background noise periodically.

The frequency-domain method of energy-based detection is based simply on the estimation of the power spectral density (periodogram) as explained earlier. The idea is simplified in Figure 10.10. One of the features of this method is that it uses the FFT algorithm. This algorithm is also used in OFDM transceivers, which is one of the potential modulation techniques proposed for cognitive radios. The problems associated with finite data and sampling on spectrum estimation have been mentioned in an earlier section.

There is one powerful method which can considerably enhance the spectrum estimation accuracy by reducing the estimation variance. This method is called the *multitaper method*. There is a wide impression that this method will become standard as energy-based spectrum detection in cognitive radios. Two reasons support this: (i) the method accuracy is considerably better than other methods such as time-domain detection and conventional periodogram methods; (ii) it can be realized using filter banks, and consequently, filter banks can be implemented using IFFT, which is the platform of OFDM transmitters which is expected to be the modulation type of cognitive radios [12].

The multitaper method is simply a windowed periodogram process of the sampled time data. The difference between multitaper and other conventional windowing is that multitaper uses Q orthogonal tapers called discrete prolate spheroidal sequence (DPSS) tapers (windows), then takes the weighting average of the result. As Q increases we reduce the error variance but sacrifice the resolution. Hence, Q represents a tradeoff between the resolution and variance. The eigenspectrum for the qth taper is computed as

$$\hat{S}_q^{mt}(f) = \frac{1}{f_s} \left| \sum_{k=0}^{N-1} h_{k,q} x_k e^{-j2\pi f k/f_s} \right|^2 \tag{10.16}$$

where $h_{k,q} \in \{h_{0,q}, h_{1,q}, \cdots, h_{N-1,q}\}$, $q = 0, 1, \cdots, Q-1$, represents the qth taper values which reduce the spectrum leakage as described before. Since we generate Q different eigenspectrum through an orthogonal process, the estimation variance can be considerably reduced by the following averaging process:

$$S^{mt}(f) = \frac{\sum_{q=0}^{Q-1} \lambda_q \hat{S}_q^{mt}(f)}{\sum_{q=0}^{Q-1} \lambda_q} \tag{10.17}$$

where λ_q represents the eigenvalue associated with each eigenvector $\mathbf{h}_q$.

There are other interesting methods that can be applied in this class of detection, such as wavelet transform. Regardless of the method applied, there are several inherent limitations, such as the inability to differentiate between the kind of signals (primary, secondary, and others). Several problems associated with energy-based detection methods have been handled by feature-based detection as will be discussed next.

10.3.2 Feature-Based Detection

As can be deduced from its name, in this class we detect the signal based on more specific features than its energy, which is common to all transmitters. There are two famous subclasses in this method. One is the matched filter which is the optimum linear method as will be shown later and the second is the cyclostationary method. We start with the matched filter method.

Assume we receive a signal $x(t)$ such as:

$$x(t) = \beta s(t) + n(t) \tag{10.18}$$

We know the shape of $s(t)$ which is time-finite signal (e.g., with duration T), $n(t)$ is a white random process with Gaussian distribution, and $\beta \in \{0, 1\}$. When $\beta = 1$, the primary user occupies the channel, otherwise not. It is not required to recover $s(t)$ form, but we are interested in detecting if $s(t)$ exists. If $x(t)$ is applied to a linear filter, what is the optimum impulse response of that filter to maximize the probability of detection whether $s(t)$ exists or not? If we call the impulse response of this optimum filter $h(t)$ then the output will be:

$$y(t) = \int_{\infty}^{\infty} x(\tau)h(t-\tau)d\tau = \int_{\infty}^{\infty} s(\tau)h(t-\tau)d\tau + \int_{\infty}^{\infty} n(\tau)h(t-\tau)d\tau \tag{10.19}$$

If $y_s(t) = \int_{-\infty}^{\infty} s(\tau)h(t-\tau)d\tau$, and $y_n(t) = \int_{-\infty}^{\infty} n(\tau)h(t-\tau)d\tau$, then we are looking for the optimum impulse response $h(t)$ which maximizes $E[y_s^2]/E[y_n^2]$. Since convolution in the time domain is equivalent to multiplication in the frequency domain then the SNR ratio can be expressed as

$$y_s(T) = \int_{-\infty}^{\infty} H(f)S(f)e^{j2\pi fT}df \Rightarrow E[y_s^2] = \left| \int_{-\infty}^{\infty} H(f)S(f)e^{j2\pi fT}df \right|^2 ;$$

$$E[y_n^2] = \sigma_n^2 = \frac{N_0}{2} \int_{-\infty}^{\infty} |H(f)|^2 df \tag{10.20}$$

So that the SNR ratio can be expressed as

$$SNR_T = \frac{\left| \int\limits_{-\infty}^{\infty} H(f)S(f)e^{j2\pi fT}\,df \right|^2}{\frac{N_0}{2} \int\limits_{-\infty}^{\infty} |H(f)|^2\,df} \tag{10.21}$$

We use time T instead of t, because in the convolution we will integrate only over the duration of $s(t)$ which is given as T. Schwartz's inequality states that for any functions $f_1(x)$ and $f_2(x)$,

$$\left| \int\limits_{-\infty}^{\infty} f_1(x)f_2(x)\,dx \right|^2 \leq \int\limits_{-\infty}^{\infty} |f_1(x)|^2\,dx \int\limits_{-\infty}^{\infty} |f_2|(x)|^2\,dx \tag{10.22}$$

the quality sign holds at $f_1(x) = kf_2^*(x)$, for any integer k, where $f_2^*(x)$ is the conjugate of $f_2(x)$. Using Schwartz's inequality with Equation (10.21), we obtain

$$SNR_T \leq \frac{\int\limits_{-\infty}^{\infty} |H(f)|^2\,df \int\limits_{-\infty}^{\infty} |S(f)|^2\,df}{\frac{N_0}{2} \int\limits_{-\infty}^{\infty} |H(f)|^2\,df} = \frac{2}{N_0} \int\limits_{-\infty}^{\infty} |S(f)|^2\,df \tag{10.23}$$

This means that the maximum achievable SNR is $\frac{2}{N_0} \int_{-\infty}^{\infty} |S(f)|^2\,df$. This maximum is achieved when the transfer function of the receiver filter is $H(f) = kS^*(f)e^{-j2\pi fT}$. From this result, the impulse response of the matched filter is the inverse Fourier transform of the transfer function, which is given by

$$h(t) = \Im^{-1}(kS^*(f)e^{-j2\pi fT}) = \begin{cases} ks(T-t) & 0 \leq t \leq T \\ 0 & \text{elsewhere} \end{cases} \tag{10.24}$$

Using this filter impulse response in the convolution integral given in Equation (10.19), we obtain:

$$y(t) = k \int\limits_0^t x(\tau)s(T-t+\tau)\,d\tau \tag{10.25}$$

and at the signal duration T, the previous result becomes

$$y(T) = k \int\limits_0^T x(\tau)s(\tau)\,d\tau \tag{10.26}$$

This important result gives the optimum linear detection algorithm. Actually it is clear that it is the correlation between the received signal and the specific signal that we want to detect.

Let's now determine the result of the correlator process presented by Equation (10.26) in the presence and absence of the primary signal as (with $k = 1$):

$$y(T) = \int_0^T [\beta s(\tau) + n(\tau)] s(\tau) d\tau = \beta \int_0^T s(\tau)^2 d\tau + \int_0^T n(\tau) s(\tau) d\tau \tag{10.27}$$

For zero average noise, which is uncorrelated with the primary user signal, the second term is $\int_0^T n(\tau)s(\tau)d\tau \simeq 0$, then Equation (10.27) becomes:

$$y(T) = \beta \int_0^T s(\tau)^2 d\tau \tag{10.28}$$

Theoretically, the output will be the energy of the received primary signal symbol if it is present ($\beta = 1$) and zero otherwise ($\beta = 0$). From this simple analysis it is clear that the matched filter or its correlation version is the optimum linear detector in the case of an uncorrelated zero mean additive stationary white noise environment. However, this matched filter is not so attractive for cognitive radio applications. There are at least three reasons for this: (i) It needs a complex receiver, because $s(t)$ is the primary user signal in baseband, hence the cognitive radio needs to have a demodulation process and then to apply the correlation algorithm to check the presence of the primary signal. This makes cognitive radio complex. Moreover, if the cognitive radio needs to check several primary users, this can be even more complex. (ii) The cognitive radio should know some of the primary signal details such as its transmitted symbols $s(t)$, this is not always available, especially for mobile cognitive radios. (iii) The assumption of stationary zero mean white Gaussian additive noise is not perfectly achieved in practice, which makes the matched filter less efficient.

The complexity of matched filters can be considerably reduced by using the cyclostationary method. This method is based on the fact that manmade signals have characteristics that are completely different than natural noise. One of these characteristics is the hidden periodicity in these signals such as the carrier frequency and the training sequence used by channel equalizers. Periodic signals appear as spectral lines in the frequency domain. The main problem here is when the spectral lines disappear because of modulation. For example, in double-side band suppressed carrier (DSB-SC) modulation the spectrum consists of the original information spectrum shifted by the carrier frequency, but with no spectral lines. It was shown that the spectral lines can be extracted by using quadratic transformation such as [13]

$$y(t) = x(t)x^*(t - \tau) \tag{10.29}$$

Now we can determine the locations of those spectral lines by a correlation process with the complex sinusoidal signal (spectral line) at frequency α which represents the line location in the spectrum as follows:

$$q(\tau, \alpha) = \int x(t)x^*(t - \tau)e^{-j2\pi\alpha t} dt \tag{10.30}$$

At $\alpha = 0$, we will have the autocorrelation function in the time domain. For this reason it is represented as a general form of the autocorrelation function:

$$R^{\alpha}(\tau) = \int x(t)x^*(t-\tau)e^{-j2\pi\alpha t}dt \tag{10.31}$$

Frequency parameters α for which $R^{\alpha}(\tau) \neq 0$ are called cyclic frequencies.

More interesting results can be obtained by treating the problem as a stochastic process utilizing an ensemble averaging process. From Equation (10.31) we can visualize the spectrum of the signal by Fourier transform such as:

$$S_x^{\alpha}(f) = \Im\{R_x^{\alpha}(\tau)\} = \int_{-\infty}^{\infty} R_x^{\alpha}(\tau)e^{-j2\pi f\tau}d\tau \tag{10.32}$$

where $S_x^{\alpha}(f)$ is called the *spectra correlation density* (SCD). At $\alpha = 0$, the SCD is reduced to the conventional power spectral density. It is clear that the SCD is a two-dimension function in frequencies f and α.

Cyclostationary is an efficient method used to detect primary user signal even when it is buried in high-level natural noise and interferences. For example, if we know that the primary user carrier frequency is 0.85 GHz, then we can multiply the received signal with a variable delayed version of itself and correlate the resultant signal with a complex sinusoidal at 0.85 GHz. If the output is greater than some threshold we decide on the presence of the primary user. The SCD can also be used to scan the spectrum to find if all transmitters use the specified band.

For a received signal associated with an additive background noise such as $r(t) = x(t) + n(t)$, it can be proved that $S_r^{\alpha}(f) = S_x^{\alpha}(f) + S_n^{\alpha}(f)$. Since background noise has no hidden periodicity, we can guess that $S_n^{\alpha}(f) = 0$ for $\alpha \neq 0$. Hence, the spectral correlation density of the received signal will depend only on the manmade signals (with periodicity) such as the primary user signal. Depending on the type of periodicity (or equivalently the SCD) we can distinguish between primary user and secondary users. For example, for certain licensed bands, we have a primary user and two different kinds of secondary users from different operators. It is not difficult to find the SCD of each signal in the band such as $S_P^{\alpha}(f), S_{S1}^{\alpha}(f)$, $S_{S2}^{\alpha}(f)$ for primary, secondary 1 and secondary 2 users, respectively. One realization for user detection using the cyclostationary method is shown in Figure 10.11. The output level of the correlator will be high if the corresponding transmitter exists. This feature is important in identifying the users of the band. This gives a large advantage of cyclostationary over energy-based detection methods. Table 10.1 shows a brief comparison between energy-based and feature-based detection methods.

Exercise 10.5 Assume the following binary signal $s(t) = \sum_{k=-\infty}^{\infty} aU(t - kT)$, where $a = \pm 1$ with equal probability. The signal is transmitted with BPSK which is the same as DSB-SC, i.e., $x(t) = s(t)\cos(2\pi f_0 t + \theta_0)$, where θ_0 is fixed phase, then

(i) Show that no spectra lines can be observed in the power spectral density.
(ii) Find the cyclostationary function.

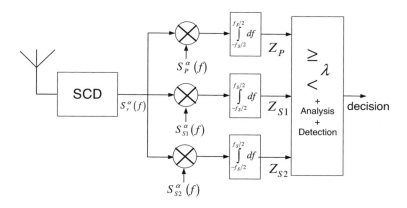

Figure 10.11   Cyclostationary method for spectrum user detection

(iii) Find the spectral correlation density.

(iv) Use MATLAB® to find the spectral lines even in the presence of large levels of additive noise, i.e., when $r(t) = s(t)\cos(2\pi f_0 t + \theta_0) + n(t)$

It is also possible to use different sensing methods within one platform. This depends on the application. For some less critical primary users such as TV channels, cognitive radio may use energy-based detection to check the presence of the user. Energy-based detection consumes less processing power. When the channel is occupied and the cognitive radio decides to jump to another critical band, e.g., the navigation band, it may start using feature-based detection to guarantee the availability of the band. Actually the decision of jumping or just reducing the transmission power is another interesting branch of research. There are several techniques proposed in the literature to handle this decision problem such as fuzzy logic, artificial intelligence, and neural networks.

Table 10.1   Comparison between energy- and feature-based detection methods

Energy-based detection	Feature-based detection
More general and can be used with any system	Only possible when some details about the primary user signal are known
High detection error probability at low SNR	Low to moderate detection error probability even at low SNR
Low sensing ability	High sensing ability
Not possible to differentiate between primary user and other secondary users	Possible to differentiate between primary user and other secondary users
Needs less processing power and lower circuit complexity	Higher processing power with more complicated circuits
Very long sensing time at low SNR	Moderate sensing time even at low SNR
Short sensing time at high SNR	Longer sensing time even at high SNR (because of the identification process)

10.4 Cognitive Radio Networks and Game Theory

Game theory is a famous analysis and optimization technique for a multi-authority with at least partial conflicted objectives. We call players the 'authority' because they should be decision makers and moreover smart and rational. In game theory we analyze the conflict between players (or authorities) and then predict the outcomes from games in different situations such as non-cooperative and cooperative cases. Furthermore, game theory can be used as a tool to design new algorithms to handle cognitive networks. It is possible to include game theory algorithms in the cognitive engine of cognitive radios to decide the best response for certain actions from other users (primary and secondary) within certain channel situations. Before we give some present applications of game theory in cognitive radios and cognitive radio networks let's start with a simple introduction to game theory. Wherever there are conflicts between different authorities to achieve their objectives, there will be a game. 'Game' is not only used as a word of fun, but all life is a big serious game in the sense that everyone plays to achieve his/her objectives, which may totally or partially conflicted with others' objectives. Maximizing our profits and properties, increasing our career levels, improving our lifestyle, minimizing our daily problems, having a happy life are some of our objectives. But these will conflict with others' objectives because we live in a limited resources environment. The relations between countries, firms, vendors, markets, people (social), non-human (animals, plants, and even viruses) can all be defined to some extent as games. Often it is possible to interpret or explain certain reactions from players in the light of game theory. However, the clearest application of game theory with remarkable success is in the area of business and economy. Eight Nobel prizes were granted to scientists for their contributions related to game theory in economy and business. Games consist of at least three units:

1. *Players:* in game theory we assume that players are rational, intelligent, and decision makers.
2. *Game rules and strategies:* every game should have defined rules. Those rules can be changed with different situations. Moreover, all sets of movements or tactics are known as game strategies.
3. *Players' objectives and payoffs:* every player plays his game with a certain objective to be achieved. The objectives depend on the payoffs of the game. If the payoff is money, then the players' objective could be maximizing their revenue or minimizing their costs or losses. In wireless communication, the payoff could be defined in terms of QoS such as throughput where players want to maximize or BER which should be minimized.

What is the difference between conventional optimization and optimum game strategies? Let us assume that we have a reward (or utility) function that depends on vector $\mathbf{x} = [x_1 \cdots x_N]$ and we want to find the optimum vector $\mathbf{x}$ which maximizes our utility such as

$$\hat{\mathbf{x}} = \arg \max_{\mathbf{x}} [\Psi_1(\mathbf{x})]$$
$$\text{s.t.} \tag{10.33}$$
$$\mathbf{x} \in \Omega_x$$

where $\Psi_1(\mathbf{x})$ is the utility function that we want to maximize and Ω_x is the feasible set of vector $\mathbf{x}$, i.e., it is not allowed to search outside this set. This set is determined by the given constraints. It is clear that Equation (10.33) is a conventional optimization problem, it is NOT a game.

Now assume that the utility function also depends on another vector of parameters ($\mathbf{y}$) which can be set by another person (or player), i.e., you do not have direct control over these new parameters. The problem now becomes

$$\hat{\mathbf{x}} = \arg\max_{\mathbf{x}} [\Psi_1(\mathbf{x}, \mathbf{y})]$$
$$\text{s.t.} \tag{10.34}$$
$$\mathbf{x} \in \Omega_x$$

and your opponent also has a similar problem, she wants to solve the following maximization problem:

$$\hat{\mathbf{y}} = \arg\max_{\mathbf{y}} [\Psi_2(\mathbf{y}, \mathbf{x})]$$
$$\text{s.t.} \tag{10.35}$$
$$\mathbf{y} \in \Omega_{\mathbf{y}}$$

Remember that, your opponent's objective can be different than maximizing her utility. For example she can select to damage your utility regardless of her payoff, in this case Equation (10.35) becomes:

$$\hat{\mathbf{y}} = \arg\min_{\mathbf{y}} [\Psi_1(\mathbf{y}, \mathbf{x})]$$
$$\text{s.t.} \tag{10.36}$$
$$\mathbf{y} \in \Omega_{\mathbf{y}}$$

Finding the optimum strategy (i.e., $\hat{\mathbf{x}}, \hat{\mathbf{y}}$) is the solution of this game formulation. Simply this is the main difference between game theory and conventional optimization. However, some researchers in communication want to formulate non-game problems in game forms. It is not thought that they will gain much from doing that. For example, if we assume channel fading as the opponent player, and the objective is to maximize the throughput or minimize transmission power, then it must be remembered that the fading channel is something random without objectives (i.e., it is not intelligent or rational), hence assuming it as another player would violate the definition of players in game theory. Therefore, most of the results obtained by this formulation are not very significant. It is alleged that there are some games that could be played randomly. This is true, but since the player is smart, if he observes better payoff from deviating from the random game then he will do. For example, assume that you play the following game with another person: both of you have two colored cards, one red and one black. In the same instant of time both of you decide to raise one card. If both cards are the same color, you win a point. Otherwise, your opponent wins a point. It is clear that the optimum way to play this game is to make your opponent unable to guess your next decision. This will be achieved when you select any card with probability 0.5! This will be the same for your opponent. But if you observe that your opponent starts to show the red card more than the black card, then your response will be to increase your probability of selecting the red card, because your objective is to have similar color cards. The channel is not able to do this. It is merely a random parameter where its statistic properties are dependent on the distance between transmitter and receiver, environment type, multipath profile, interference structure, and so on.

How do we play the game (i.e., set the optimum parameters) in the simple game formulation given in Equations (10.34) and (10.35)? The method depends on the game type. There are two types of games – cooperative and non-cooperative – and several degrees between them. If we play the non-cooperative game, then there is an equilibrium solution (when certain required conditions achieved). This is called the Nash equilibrium solution. The Nash equilibrium is not necessarily an efficient solution; actually it is not usually efficient for all players. Moreover, in several cases it is not a Pareto optimal solution, i.e., there is room for at least one player to improve their payoff without violating the others. However, the Nash equilibrium explains the solution that players will approach to if they play a non-cooperative and selfish game. Mathematically, the Nash equilibrium solution for two players $(\mathbf{x}^*, \mathbf{y}^*)$ can be described as:

$$\Psi_1(\mathbf{x}, \mathbf{y}^*) \leq \Psi_1(\mathbf{x}^*, \mathbf{y}^*) \ \forall \mathbf{x} \in \Omega_{\mathbf{x}}$$
$$\Psi_2(\mathbf{x}^*, \mathbf{y}) \leq \Psi_2(\mathbf{x}^*, \mathbf{y}^*) \ \forall \mathbf{y} \in \Omega_{\mathbf{y}} \tag{10.37}$$

Hence, it is clear that no player can improve their utility if they deviate from the Nash equilibrium point given that the other player will stick with the Nash equilibrium point. Many games are played in a sequential manner, so that your response is based on your opponent's action. This can be described as a closed loop control system. Let's describe the types of games based on strategies. From a strategy point of view we have two different games either with finite or infinite strategies. For example, if a player plays with a set of transmitted data rates (e.g., 1 kbps, 10 kbps, 100 kbps) then we have finite discrete strategies. On the other hand if the strategy is played by adjusting the transmission power to be any value between 0 and P_{max}, then we have infinite and continues type of strategies. For finite strategy games we may have solutions in the pure strategies, i.e., when at least one player guarantees achieving a satisfied payoff if he plays with specific strategy regardless of the played strategies of others. If not, then all players would play with mixed strategies where every strategy is played with certain probability. For games with continuous strategies, there are solid mathematical theories to find the equilibrium solutions. Games can also be classified as zero-sum and nonzero-sum games. Zero-sum games refer to games where if player gets more payoff then at least one other player will lose the same amount. Nonzero-sum games refer to games where there is a chance that some players receive more payoff without degrading other player payoffs.

Example 10.2

Two transmitters compete to use the available resources from a single access point as shown in Figure 10.12. Every transmitter has only three possible transmission powers: 0, P_{ei}, and P_{max}. Where $P_{e1} = \alpha/G_1$, and $P_{e2} = \alpha/G_2$, α is constant, and G_i is the average channel gain for transmitter i. The payoff of every terminal is the obtained capacity which can be formulated as the upper capacity as

$$C_i = B \log_2 \left(1 + \frac{P_i G_i}{P_j G_j + \sigma_n^2}\right), \ i \neq j \in \{1, 2\} \tag{10.38}$$

[where σ_n^2 is the average power of the background noise. Assume also that $G_1 = 100G_2$, $P_{max} = 1, B = 1, \sigma_n^2 = 0.001 G_2 P_{max}$, and $\alpha = G_2 P_{max}$. Observe that we consider a static system or a snapshot of the channel. In real systems the channel gains are time varying.

Study this game between both transmitters and find the optimum as well as the equilibrium solution.

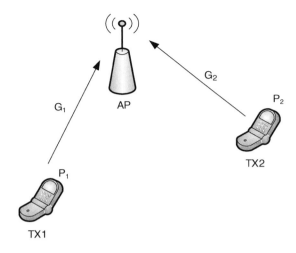

Figure 10.12   Two transmitters compete for the available resources

Solution

It is clear that the transmitted power value is the game strategy and the upper capacity is the payoff in this example. Since this game has a discrete and finite number of strategies to be played by each player, it is easy to represent it in matrix form as shown in Table 10.2.

The first element in each pair represents the payoff of terminal one, and the second is for terminal two. For each strategy played by terminal two, terminal one will select the strategy which maximizes its payoff. This means that we check the maximum of the first element in each column and mark it with a star. We should do the same for the second element in each row, because it represents second player action. The resultant stars are shown in Table 10.3.

As a non-cooperative game, it is clear that player 1 (or TX1) will play with strategy $P_1 = P_{max}$. With this strategy, player 2 has no other option other than to play also with $P_2 = P_{max}$. Based on the values given in the example, the second and third strategies of player 2 are the same. The payoffs pair with stars on both values represent the Nash equilibrium. This solution is the actual result if both terminals play a non-cooperative game without central control. Is this solution efficient? Actually, it is not so efficient because both players can do better as we will see later. But if both terminals belong to two different competitive operators, then this will be the solution. It is clear that player 1 is the winner because its channel gain is

Table 10.2   Game payoff matrix of Example 10.2

TX1 \ TX2	0	α/G_2	P_{max}
0	(0,0)	(0,9.97)	(0,9.97)
α/G_1	(9.97,0)	(1,1)	(1,1)
P_{max}	(16.61,0)	(6.66,0.01)	(6.66,0.01)

Table 10.3 Nash equilibrium solution of Example 10.2

TX1 \ TX2	0	α/G_2	P_{max}
0	(0,0)	(0,9.97*)	(0,9.97*)
α/G_1	(9.97,0)	(1,1*)	(1,1*)
P_{max}	(16.61*,0)	(6.66*,0.01*)	(6.66*,0.01*)

much better. The question now is what is the optimum solution if there is cooperation between both players or if they are controlled by the same base station? To answer this question we need to define two different types of services in telecommunications. The first has a fixed target QoS. In this case, when a certain QoS is achieved then we will have successful communication. If we increase the QoS parameters beyond this target, no significant improvement can be observed. An example is voice communication, where we need a certain SNR and data rate to have good voice quality. The second class is elastic services where a higher QoS obtains better services. An example is large files up/down loads. There is another class which also needs some target QoS to work, but as the QoS parameters are increased, better service is achieved such as video streaming with MPEG. Now let's check the optimum solution in the light of the above services. What is the optimum solution if both terminals are used for voice communication which needs minimum capacity of 0.5? The optimum solution in this case is (1,1), i.e., both terminals transmit with a power value related to the reverse of the channel gain to make the received power equal from both. This is the main concept of conventional power control which is used to mitigate the near-far problem in non-perfect orthogonal multiple access systems such as CDMA. Observe that this solution is not a Nash equilibrium, i.e., it is not possible to achieve in pure non-cooperative games.

What is the optimum solution if both terminals belong to same operator and the objective is to maximize the total capacity? We have an evident solution for this case which is (16.61,0). This means that the decision maker in the operator will command the second terminal to switch off and make all the spectrum available to the first terminal. What is the problem associated with such a decision? Simply it is unfair! The second user will always be off because its channel is the worst. If you can select any power values, the optimum solution of the power allocation could be based on the famous water-filling algorithm. Are there other solutions which take into account high capacity and greater fairness? There are several algorithms including the asymptotic fair algorithm. One simple and logical way is to divide the time between both terminals. For example if we allocate 70% of the time slot to terminal 1 and 30% to terminal 2, then on average the capacity of terminal 1 is 11.6 and for terminal 2 is 3.0. Now terminal 1 gets better than its Nash equilibrium point (it was 6.66) and terminal 2 will also obtain much better its equilibrium point (it was only 0.01!). Also the average total is better. This indicates that the Nash equilibrium for non-cooperative games is not necessarily Pareto optimal. As we saw both players can improve their payoffs if they cooperate.

However, since cognitive radios are unlicensed terminals, then it is possible to be independent, i.e., use different operators. In such situations it is more likely to observe the non-cooperation situation. The intelligent system in the cognitive radios will assist in deciding which is the better situation: playing cooperatively or with selfish behavior! In the previous simple example we showed the case when the game is played with finite strategies. Let's consider the case where we have N players and each player has a utility function they want to maximize. The utility function of each player is formulated as

$$u_i = \Psi_i(\mathbf{x}_1, \mathbf{x}_2, \cdots, \mathbf{x}_i, \cdots, \mathbf{x}_N) \tag{10.39}$$

where $\mathbf{x}_i = [x_{i1}, x_{i2}, \cdots, x_{im_i}]'$ is the strategy vector of player i, and m_i is the number of parameters of player i. Please note that m_i is not the number of strategies. For example $\mathbf{x}_i = [P_i, f_{0i}, R_i]'$ can be a strategy vector consisting of the transmitted power P_i (e.g., from 0 to $P_{\max}$), carrier frequency f_{0i} (e.g., a set of finite possible transmission carriers), and data rate R_i (e.g., a finite set of possible rates).

The Nash equilibrium solution $[\mathbf{x}_1^*, \mathbf{x}_2^*, \cdots, \mathbf{x}_i^*, \cdots, \mathbf{x}_N^*]$ for N players' nonzero-sum game should achieve the following condition:

$$\Psi_i(\mathbf{x}_1^*, \mathbf{x}_2^*, \cdots, \mathbf{x}_i^*, \cdots, \mathbf{x}_N^*) \geq \Psi_i(\mathbf{x}_1^*, \mathbf{x}_2^*, \cdots, \mathbf{x}_i, \cdots, \mathbf{x}_N^*) \, \forall i = 1, ..., N \tag{10.40}$$

This means that no player may improve his utility if he deviates from the Nash equilibrium point if no one else deviates. The sufficient conditions for the Nash equilibrium are:

$$\frac{\partial \Psi_i(\mathbf{x}_1, \cdots, \mathbf{x}_N)}{\partial \mathbf{x}_i} = \mathbf{0}, \quad i = 1, \cdots, N \tag{10.41}$$

$$\mathbf{x} \mapsto \Psi_i(\mathbf{x}_1^*, \cdots, \mathbf{x}_{i-1}^*, \mathbf{x}, \mathbf{x}_{i+1}^*, \cdots, \mathbf{x}_N^*) \forall \mathbf{x}_i \in Q_i \tag{10.42}$$

$$\frac{\partial^2 \Psi_i(\mathbf{x}_1, \cdots \mathbf{x}_N)}{\partial \mathbf{x}_i^2} < 0, \quad i = 1, \cdots, N \tag{10.43}$$

Condition (10.41) ensures that the point is an extreme point, condition (10.42) means that this point is unique, and condition (10.43) guarantees that the extreme point is a local maximum. More interesting analysis and proofs can be found in [14].

The cooperation between cognitive radios to form ad hoc cooperated radio networks can significantly enhance the performance of the secondary utilization of the spectrum. The performance can be achieved over more than one dimension such as:

1. *Cooperation in exchanging spectrum-sensing information:* This can considerably enhance the ability of detecting primary users without the necessity of using very high sensitive detectors at cognitive radios. Because cognitive radios are usually spatially distributed they experience different channel characteristics (in terms of fading and interference), hence every cognitive radio will have a different vision about the environment situation (e.g., about primary users). Exchanging these data and using proper data fusion algorithms, cognitive radios may make more robust decisions in terms of transmission power and so on. What happens if one cognitive radio terminal decides to cheat by sending false information that the primary user is present to enforce other terminals to shut down or minimize their transmission power? Therefore, the spectrum becomes free for this cheater alone! This situation and others are analyzed in literature, and several interesting proposals have been given.

2. *Cooperation in terms of multi-hope relaying:* The energy consumption of cognitive radios can be reduced if data transmission is done through shorter paths. Of course, there are several constraints such as maximum allowed latency or maximum number of hops.

3. *Cooperation in dividing the available resources:* As we show in Example 10.2, if two cognitive radios divide their utilization of the available spectrum in terms of time, both of them can achieve high capacity. But again, terminals should be honest and not selfish in order to achieve this task. Moreover, there should be a gain or a benefit for this cooperation in order to play it. Assume in Example 10.2, that there are many different users (not just two) competing for the transmission, in this case, it is less likely that the excellent channel user (if any) would be eager to play a cooperative game.

As usual there are no free benefits. Two major problems may reduce the chance of such cooperation.

1. Cooperated terminals mean predefined control channel is used to exchange spectrum information, resources scheduling, etc. The control channel is usually part of the available spectrum which reduces the spectrum efficiency. This can be mitigated by using very low data rates in the control channel. Moreover, using cooperation means more complicated terminals and standard cooperation algorithms to facilitate the cooperation between different cognitive radio types. These cooperation algorithms can be integrated in the MAC layer.
2. Since cognitive radios are unlicensed and every party could distribute them, it is not feasible to have any real desire of cooperation between competitor operators. Moreover, as the number of cognitive radios increases, it will be difficult to find a good context for cooperation, e.g., a standard common talking language.

From the above discussion we can see a real opportunity to apply game theory algorithms in the cognition engine of cognitive radios. These will set the rules of the optimum response of cognitive radio terminals for the actions of other cognitive radios, primary users, and channel fluctuations.

10.5 Systems Engineering and Cognitive Radios

Systems engineering is inherent in cognitive radio terminals and cognitive networks. On a single terminal level, the terminal parameters adaptation is based on closed and open control loops. For example, the cognitive radio can adjust its transmit power in order to avoid any harmful interference with the primary user, but usually there is no direct channel between primary and secondary users. Therefore the secondary users need to sense the spectrum and decide the optimum transmission power value. This is one version of open loop control. Moreover, cognitive radios utilize the closed loop control between transmitter and receiver in order to maximize the channel efficiency and avoid harmful interference with primary users close to the receiver. Figure 10.13 shows those concepts.

In Figure 10.13, we show one cognitive radio transmitter and receiver pair. It is also possible to reverse the situation, i.e., the transmitter becomes receiver and vice versa.

The cognitive radio network can also be studied and analyzed as a distributed control system with multiple inputs and multiple outputs (MIMO). Every cognitive transmitter can be modeled as a controller + actuator. The cognitive radio receiver can be modeled as a sensor which sends the environment state to the transmitter to adapt its transmission parameters. Moreover, every transmitter also has a sensor to check the primary user signal. In the case of a non-cooperative

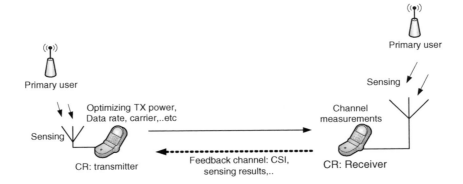

Figure 10.13   Open and closed loop control applications in cognitive radios

network, all other pairs can be modeled as a time-varying disturbance. This is a trivial situation. The interesting point is that the optimum transmission parameters of every cognitive radio are based on MIMO control theory. Figure 10.14 shows a simple schematic for the cognitive radio network. Dashed lines show possible cooperation between different cognitive terminals such as exchanging sensing information or resource scheduling proposals.

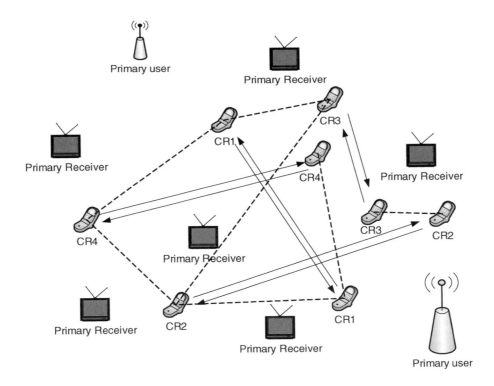

Figure 10.14   Cognitive radio network with primary users' transmitters and receivers

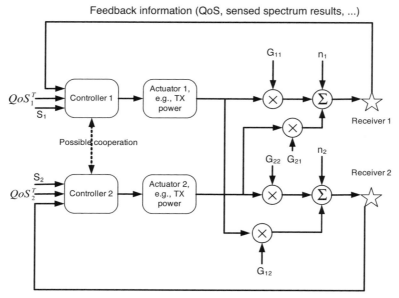

Figure 10.15   Distributed control system

From a systems engineering perspective, the above network can be described as a distributed MIMO control system. The idea can be illustrated by considering only two cognitive transmitter/receiver pairs and a single primary user as shown in Figure 10.15. The controller of each transmitter decides the output level based on the QoS at its receiver, the required QoS target (QoS_i^T), and the sensed spectrum at the receiver as well as at the transmitter side (S_i). Based on this structure we can test several MIMO centralized and distributed control system algorithms. Dashed lines between controllers may refer to information exchange between controllers such as sensed information, measured QoS, channel gain, and so on. Moreover, it may also refer to a centralized controller where the optimum controllers' response is done jointly in a central controller.

Exercise 10.5   Propose an MIMO PI controller for two cognitive radios pairs as shown in Figure 10.15. Use MATLAB$^{\circledR}$ to simulate the controller's response with static as well as dynamic channels.

References

[1] A. Ghasemi and E. Sousa, 'Spectrum sensing in cognitive radio networks: requirements, challenges, and design trade-offs', *IEEE Communications Magazine*, April, 2008.

[2] S. Haykin, 'Cognitive radio: brain-empowered wireless communications', *IEEE Journal on Selected Areas in Communications*, **23**(2), 2005.

[3] B. Wang, Y. Wu, K. Liu, and T. Clancy, 'Game theoretical mechanism design methods', *IEEE Signal Processing Magazine*, November, 2008.

[4] D. Niyato and E. Hossain, 'Market equilibrium, competitive, and cooperative pricing for spectrum sharing in cognitive radio networks: analysis and comparison', *IEEE Transactions on Wireless Communications*, November, 2008.

[5] J. Mitola, Cognitive radio: an integrated agent architecture for software defined radio, Ph.D. thesis, Computer Communication System Laboratory, Department of Teleinformatics, Royal Institute of Technology (KTH), May, 2000.

[6] H. Arslan, *Cognitive Radio, Software Defined Radio, and Adaptive Wireless Systems*, Springer, 2007.

[7] C. Stevenson, G. Chouinard, Z. Lei, W. Hu, S. Shellhammer, and W. Caldwell, 'IEEE 802.22: the first cognitive radio wireless regional area network standard', *IEEE Communications Magazine*, January, 2009.

[8] C. Wang, H. Chen, X. Hong, and M. Guizani, 'Cognitive radio network management', *IEEE Vehicular Technology Magazine*, March, 2008.

[9] A. Papoulis and S. Pillai, *Probability, Random Variables and Stochastic processes*, 4th edition, McGraw-Hill, 2002.

[10] D. Cabric, 'Addressing the feasibility of cognitive radios', *IEEE Signal Processing Magazine*, November, 2008.

[11] Z. Ye, G. Memik, and J. Grosspietsch, 'Energy detection using estimated noise variance for spectrum sensing in cognitive radio networks', *IEEE Wireless Communications and Networking Conference*, 2008.

[12] B. Farhang-Boroujeny and R. Kempter, 'Multicarrier communication techniques for spectrum sensing and communication in cognitive radios', *IEEE Communications Magazine*, April, 2008.

[13] W.A. Gardner, 'Exploitation of spectral redundancy in cyclostationary signals', *IEEE Signal Processing Magazine*, April, 1991.

[14] E.N. Barron, *Game Theory: An Introduction*, John Wiley & Sons, 2008.

Bibliography

Ahlin, L., Zander, J. and Slimane, B. (2006). *Principles of Wireless Communications*. Lund: Studentlitteratur.

Åström, K.J. and Hägglund, T. (2006). *Advanced PID Control*. ISA: Research Triangle Park.

Åström, K.J. and Wittenmark, B. (1989). *Adaptive Control*. Reading, MA: Addison Wesley.

Åström, K.J.And Wittenmark, B. (1998). *Computer-Controlled Systems: Theory and Design*. 3rd Edition. Englewood Cliffs: Prentice Hall.

Chen, C.-T. (1999). *Linear System Theory and Design*. Oxford: Oxford University Press.

Communications Blockset User's Guide (2009), http://www.mathworks.com/access/helpdesk/help/pdf_doc/commblks/usersguide.pdf.

Demuth, H., Beale, M. and Hagan, M. (2009). *Neural Network Toolbox, User's Guide*. www.mathworks.com, The Math-Works.

Dorf, R.C. and Bishop, R.H. (2008). *Modern Control Systems*, 13th edition. Upper Saddle River, NJ: Pearson Prentice Hall.

Egan, W.F. (1998). *Phase-Lock Basics*, New York: John Wiley & Sons, Inc.

Ekman, T. (2002). Prediction of mobile radio channels – modeling and design, PhD thesis. Uppsala University, Sweden.

Franklin, G.F., Powell, J.D. and Workman, M. (2006). *Digital Control of Dynamic Systems*, 3rd edition. Upper Saddle River, NJ: Pearson Prentice Hall.

Garg, V.K. and Wilkes, J.E. (1996). *Wireless and Personal Communications Systems*. London: Prentice Hall.

Goldsmith, A. (2005). *Wireless Communications*. Cambridge: Cambridge University Press.

Gran, R.J. (2007). *Numerical Computing with Simulink, Volume I, Creating Simulations*. Philadelphia: SIAM, Society for Industrial and Applied Mathematics.

Grewal, M.S. and Andrews, A.P. (2001). *Kalman Filtering: Theory and Practice*, 2nd edition. New York: John Wiley & Sons, Inc.

Haykin, S. (2001). *Communication Systems*, 4th edition. New York: John Wiley & Sons, Inc.

Haykin, S. (2009). *Neural Networks and Learning Machines: A Comprehensive Foundation*. London: Pearson.

Haykin, S. and Moher, M. (2005). *Modern Wireless Communications*. London: Prentice Hall.

http://en.wikipedia.org/wiki/Rayleigh_fading. http://www.mathworks.com/matlabcentral/ – provides many Rayleigh fading MATLAB models including the Jakes' model.

Jakes, W.J. (1974). *Microwave Mobile Communications*. Chichester: John Wiley & Sons, Ltd.

Jeruchim, M.C., Balaban, P. and Shanmugan, K.S. (2000). *Simulation of Communication Systems. Modeling, Methodology, and Techniques*. New York: Kluwer Academic/Plenum.

Ljung, L. (1999). *System Identification – Theory for the User*, 2nd edition. London: Prentice Hall.

Ljung, L. (2009). *System Identification Toolbox User's Guide*. http://www.mathworks.com/matlabcentral/.

Mark, J.W. and Zhuang, W. (2003). *Wireless Communications and Networking*. Upper Saddle River, NJ: Prentice Hall.

Norgaard, M. (2000). NNSYSID. http://www.mathworks.com/matlabcentral/fileexchange/87.

Paulraj, A., Nabar, R. and Gore, D. (2003). *Introduction to Space-Time Wireless Communications*. Cambridge: Cambridge University Press.

Rappaport, T.S. (1996). *Wireless Communications – Principles and Practice*. London: Prentice Hall.

Rappaport, T.S. (2001). *Wireless Communications – Principles and Practice*, 2nd edition. London: Prentice Hall.

Rintamäki, M. (2005). Adaptive power control in CDMA cellular communication systems, Doctoral dissertation, Helsinki University of Technology, Finland.

Sklar, B. (1997). Rayleigh fading channels in mobile digital communication systems, Part 1: characterization. *IEEE Communications Magazine*, July, 90–100.

Stüber, G.L. (2000). *Principles of Mobile Communication*, 2nd edition. Boston, MA: Kluwer.

Tse, D. and Viswanath, P. (2005). *Fundamentals of Wireless Communication*. Cambridge: Cambridge University Press.

Wellstead, P.E. and Zarrop, M.B. (1991). *Self-Tuning Systems – Control and Signal Processing*. Chichester: John Wiley & Sons, Ltd.

Yen, J. and Langari, R. (1999). *Fuzzy Logic – Intelligence, Control, and Information*. London: Prentice Hall.

Index